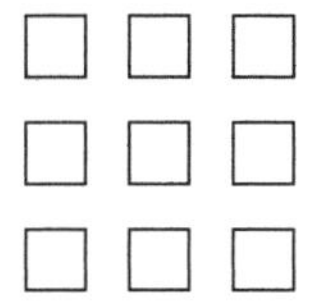

Active Learning Guide for College Physics

Eugenia Etkina

RUTGERS UNIVERSITY

Michael Gentile

RUTGERS UNIVERSITY

Alan Van Heuvelen

RUTGERS UNIVERSITY

PEARSON

BOSTON COLUMBUS INDIANAPOLIS NEW YORK SAN FRANCISCO UPPER SADDLE RIVER
AMSTERDAM CAPE TOWN DUBAI LONDON MADRID MILAN MUNICH PARIS MONTRÉAL TORONTO
DELHI MEXICO CITY SÃO PAULO SYDNEY HONG KONG SEOUL SINGAPORE TAIPEI TOKYO

Publisher: Jim Smith

Project Managers: Katie Conley and Beth Collins

Director of Development: Laura Kenney

Managing Development Editor: Cathy Murphy

Editorial Assistant: Kyle Doctor

Team Lead, Program Management, Physical Sciences: Corinne Benson

Director of Production: Erin Gregg

Full-Service Production and Composition: PreMediaGlobal

Illustrators: PreMediaGlobal

Manufacturing Buyer: Jeff Sargent

Marketing Manager: Will Moore

Cover Designer: Riezebos-Holzbaur Design Group and Seventeenth Street Studios

Cover Photo Credit: © Markus Altmann/Corbis

---

1 2 3 4 5 6 7 8 9 10—EBM—17 16 15 14 13

www.pearsonhighered.com

ISBN 10: 0-321-86445-X
ISBN 13: 978-0-321-86445-1

CONTENTS

PREFACE TO THE STUDENT

Welcome to college physics! If you are reading this preface, it is probably because your physics instructor has assigned the ***Active Learning Guide,*** a workbook to accompany the textbook ***College Physics,*** by Eugenia Etkina, Michael Gentile, and Alan Van Heuvelen. This workbook consists of carefully-crafted activities that supplement the textbook and provide an opportunity for further observation, sketching, analysis, and testing.

The chapters in this workbook correlate to those in the text. You will find marginal "Active Learning Guide" icons throughout the text that indicate content for which a workbook activity is available. Whether assigned to you or not, you can always use this workbook to reinforce the concepts you have read about in the text, to practice applying the concepts to real-world scenarios, or to work with sketches, diagrams, and graphs that help you visualize the physics.

The exercises in the ***Active Learning Guide*** are designed to help you learn to connect your intuition and everyday experiences with the words, graphs, and equations used in physics. In addition, we recommend that you apply these tips for success:

1. Actively participate in your own learning. Research shows that knowledge cannot be transmitted from one individual to another. It has to be actively constructed by the learner. This construction happens when you connect new ideas to what you already know.

2. Learn to think and act like a scientist. Be critical. Evaluate every piece of information and every new idea. Understand what is a result of observation and what is a result of reasoning. If you developed a new idea through reasoning, think of how you can test it in new experiments.

3. Reflect on the processes you use to construct and apply physics concepts. The logic you develop in physics will help you in your other courses and throughout your career.

4. Connect what you learn to the world around you. Sometimes what you learn may contradict what you thought you knew. Do not be discouraged. Great scientists such as Newton and Galileo also had to look at the world from a new point of view. This new perspective can make your study of physics more interesting and give you a deeper understanding of the world around you.

5. Learn to represent concepts in different ways to deepen your understanding. Physics uses many different ways to represent knowledge—pictures, diagrams, words, graphs, and equations. Memorizing formulas is not the key to success.

6. Learn to explain your ideas in simple language. The language that you use shapes your thinking; thus, being clear in your language will help you be clear in your thinking.

7. Pay attention to *how* you learn. You may understand better if you draw a picture or represent a process with a diagram. Perhaps you need to go over things several times. Maybe you need to check your algebra to avoid mistakes. Because people learn in different ways, this workbook employs many ways to construct physics knowledge. Knowing what helps you learn and apply knowledge most effectively will save you time and improve your work in physics and other classes.

8. Work collaboratively with friends. Research shows that students working together in study groups can solve problems that none of them can solve on their own. Through this collaboration, they can increase their skills individually.

9. Be patient. Learning physics requires focus and practice, just like learning a new sport or a musical instrument. As you become more proficient in physics, you will enjoy the learning process more and become more adept at applying your knowledge. Eventually, you will see physics all around you: in the bubbles rising from a pot of boiling water, in the behavior of a shower curtain when you take a shower, and in the glow of a filament in a lightbulb. Learning physics is empowering: it allows you to explain the world!

PREFACE TO THE INSTRUCTOR

Learning physics is easier for students when they actively observe, explain, test, represent, and evaluate explanations of physical phenomena they encounter in their everyday experiences. That understanding is the basis for the textbook ***College Physics,*** by Eugenia Etkina, Michael Gentile, and Alan Van Heuvelen, as well as this workbook, which accompanies that textbook.

Both the textbook and the ***Active Learning Guide*** are designed to encourage students in their study of physics by actively engaging them in the discovery process. The textbook includes Observational Experiment Tables and Testing Experiment Tables, as well as a problem-solving approach that will help students develop the skills they need to work through the assigned homework.

This workbook includes a set of activities for use in a variety of settings—large enrollment lectures and recitations, small classes, labs, and homework. You can use the activities in conjunction with the end-of-chapter problems in the textbook and/or ***MasteringPhysics***, or on their own. Each chapter includes many activities. You do not have to assign them all. The ***Instructor's Guide*** that accompanies ***College Physics*** describes options for assigning the various activities. Each chapter of the textbook includes icons in the margins identifying where ***Active Learning Guide*** activities correspond to chapter content.

Although the ***Active Learning Guide*** is based on an active, process-oriented learning system, the activities will fit easily into any introductory physics course, regardless of the teaching method.

Within the ***Active Learning Guide*** are four categories of activities, each subdivided into activity types that help students develop specific cognitive and science-process abilities for learning and applying physics concepts. Like the textbook, each workbook chapter begins with qualitative concept building before introducing quantitative reasoning. Readers familiar with the original edition of the ***Active Learning Guide*** will recognize some activities and find many new ones.

1. Qualitative Concept Building and Testing activities help students construct new qualitative concepts by observing phenomena, recording their observations, devising qualitative explanations for their observations, and testing these explanations by using them to predict the outcome of testing experiments. These activities involve quantitative reasoning but no formal mathematics.

2. Conceptual Reasoning activities help students learn to reason about the physical world using the qualitative explanations that they have already tested and accepted. The students learn to represent phenomena in a number of different ways—through motion diagrams, force diagrams, work-energy bar charts, sketches, ray diagrams, and so forth—and in this way create referents that enhance their understanding of the more abstract physics quantities and concepts. These qualitative representations improve students' ability to reason about the world without using mathematics. Even instructors who prefer to introduce concepts differently from the method used in the Qualitative Concept Building and Testing activities will find these Conceptual Reasoning activities useful in helping students further their conceptual reasoning skills.

3. Quantitative Concept Building and Testing activities help students devise relationships between physical quantities based on the data that they collect or that are provided with the activity. Students then use the relationship to predict the outcome of a new experiment, which we call a *testing experiment.* It is important that students both participate in and reflect on this process. Reflection will improve their retention of ideas and allow them to see patterns in how they construct knowledge.

4. Quantitative Reasoning activities help students learn to use different types of representations to describe physical processes and to solve problems. Students use verbal, pictorial, diagrammatic, graphical, and mathematical representations to solve problems. Special evaluation activities help students learn to check for the consistency of their representations and to evaluate the correctness of their work. Examples of more complex problems and experimental design activities that can be used in labs are also provided. All activities within this section can be used effectively with various instructional methods.

The ***Instructor's Guide*** that accompanies ***College Physics*** will help you make the transition to this approach from the materials you have in prior classes. That guide includes suggestions for how, when, and why to implement activities, as well as advice about helping students avoid common pitfalls.

We hope that our approach to helping students learn physics will enhance your teaching experience, as it has our own and those of many colleagues who have used these materials in their classrooms.

1 Kinematics: Motion in One Dimension

1.1 Qualitative Concept Building and Testing

1.1.1 Describe A person sits in the passenger seat of a car that is traveling along a street. Describe the person as seen by each of the following observers: (a) a person sitting in the backseat of the car; (b) a pedestrian standing on the sidewalk as the car passes; and (c) the driver of a second car moving in the same direction and passing the first car.

1.1.2 Describe A person stands near a bus stop. Describe the standing person's motion as seen by the following observers: (a) a person sitting in an approaching bus; (b) a person riding in a car moving away from the bus stop; and (c) another person standing at the bus stop.

1.1.3 Explain Review your analyses for Activities 1.1.1 and 1.1.2 and answer the questions that follow.

a. Do any observers say that the person sitting in the passenger seat of the car in 1.1.1 was moving? Explain.

b. Do any observers say that the person sitting in the passenger seat of the car in 1.1.1 was not moving? Explain.

c. Do any observers say that the person standing near a bus stop in 1.1.2 was moving?

d. Do any observers say that the person standing near a bus stop in 1.1.2 was not moving?

e. Based on your answers in parts (a) through (d), explain what it means when someone says an object is "moving."

1.1.4 Observe Set a metronome at about one beat per second. Turn on a battery-operated toy car, and release it to roll across the floor when the metronome makes a flash or blip.

When you feel comfortable with the sequence of events, repeat the experiment but this time have a friend place sugar packets on the floor at the point where the car is located at every blip of the metronome. After about 5 blips, stop the car and draw a sketch showing the locations of the sugar packets as dots. Discuss how you can use the dots to describe the motion of the car. If you do not have the car, you can use a hard ball such as a billiard ball or bowling ball that you roll on a smooth floor.

1.1.5 Represent and reason You have two battery-operated toy cars that you can release simultaneously on a smooth floor and a metronome set to 1-second intervals. You and a friend each walk next to one of the cars, and at every blip of the metronome, you place sugar packets at the cars' location. The dots in the figure below represent the locations of the packets for the two cars. The cars start simultaneously at the dot on the left and move to the right.

Describe the motion of each car as fully as possible by answering the following questions.

a. Were the cars ever next to each other? If so, where?

b. If there were a passenger in car 1, how would the passenger describe the motion of car 2?

c. If there were a passenger in car 2, how would the passenger describe the motion of car 1?

1.1.6 Observe Place a flexible ball such as a hollow rubber ball at rest on the floor (The ball should be flexible enough to change shape a little when on a surface.) Set a metronome to 1-second intervals. Push abruptly one time on the ball with a ruler so that the ball leaves the ruler rolling with considerable speed. Have a friend move beside the ball and place sugar packets on the floor to indicate positions of the ball every second. Draw a sketch representing the sugar packets with dots and describe the relative distance between the packets. How does the distance between the packets correspond to the observed motion of the ball?

1.2 Conceptual Reasoning

1.2.1 Represent and reason The illustration below relates to the experiment you performed with the flexible ball in Activity 1.1.6. The dots represent the locations of the ball measured each second. The arrows represent the direction of motion and how fast the ball was moving (we call them $\vec{v}$ arrows). Consider $\vec{v}$ arrows 1 and 2. Move them side by side with their tails at the same horizontal position. Decide what change arrow $\Delta\vec{v}_{1\text{ to }2}$ you would have to add to arrow 1 to make it the same length as arrow 2. Repeat for arrow 2—what change arrow is needed to change it into arrow 3, and what change arrow is needed to change arrow 3 into arrow 4? We call these *velocity change arrows.*

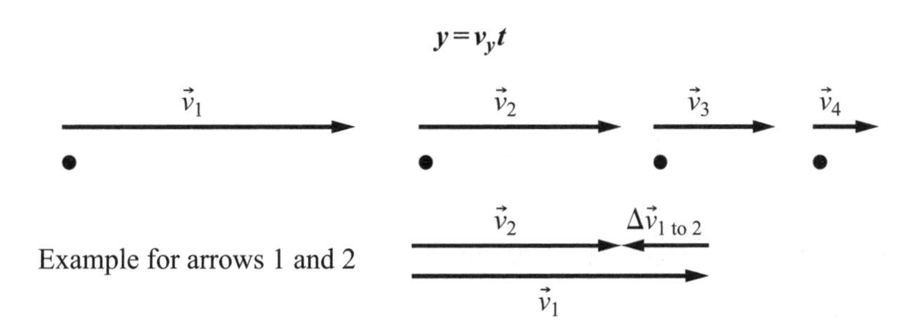

Use the Reasoning Skill box "Constructing a motion diagram" in chapter 1 of *College Physics* to learn how to represent motion using qualitative motion diagrams.

1.2.2 Represent and reason The illustration below is a motion diagram for an object. The dots represent the object's position at even time intervals. Describe the object's motion in words by devising a story that is consistent with this diagram. Note that the process has three distinct parts: vertical dashed lines separate the parts.

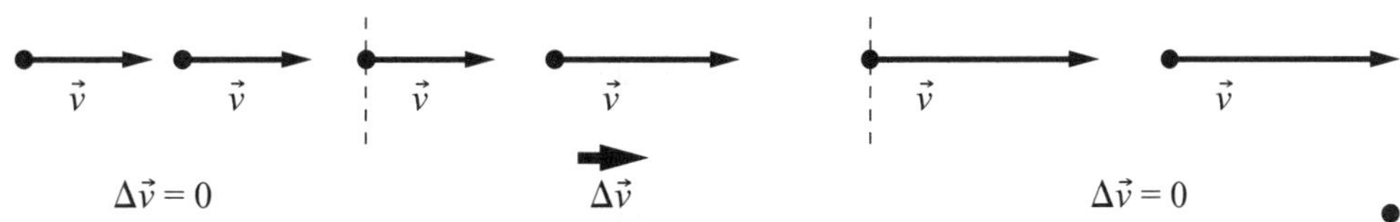

1.2.3 Represent and reason The illustration to the right is a motion diagram for an object. Devise a story that is consistent with this diagram. Note that the process has two distinct parts: the horizontal dashed line separates the parts.

1.2.4 Represent and reason A car stops for a red light. When the light turns green, the car moves forward for 3 s at a steadily increasing speed. The car then travels at constant speed for another 3 s. Finally, when approaching another red light, the car steadily slows to a stop during the next 3 s. Draw a motion diagram that describes this process.

1.2.5 Represent and reason Mike is traveling to a store and then to his friend's house. His trip is represented on a sketch below.

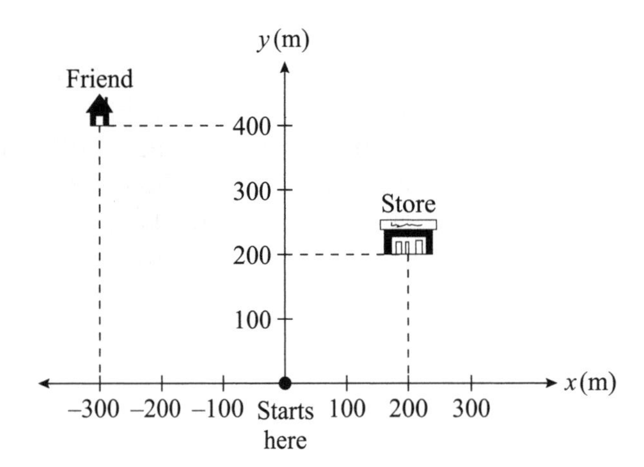

a. What is Mike's original position? Final position? What is the position of the store?

b. Draw a vector to represent his total displacement, determine the scalar x component of the displacement, total distance traveled and path length.

c. Choose a different origin for the coordinate system and repeat a–b. What quantities changed? What quantities remained the same?

1.2.6 Represent and reason Tell a story and draw pictures representing Heather's trips A and B, as described below. On the pictures show the coordinate axis and the displacement vector. Think of where you will choose the origin.

a. Trip A: $d_x = 0.7$ mi; $l = 2.4$ mi.

b. Trip B: $d_x = -3.7$ mi; $l = 4.4$ mi.

c. What are Heather's original and final positions for each trip? How do those depend on the choice of the direction of the coordinate axis and the location of the origin?

d. Who is the observer for the trips described above? Find an observer for whom during the same trips Heather's displacements and path length traveled are zero.

1.2.7 Pose your own problem Imagine any motion you participate in every day, such as going to classes or to a movie and dinner afterward. Pose a problem to solve about this motion in which one needs to understand the difference between the physical quantities position, displacement, distance and path length, and the difference between a vector and the scalar component of a vector.

1.3 Quantitative Concept Building and Testing

1.3.1 Observe and describe Imagine that you and your friend ride bicycles along a straight path beside a river. A coordinate axis is shown above the path.

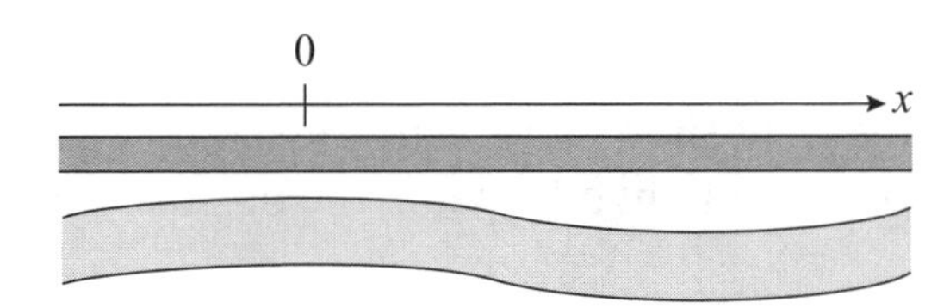

The table indicates your position along the path at different clock readings.

a. Write everything you can about the bike rides and indicate any pattern in the data. What was happening at the clock reading of zero?

Clock reading t (s)	Your position x (m)	Your friend's position
$t_0 = 0$	$x_0 = 640$	$x_0 = 640$
$t_1 = 20$	$x_1 = 500$	$x_1 = 490$
$t_2 = 40$	$x_2 = 360$	$x_2 = 340$
$t_3 = 60$	$x_3 = 220$	$x_3 = 190$
$t_4 = 80$	$x_4 = 80$	$x_4 = 40$
$t_5 = 100$	$x_5 = -60$	$x_5 = -110$
$t_6 = 120$	$x_6 = -200$	$x_6 = -260$

b. Construct position-versus-clock reading graphs for both bike trips using the same coordinate axes in which x is a dependent variable and t is an independent variable. Compare and contrast the graphs—how do the graph lines represent the differences in the bikes' motions?

c. Write two function $x(t)$ expressions for the graphs in part b. Think of your labeling system. How would one distinguish the function for your bike from the function for your friend's bike?

d. What is the physical meaning of the slope of each function and the intercepts? What common name could you use for the slope? Explain the meanings of positive or negative values for these quantities.

e. Compare and contrast how one writes linear functions in mathematics and how you just wrote position-versus-time functions for motion. What is the same between them? What is different?

1.3.2 Reason You are sitting on a park bench. A jogger is passing the bench jogging at a relatively constant speed of 6 mph when you start observing her. Write a position-versus-time function for the jogger that will allow you to predict her position at any clock reading. What assumptions do you need to make to write this function?

How many "correct" functions can you write?

1.3.3 Analyze The figure at the right shows a *velocity*-versus-time graph that represents the motion of a bicycle moving along a straight bike path. The positive direction of the coordinate axis is toward the east.

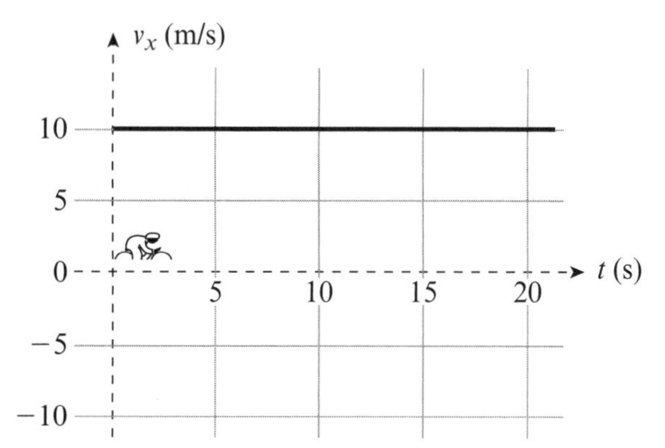

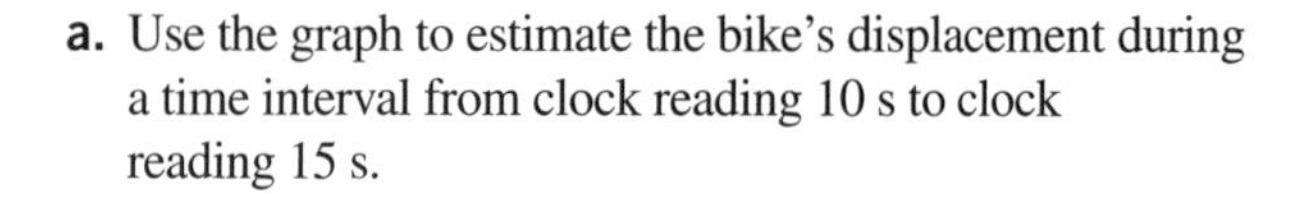

a. Use the graph to estimate the bike's displacement during a time interval from clock reading 10 s to clock reading 15 s.

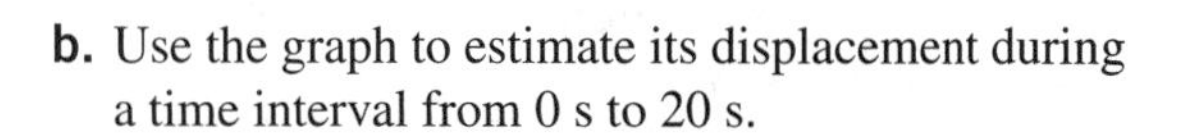

b. Use the graph to estimate its displacement during a time interval from 0 s to 20 s.

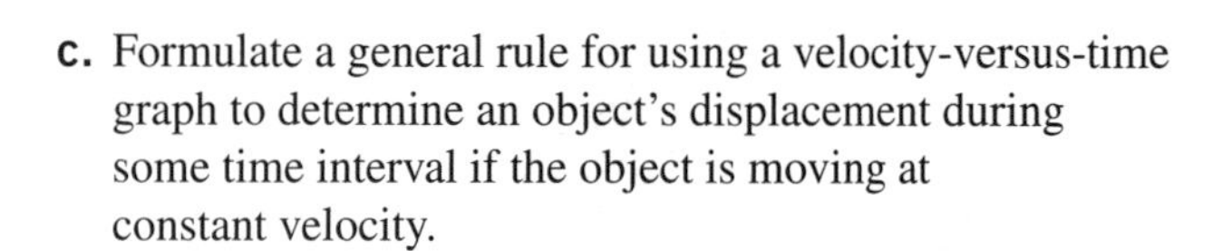

c. Formulate a general rule for using a velocity-versus-time graph to determine an object's displacement during some time interval if the object is moving at constant velocity.

1.3.4 Test your idea You have two motorized toy cars, sugar packets, a meter stick, and a watch with a secondhand.

a. For car *A* design an experiment to decide if the car moves with constant velocity. If it does, determine the magnitude of the velocity (the car's speed).

b. For car *B*, use the same equipment and method to decide if this car moves with constant velocity. If it does, determine the magnitude of the velocity (the car's speed).

c. Predict where the cars will meet if you release them from 2.0 m apart moving straight toward each other. List all assumptions that you made about how the cars move. If the assumptions were not valid, how would your prediction change?

d. Perform the experiment and collect data. Did the outcome match your prediction? How many times do you need to conduct the experiment to be able to say for sure that the outcome of the experiment matches or does not match the prediction? Write the result of the experiment accounting for the discrepancies in the meeting location in different repetitions of the experiment.

1.3.5 Test an idea Design an experiment to test whether the motion of a dropped golf ball can be described mathematically as $y = v_y t$. Consider the location from which the ball is dropped to be the origin. Before you perform the experiment, think about how you will take measurements at time intervals, and make a prediction of expected results based on your idea. Be sure to write the prediction down. Then perform the experiment as many times as necessary for you to be convinced that the data you collected are sufficient to support or reject the idea.

1.3.6 Observe and analyze You place a cart at the top of a smooth metal track tilted at a 10° horizontal angle. The data table provides records of the position of the front of the cart at different times. The x-axis points down and to the left along the track.

Clock reading t (s)	Position x (m)
$t_0 = 0.00$	$x_0 = 0.00$
$t_1 = 0.50$	$x_1 = 0.21$
$t_2 = 1.00$	$x_2 = 0.85$
$t_3 = 1.50$	$x_3 = 1.91$
$t_4 = 2.00$	$x_4 = 3.40$
$t_5 = 2.50$	$x_5 = 5.31$

a. Draw a motion diagram for the cart.

b. Draw a position-versus-time graph for the cart. Discuss whether the graph resembles a position-versus-time graph for an object moving at constant velocity.

c. Determine the scalar component of the average velocity for the cart for each time interval by completing the table that follows.

Time interval $\Delta t = t_n - t_{n-1}$	Displacement $\Delta x = x_n - x_{n-1}$	Average time $(t_n + t_{n-1})/2$	Average velocity $\frac{\Delta x}{\Delta t}$

d. Plot this average velocity v_x on a velocity-versus-time graph. The time coordinate for each average velocity coordinate should be in the middle of the corresponding time interval (the average time for that time interval). Make a best-fit curve for your graph line.

e. Discuss the shape of the graph: How does the speed change as time elapses? Suggest a name for the slope of the graph.

f. Write an equation for the velocity as a function of time that is consistent with the graph line. Discuss how your equation would change if you started observing the cart when it was already moving down the same slope. Discuss how the equation would change if the cart were slowing down instead of speeding up.

1.3.7 Derive Use the method developed in Activity 1.3.6 to find a relationship between the displacement of the cart during some time interval, its velocity at the beginning of this interval, its acceleration, and the length of the time interval. There are many ways of doing this. Think of average velocity or a velocity-versus-time graph.

1.3.8 Analyze The data recorded in the table at the right are a record of the up and down motion of the center of a ball thrown upward (the y axis points up). Fill in the table that follows.

Clock reading t (s)	Position y (m)
0.000	0.00
0.133	0.44
0.267	0.71
0.400	0.80
0.533	0.71
0.667	0.42
0.800	−0.04

a. Sketch a motion diagram for the ball modeled as a point object.

b. Draw a position-versus-time graph.

c. Draw a velocity-versus-time graph. Find its slope. What do you call this slope?

d. Use the velocity-versus-time graph to determine the ball's acceleration at the very top of its trajectory.

e. What is the ball's velocity at the top?

f. Can you reconcile these two answers? Explain.

g. Use the velocity-versus-time graph to determine the distance that the ball traveled during the trip from clock reading 0.000 s to 0.800 s.

1.3.9 Design an experiment Design an experiment to investigate the motion of a cart moving up and down an inclined plane. You push it forcefully at the bottom of the plane so it moves up, and then it stops and rolls down. You can use a motion detector, a stopwatch, a meter stick, and other equipment available in the lab. Describe your experiment in words and draw the setup. What data will you collect?

a. How will you organize your data?

b. After you made all the decisions, conduct the experiment and write your findings. What can you say about the motion of the cart?

1.3.10 Summarize Use different representations of the two types of motion we have studied to fill in the empty cells in the table. Some cells are completed to give you an idea of the motions and the direction of the coordinate axis for each case. Your responses should relate to the motion already described. Completing the table will help you summarize everything you have learned about motion description.

Motion with constant velocity	**Motion with constant acceleration**
Describe the motion in words, providing an example.	Describe the motion in words, providing an example. The object's velocity is decreasing by the same amount every second—for example a cart going up a smooth track tilted at an angle.
Provide a motion diagram that describes this type of motion.	Provide a motion diagram that describes this type of motion.
Provide a position-versus-time graph that describes this type of motion. 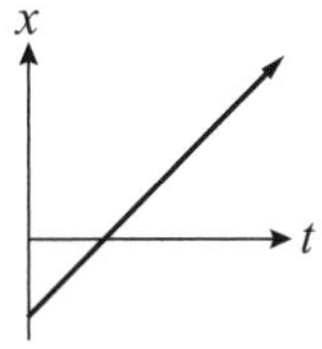	Provide a position-versus-time graph that describes this type of motion.
Describe motion mathematically as $x(t)$.	Describe motion mathematically as $x(t)$. $x = -v_0 t + \frac{1}{2} a_x t^2$
Provide a velocity-versus-time graph that describes this type of motion.	Provide a velocity-versus-time graph that describes this type of motion. 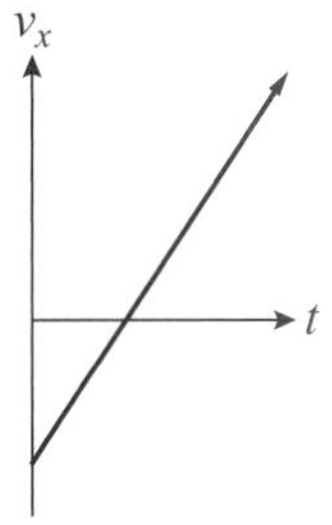
Describe motion mathematically as $v_x(t)$. $v_x(t) = const$	Describe motion mathematically as $v_x(t)$.
Provide an acceleration-versus-time graph that describes this type of motion. 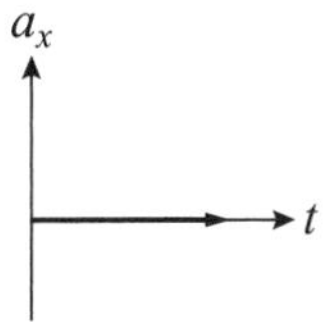	Provide an acceleration-versus-time graph that describes this type of motion.
Describe motion mathematically as $a_x(t)$.	Describe motion mathematically as $a_x(t)$. $a_x(t) = const$

1.4 Quantitative Reasoning

1.4.1 Represent and reason You have two identical billiard balls on the top of an inclined track. Assume that the balls roll the same way a small cart does (constant acceleration). Now imagine you released one ball and it is rolling down the track. When it is about 10 cm down the track, you release the second ball. Draw a picture of the situation and describe it using motion diagrams, graphs, and mathematical equations. Use those representations to predict what will happen to the distance between the two balls—will it increase, decrease, or stay the same at 10 cm?

1.4.2 Represent and reason The motion diagrams in the illustrations represent the motion of different objects. The arrows are velocity arrows.

a.
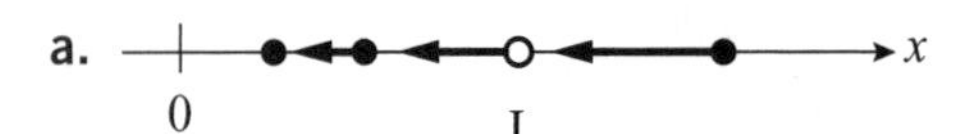

b.

I

x

0

II

c.
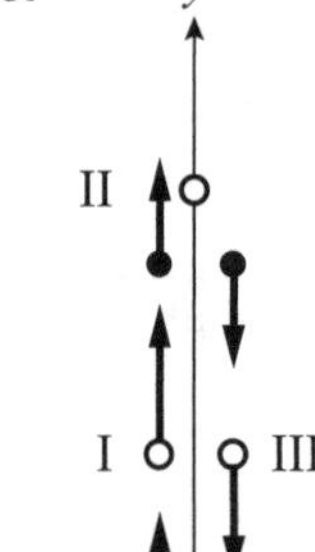

A different coordinate axis is provided for each of the three motion diagrams. An open circle(s) indicates a location of interest. Add a single velocity change arrow for each diagram. Then, determine the signs of the position, velocity component, velocity change component, and acceleration component at the position(s) of the open dots for each diagram. Note: what assumptions about motion do you need to make to use only *one* velocity change arrow for part (b) and one for part (c).

1.4.3 Represent and reason A stoplight turns yellow when you are 20 m from the edge of the intersection. Your car is traveling at 12 m/s. After you hit the brakes, your car's speed decreases at a rate of 6.0 m/s each second until the car stops. Ignore the reaction time needed to bring your foot from the floor to the brake pedal.

a. Sketch the process.

b. Draw a motion diagram representing the process. What are the signs of the v_x, Δv_x and a_x?

c. Construct an $x(t)$ graph for the process.

d. Construct a $v_x(t)$ graph for the process.

e. Write $x(t)$ and $v_x(t)$ expressions representing the process.

1.4.4 Regular problem You ride your bike west at speed 8.0 m/s. Your friend, 400 m east of you, is riding her bike west at speed 12 m/s. Complete the following steps to determine when your friend passes you.

a. *Sketch and Translate* Draw a sketch of the initial situation and choose a coordinate system to describe the motion of both bikes. Put all given information on the sketch; identify the unknown.

b. *Simplify and Diagram* Draw a motion diagram for each bike. Sketch a position-versus-time graph for each bike using the same coordinate axes.

c. *Represent Mathematically* Construct equations that describe the positions of each bicycle as a function of time relative to the chosen coordinate system.

d. *Solve and Evaluate* Use the equations to determine when the bicycles are at the same position. Does your result make intuitive sense? How do you know?

1.4.5 Represent and reason An imaginary object moves horizontally. The position-versus-time function represents the object's motion mathematically. Describe in different ways a process that the equation below might describe. The equation could describe many different processes.

$$x(t) = (-200.0\text{ m}) + (-20.0\text{ m/s})t + (1.0\text{ m/s}^2)t^2$$

a. Describe the motion in words. Note that it is important to focus on what was happening at $t = 0$. Use physical quantities to write down all of the information that you can "extract" from the function. If you can write other functional dependencies—do it!

b. Draw a motion diagram that represents the process.

c. Draw a position-versus-time graph that represents the process.

d. Draw a velocity-versus-time graph that represents the process.

e. Determine when and where the object for your chosen process stops.

1.4.6 Represent The motion of two objects is represented below. Study the motions and act them out with your classmate. Note that it is important to focus on what was happening at $t = 0$. What are your assumptions about the observer?

$$x_A = (-7.5\text{ m}) + (1.7\text{ m/s})t$$

$$x_B = (5.2\text{ m}) + (-2.2\text{ m/s})t$$

1.4.7 Evaluate You learned that the equation describing position-versus-time of an object moving at constant acceleration is $x = x_0 + v_0 t + \frac{1}{2} a_x t^2$. Use both algebraic and graphical approaches to show that in a limiting case of $\vec{a} = 0$ this equation describes the motion of an object that is traveling at constant velocity.

1.4.8 Evaluate You learned that in the equation describing an object moving at constant acceleration, the position as a function of time is $x = x_0 + v_{0x} t + \frac{1}{2} a_x t^2$. Use algebraic and graphical approaches to show that in a limiting case of $x_0 = 0$ $\vec{v}_0 = 0$ the displacements of the object change in proportion as the integers squared: 1, 4, 9, 25, etc.

1.4.9 Regular problem While traveling in your car at 24 m/s, you find that traffic has stopped 30 m in front of you. Will you smash into the back of the car stopped in front of you? Your reaction time is 0.80 s and the magnitude of your car's acceleration is 8.0 m/s^2 after the brakes have been applied. List all assumptions you make.

1.4.10 Evaluate the solution Identify any errors in the proposed solution to the following problem and provide a corrected solution if there are errors.

The problem: You are driving at 20 m/s and slam on the brakes to avoid a goose walking across the road. You stop in 1.2 s. How far did you travel after hitting the brakes?

Proposed solution:

$$(x - x_0) = vt = (20\text{ m/s})(1.2\text{ s}) = 24\text{ m}.$$

1.4.11 Evaluate the solution Identify any errors in the proposed solution to the following problem and provide a corrected solution if there are errors.

The problem: Use the graphical representation of motion to determine how far the object travels until it stops.

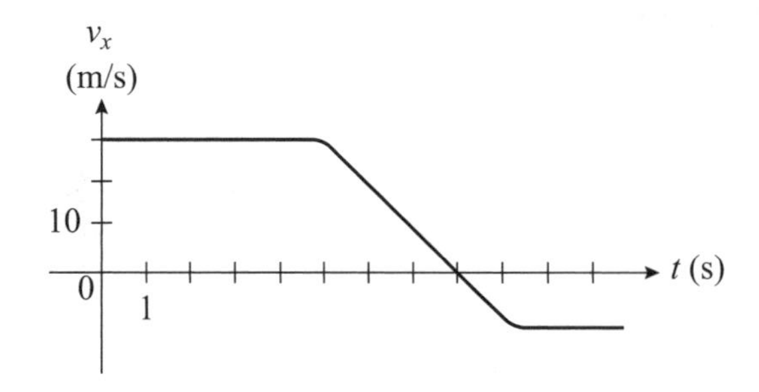

Proposed solution: The object was at rest for about 5 seconds, then started moving in the negative direction and stopped after about 9 seconds. During this time its position changed from 30 m to –10 m, so the total distance that it traveled was 40 m.

1.4.12 Represent A small cart moves along a track shown below. Sketch position, velocity and acceleration-versus-time graphs for its motion. Position the graphs under each other so one can see all three quantities for the same clock reading. List all assumptions that you made.

1.4.13 Design an experiment You have a cotton ball, a stopwatch, and a meter stick.

a. Describe in detail (including a sketch) an experiment that you can perform to determine whether the cotton ball falls with constant speed, constant acceleration or changing acceleration.

b. Write the physical quantities that you will measure and the quantities that you will calculate.

c. List experimental uncertainties and how you will minimize them.

d. Perform the experiment; record the data in a table and use a best-fit function for the data to make a judgment.

e. Write your analysis and conclusion about the ball's motion.

1.4.14 Evaluate the solution Allison proposed the following experiment to investigate the motion of the cotton balls: Drop one ball from 1 m above a surface and record the time during which it falls. If it falls at constant speed, then when it is dropped from the height of 2 meters, the time in flight should double. If it falls at constant acceleration, the time should be equal to the square root of 2 times the time it took the ball to fall from 1 meter. Now that the predictions for the two models are made, perform the experiment dropping the ball from 2 meters and record the time. Depending on the outcome—discard one or both models of motion. What do you think of Allison's design, her description of the experiment, and the mathematical analysis?

2 Newtonian Mechanics

2.1 Qualitative Concept Building and Testing

2.1.1 Observe and represent While standing: (I) first hold a basketball and then (II) put down the basketball and hold a medicine ball.

a. Choose the ball as the object of interest (the system). For each situation, construct a sketch; circle in the sketch the system (the ball); and below each sketch start the construction of a diagram where you model the system as a point object.

b. On that diagram draw an arrow to show how your hand pushes on the ball. *Connect the tail of the arrow to the dot.* This arrow represents the **force** that your ***hand exerts on the ball***. How could you label this force arrow to show that it is the force your hand exerts on the ball? Add this label to your representation.

c. What do you think would happen to the ball if your hand were the only object interacting with it? What does this tell you about other objects interacting with the ball?

d. What other objects are interacting with the ball? List each object and the direction of the push or pull. Represent these interactions on the diagram. Try to make the lengths of the force arrows in the two diagrams representative of the relative magnitudes of the forces. The diagrams are called force diagrams.

e. The word **"force"** is used in physics for a physical quantity that characterizes ***the interaction between two objects***. A single object does not have a force because the force is defined as the interaction of two objects. Using the definition of force in physics, give three examples from everyday life when the use of the term force does not match the meaning of this word in physics.

2.1.2 Test your idea

a. In the previous activity, did you say that air interacts with the ball?

b. Do you think that the total force that the air exerts on the ball points up or down?

c. What experiment can you perform to test your idea about whether the air pushes up or down on the ball? Describe the experiment and state the predictions of what should happen if the air pushes up/down on the ball.

Before working through the problems in this section read Reasoning Skill box "Constructing a force diagram" in Chapter 2 of *College Physics*.

2.1.3 Observe and find a pattern Perform or observe each experiment described below. After each experiment, describe what happened and construct a motion diagram and several force diagrams for the ball's motion during that experiment.

a. Steadily push a bowling ball along a smooth surface (using a meter stick will work here). How did you have to move in order to keep pushing steadily?

b. Stop pushing the ball and let it roll.

c. Set the ball in motion and push gently on the front.

d. Is there a pattern in the directions of the *vector sum of the forces* that other objects exert on the ball and in the directions of the $\vec{v}$ arrows? Is there a pattern in the directions of the *vector sum of the forces* that other objects exert on the cart and the directions of the $\Delta\vec{v}$ arrows on the motion diagrams? If so, describe the patterns.

e. Use the pattern that you found to formulate a statement relating the direction of the sum of the forces exerted on an object by other objects and one or more of the kinematics quantities that describe its motion.

2.1.4 Observe and find a pattern Perform or imagine performing each experiment described below. Then construct a motion diagram and several force diagrams for the ball's motion during that experiment. Based on our investigation in Activity 2.1.2, you can ignore any force or forces that the air might exert on the ball.

a. Throw a golf ball upward. Observe its motion AFTER it leaves your hand until it reaches the top of the flight.

b. Hold a golf ball above your head and drop it. Observe its motion AFTER it leaves your hand until right before it hits the floor.

c. Examine the results of both experiments. Is there the same pattern in the directions of *sum of the forces* that other objects exert on the ball and in the directions of the $\vec{v}$ arrows as in the previous activity? Is there the same pattern in the directions of the sum of the forces and the directions of the $\Delta\vec{v}$ arrows on the motion diagrams? If so, describe the patterns.

d. Use the patterns that you found in Activity 2.1.3 and in this activity to formulate a statement relating the direction of the sum of the forces exerted on an object by other objects and one or more of the kinematics quantities that describe its motion.

Testing a relationship

You have now devised a relationship between the sum of all external forces that other objects exert on an object of interest and some kinematics quantity used to describe its motion in a motion diagram. Use that proposed relationship to predict the outcome of a new experiment. If the outcome of the experiment does not match the prediction, you may have rejected the relationship. If the outcome of the experiment does match the prediction, support builds for the relationship.

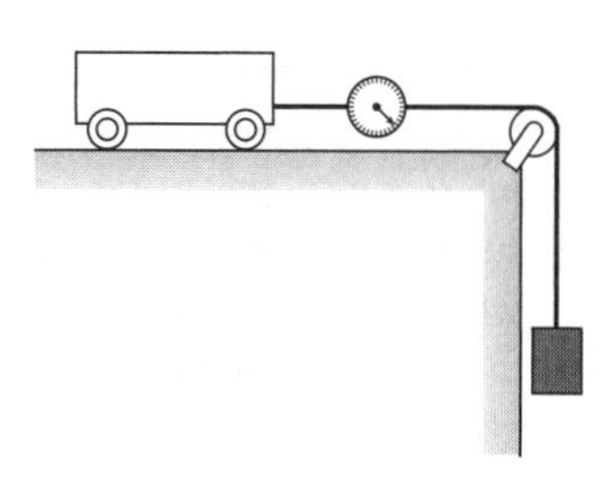

2.1.5 Test your idea You have a cart on a smooth track connected to a scale, which is then connected by a string that passes over a pulley to a hanging object. At first you are supporting the hanging object with your hand and then you let it go. Use the relationship between force and motion that you constructed in the activities above to predict what will happen to the reading on the scale as you remove your hand from the hanging object. Think of what object you need to choose as your system for analysis: the cart or the hanging object.

Did the outcome of the experiment match your prediction? If not, do you need to revise the analysis or revise the idea?

2.1.6 Test an idea Aaron says that an object always moves in the direction of the sum of the forces exerted on it by other objects. Design an experiment to test his idea. Make a prediction of the outcome of the experiment based on Aaron's idea. What additional assumptions are you making? (An assumption is something that you accept as true.) Perform the experiment and record the outcome. What can you say about the idea? What can you say about assumptions?

In 2.1.6 think of a force as a continuous interaction, and thus in your experiments try to find a way to exert continuous force for a while to see the changes in motion.

2.1.7 Design an experiment Design an experiment to test the following idea: "When you place a heavy object on a table, the table does not exert any forces on that object, it is merely in the way." You have a spring scale, a platform scale, a foam cushion, and a small heavy object. You can use other materials if you wish, such a laser pointers, scotch tape, sheets of cardboard, etc. Decide how you will know whether the support object exerts any forces on the object of interest in your experiments. Draw pictures of the experimental setup and predict the outcomes of your experiments based on the idea under test before you perform the experiments. Then conduct the experiments, record the outcome and write your judgment about the idea.

2.2 Conceptual Reasoning

2.2.1 Reason You stack three identical books on a table T. Book A is on the top, book B in the middle, and book C on the bottom.

a. Sketch the situation.

b. Choose book B as the system object of interest for a force diagram. Carefully draw a closed dashed line around book B indicating that it is the system—your whole focus of attention. The top of the line should go between the top of book B and the bottom of book A. The bottom of the line should go between the bottom of book B and the top of book C. This is very important as it allows you to visualize what external forces to include in a force diagram for the system object (book B in this case).

c. Below the sketch, construct a force diagram for book B. Remember in the force diagram to model the system object as a point object, to include arrows representing each object outside the system (external objects) that interact with the system object, and to label each arrow in the force diagram with a force symbol that includes two subscripts (the first for the object that exerts the force on the system object and the second for the system object on which that force is exerted). For example, the force that the book A exerts on book B can be written as $\vec{F}_{\text{A on B}}$. Try to make the lengths of the force arrows representative of the relative magnitudes of the forces.

2.2.2 Represent and reason Below you see four different situations and force diagrams for a circled system object for each situation. Draw a motion diagram for each system object. Describe the motion of the object in words. Then add the force arrows graphically in each force diagram to determine the direction of sum of the forces that other objects exert on each system object. Explain how the motion diagram and the sum of the forces are related.

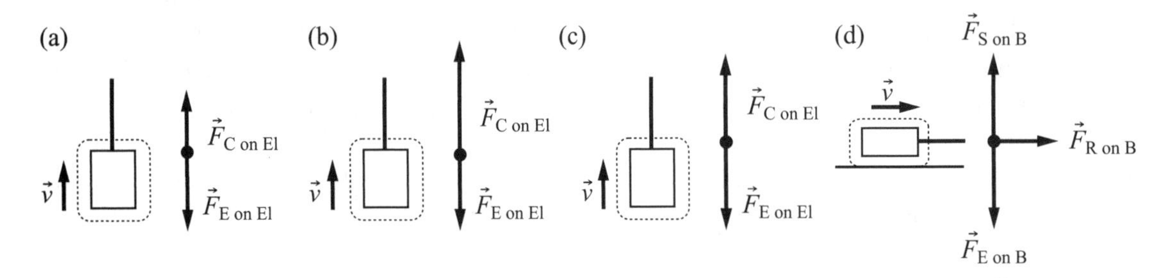

2.2.3 Reasoning in a worked example A paratrooper whose parachute did not open lands in a deep snow bank. He sinks about one meter into the snow while stopping. How does the magnitude of the force that the snow exerts on the paratrooper compare to the magnitude of the force that Earth exerts on him as he is sinking into the snow?

a. Make an enlarged sketch of the process (from the initial instant he first contacts the snow to the final instant he stops). Choose the trooper as the system object of interest for a force diagram, indicated by a closed dashed line around him in the sketch halfway through the process.

b. Construct a motion diagram for the trooper for this process. The motion diagram shows one velocity change arrow for simplicity (in fact the velocity change arrow might vary through the process).

c. Next to the motion diagram construct a force diagram for the trooper when he is about half way through the process. Model him as a point object, include arrows representing each external object that interacts with the trooper, and label each arrow in the force diagram with a force symbol that includes two subscripts. Ignore the force the air exerts on him.

d. Check to be sure the force diagram is consistent with the motion diagram—does the sum of the forces point in the same direction as the change in velocity and acceleration?

The $\Delta\vec{v}$ arrow points up and the sum of the forces $\Sigma\vec{F}_{\text{External on Trooper}}$ points up.

e. Finally, answer the question—compare the force magnitudes. Notice that the sum of the forces points upward, although the paratrooper moves downward—the direction of the sum of the forces and of the velocity are not the same!

Tip! The goal of all of the activities in this section is to help you draw both motion and force diagrams for the system of interest and make sure they are consistent with each other.

2.2.4 Represent and reason A friend drops a medicine ball so that it falls straight down. You catch it. Consider the motion of the ball from the instant it touches your hands (the ball's downward speed starts to decrease) until the instant it stops (your hands have moved lower with the ball while stopping it).

a. Sketch the process from the instant the ball touches your hands until the instant it stops. Choose the ball as the system object of interest.

b. Construct a motion diagram for the stopping process.

c. Construct three force diagrams for the ball for three different instants during the motion. Be sure each force diagram is consistent with the motion diagram.

d. What object exerts a bigger force on the ball as you are stopping it—you or Earth?

2.2.5 Represent and reason A hermit hangs from a rope that passes over pulleys and is attached to a wagon filled with supplies. As the wagon moves at increasing speed up the hill, the hermit moves with increasing speed downward (to get the wagon to his cabin at the top of the hill).

a. Sketch the process from the instant the hermit starts to fall until the instant just before the wagon gets up the hill and the hermit is somewhat lower. Choose the hermit as the system.

b. Construct a motion diagram for the hermit while his speed is increasing. The hermit's acceleration is constant.

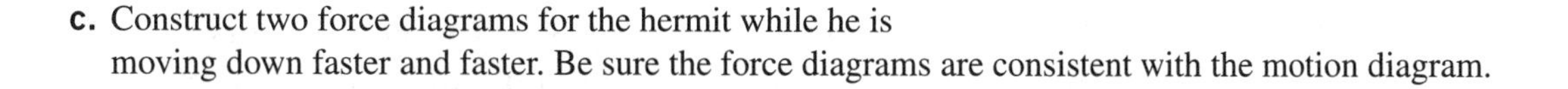

c. Construct two force diagrams for the hermit while he is moving down faster and faster. Be sure the force diagrams are consistent with the motion diagram.

d. How does the magnitude of the force that the rope exerts on the hermit $(F_{\text{R on H}})$ compare to the magnitude of the force that Earth exerts on the hermit $(F_{\text{E on H}})$?

2.2.6 Represent and reason An elevator starts at rest on the ground floor of a building and stops at the top floor. The elevator then returns to the ground floor. Construct a motion diagram and a consistent force diagram for each part of the downward trip to determine the relative magnitude of the force that the supporting cable exerts on the elevator $(F_{\text{C on El}})$ compared to the force that Earth exerts on the elevator $(F_{\text{E on El}})$. The motion diagram and the force diagram for each part should be consistent with each other and with the rule relating motion and forces developed in Activities 2.1.3 through 2.1.6. In short, for each part of the trip determine if: $F_{\text{C on El}} > F_{\text{E on El}}$, $F_{\text{C on El}} = F_{\text{E on El}}$, or $F_{\text{C on El}} < F_{\text{E on El}}$.

a. The elevator hangs at rest at the top floor.

b. Elevator starts moving downward at increasing speed.

c. Elevator reaches a constant downward speed.

d. Elevator moves at decreasing speed as it approaches the ground floor.

2.2.7 Represent and reason A person pulls a rope, which in turn pulls a crate across a horizontal rough surface (the surface exerts an opposing friction force). Three motion diagrams are shown below for the crate (with velocity arrows only). Construct two force diagrams for each type of crate motion and be sure to make the horizontal force arrows the correct relative lengths.

2.2.8 Evaluate the solution A crate sits on the flatbed of a truck that moves right at increasing speed. The crate does not slide but instead moves right with the truck. A friend draws the force diagram shown at the right for the crate. Is anything wrong with this diagram? Explain.

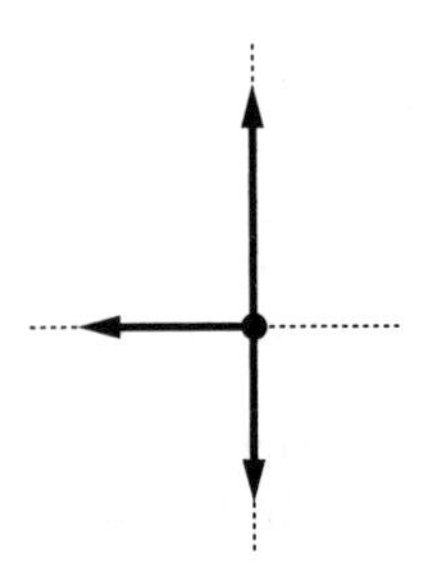

2.3 Quantitative Concept Building and Testing

2.3.1 Observe and find a pattern Consider the three experiments below, each observed by two different persons. For each experiment, draw a motion diagram and a force diagram on behalf of the specified observers and indicate which observer thinks that the rule relating motion diagram and force diagram holds and which observer finds that the rule does not apply. Based on this analysis, devise a new rule that restricts the type of observer that can use the rule without conflict.

a. The system object is a cart resting on a frame with small wheels; the cart sits on the smooth surface of a flatbed truck. Observer 1 is the passenger in the truck and observer 2 is the person on the street. The driver of the truck pushes the gas pedal. A passenger in the truck sees the cart start to roll back and fall off the back of the truck. A person on the street sees the cart remain stationary while the truck moves faster and faster to the right out from under it.

Motion and force diagrams for the cart

As seen by the passenger in the truck **As seen by the person on the street**

b. The system object is a light box with potted plants sitting on a flatbed cart that you are pushing relatively fast at a home supply store. Observer 1 is you and observer 2 is a person standing on the sidewalk. All of a sudden you see the plants slide forward off the cart. The person standing on the sidewalk sees the plants continue forward a constant speed as the cart is stopping.

Motion and force diagrams for the box of plants

As seen by you **As seen by the person on the sidewalk**

c. The system object is a small ball held by a person standing on a high diving board. Observer 1 is the diver and observer 2 is a spectator standing below the diving board. The diver steps off the board and simultaneously releases the ball. She observes that the ball stays at rest beside her as she falls toward the water. Observer 2 sees the ball moving faster and faster toward the water.

Motion and force diagrams for the ball

As seen by the diver **As seen by the spectator**

We learned in Section 2.1 that the sum of forces that other objects exert on a system object is in the same direction as the object's velocity change. Now we understand that this is only true if the observer is in the inertial reference frame. Newton's first law is a formal statement of that idea for the case where the sum of the forces is zero. We now see that the rule works only for observers in inertial reference frames. Next, we want to construct a quantitative relationship that can be used to analyze processes in which the sum of the forces exerted on a system object by other objects is not zero.

2.3.2 Observe and find a pattern Student *A* is wearing rollerblades and stands in front of a motion detector. The detector produces velocity-versus-time graphs. Student *B* (not wearing rollerblades) stands behind Student *A* and pushes her forward. Student *A* starts moving. Use the graph below to find a qualitative pattern between the change in Student *A*'s motion and the force exerted on her by Student *B*.

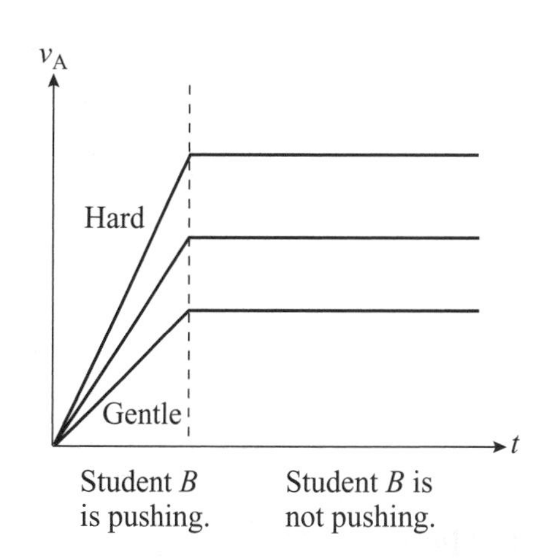

2.3.3 Observe and find a pattern Student *A* is wearing rollerblades and stands in front of a motion detector. The detector produces velocity-versus-time graphs for Student *A*.

Student B (not on rollerblades) stands behind Student A and pushes her forward. Student A wears a backpack that can hold textbooks. Student B pushes Student A several times; each time student A adds three more books to the backpack. Student B pushes with the same force each time.

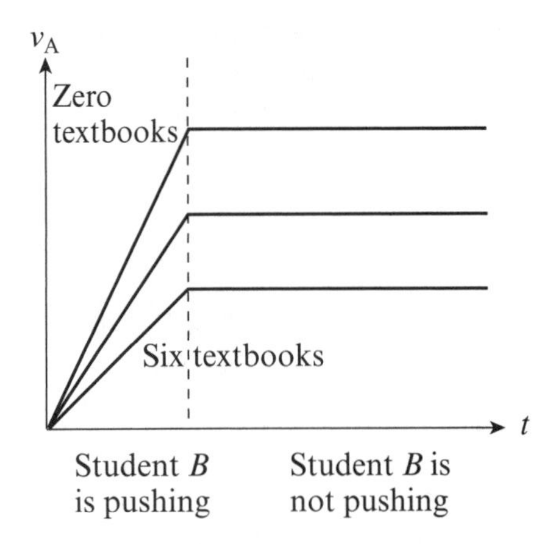

a. Use the graph to find a qualitative pattern between the change in Student A's velocity and the amount of material in her backpack.

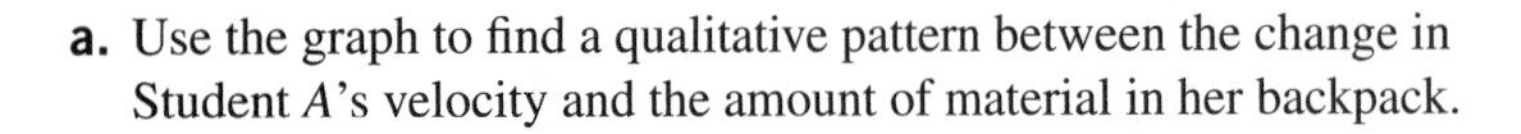

b. A physical quantity that is used to quantify "the amount of material" is called *mass*. How does the mass affect the change in velocity of an object if the external interaction is the same?

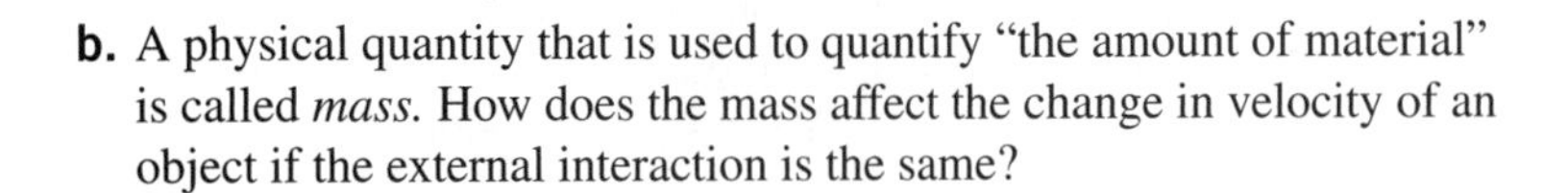

Next consider how the magnitudes and directions of the sum of external forces exerted on a system object affect the acceleration of that object.

2.3.4 Find a pattern Imagine an experiment in which one or more identical springs pull in the same direction on one or more identical carts on a smooth horizontal track. The springs are stretched the same amount so that each spring exerts the same force on the cart.

Experiment 1 cart FD	Experiment 2 cart FD

a. Draw a force diagram for the cart in Experiment 1 and another for the cart in Experiment 2 described below.

b. Then use the data in the table that follows to devise a relation that shows how the carts' acceleration depends on the carts' mass and on the sum of the forces exerted on the carts by the springs, Earth, and the track. Note: When doing such analysis, devise a relation for each independent variable one at a time and for the dependent variable (for example, use some of the data to see how the acceleration depends on the number of springs and then use other parts of the data to see how the acceleration depends on the number of carts). Then combine these relations to get a final relation.

Experiment number	Number of springs	Number of carts	Acceleration
1	0	1	0
2	1	1	1.03 m/s^2
3	2	1	1.98 m/s^2
4	3	1	3.03 m/s^2
5	4	1	3.95 m/s^2
6	1	1	1.03 m/s^2
7	1	2	0.51 m/s^2
8	1	3	0.32 m/s^2
9	2	2	1.02 m/s^2
10	2	3	0.66 m/s^2

The previous activity helped us develop an expression for the magnitude of the sum of all forces exerted on a system object, its mass, and the acceleration caused by the net force. How does the direction of the different external forces affect the acceleration?

2.3.5 Find a pattern Imagine springs attached to either end of a cart. The springs can pull the cart left and right. Each spring pulls with the same strength, but the number of springs on either side of the cart can vary.

a. Examine the data in the table that follows and draw a force diagram for the cart for each experiment (show the horizontal forces only—the upward force of the surface on the cart's wheels and downward force of Earth on the cart cancel).

Number of springs pulling to the right	Number of springs pulling to the left	Acceleration of the cart	Draw a force diagram; show horizontal forces only.
3	3	0	
1	2	$-1.03\ \text{m/s}^2$	
3	1	$1.98\ \text{m/s}^2$	
4	1	$3.03\ \text{m/s}^2$	
2	6	$-3.95\ \text{m/s}^2$	

b. Explain why we use negative signs in the acceleration column of the table.

c. Represent mathematically the relation between the object's acceleration, the sum of the forces exerted on it by other objects, and its mass.

d. Is this relation consistent with the relationship you devised in Activity 2.3.4? Explain.

2.3.6 Represent and reason Several force diagrams are shown in the table below. Apply the component form of Newton's second law in symbols for each diagram. The first two cells on the top left show an example of what to do.

Sketch and force diagram	Newton's second law in component form	Sketch and force diagram	Newton's second law in component form
$\vec{v}$ B m y, x, $\vec{N}_{\text{S on B}}$, $\vec{f}_{\text{S on B}}$, $\vec{F}_{\text{R on B}}$, $\vec{F}_{\text{E on B}}$	$a_x = \dfrac{+f_{\text{S on B}} + (-F_{\text{R on B}})}{m}$ $a_y = 0 = \dfrac{+N_{\text{S on B}} + (-F_{\text{E on B}})}{m}$	$\vec{v}$ B y, x, $\vec{N}_{\text{S on B}}$, $\vec{f}_{\text{S on B}}$, $\vec{F}_{\text{R on B}}$, $\vec{F}_{\text{E on B}}$	

(continued)

Sketch and force diagram	Newton's second law in component form	Sketch and force diagram	Newton's second law in component form
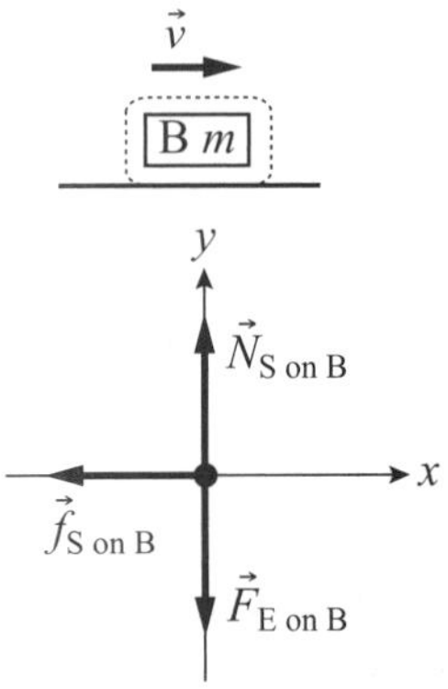 $\vec{v}$ B m y $\vec{N}_{\text{S on B}}$ x $\vec{f}_{\text{S on B}}$ $\vec{F}_{\text{E on B}}$		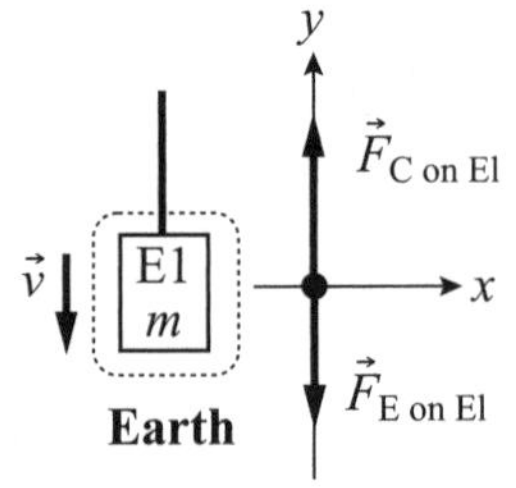 y $\vec{F}_{\text{C on El}}$ $\vec{v}$ El m x Earth $\vec{F}_{\text{E on El}}$	

2.3.7 Reason When using the component form of Newton's second law for a vertical motion problem (and for most problems), we need a method to determine the magnitude of the gravitational force that Earth exerts on an object of known mass or to determine the mass of an object for which we know the gravitational force that Earth exerts. In Chapter 1 you learned that all objects near Earth, when dropped or thrown, move with the same acceleration pointed downward, independently of their mass. Think of how this observational fact helps you determine the relation between the mass of the object and the force that Earth exerts on it.

2.3.8 Test your idea Design and carry out an experiment to test the relation between the mass of the object and the force that Earth exerts on it, which you devised in Activity 2.3.6. Summarize the data and conclusions.

2.3.9 Observe and explain Students *A* and *B* are wearing rollerblades. Student *B* pushes Student *A* abruptly. This causes Student *A*'s velocity to change (from zero to a final velocity toward the right). While pushing, Student *B* starts rolling in the opposite direction toward the left. Student *B*'s velocity change is always less in magnitude than Student *A*'s velocity change (Student *B* is a tall and muscular guy and Student *A* is a petite woman). How can you explain these observations?

2.3.10 Observe and explain You and your friend each hold a spring scale and you hook the scales to each other.

a. Let you and your friend take turns pulling on the scales. What do the scales read in each case? Make a table of data and record the results. For example, what does your friend's spring scale read when you pull yours to exert a force of 3 units? What does your scale read when she pulls with a force of 5 units?

b. Can you arrange it so that one of you pulls with 3 units and the other pulls with 5 units?

c. What is the pattern in the readings of the scales that you can infer from the experiments?

2.3.11 Observe and explain The graphs shown below were drawn from data gathered when two different carts collided. The carts have different masses and move toward each other at different speeds. A force probe on each cart measures the force that the other cart exerts on it during the collision.

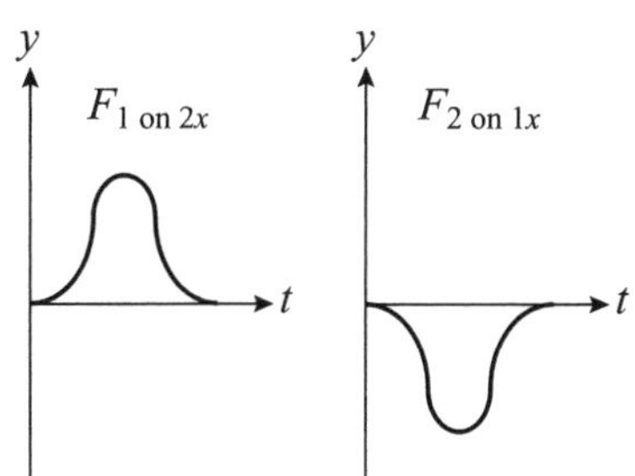

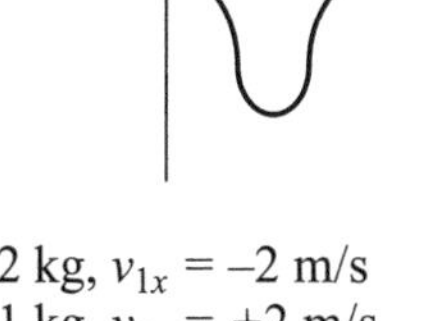

$m_1 = 2$ kg, $v_{1x} = -2$ m/s
$m_2 = 1$ kg, $v_{2x} = +2$ m/s

a.

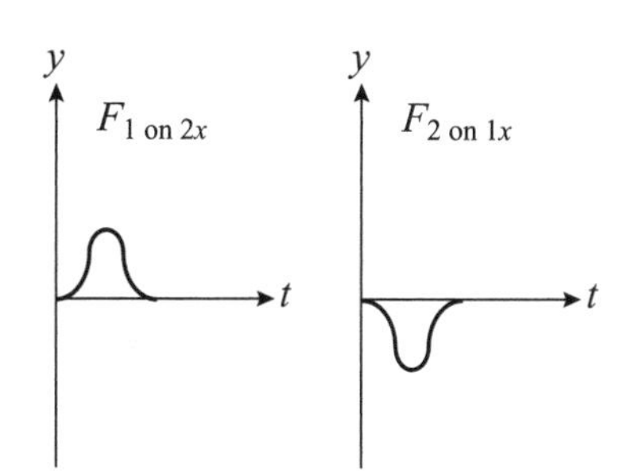

$m_1 = 1$ kg, $v_{1x} = -1$ m/s
$m_2 = 1$ kg, $v_{2x} = +3$ m/s

b.

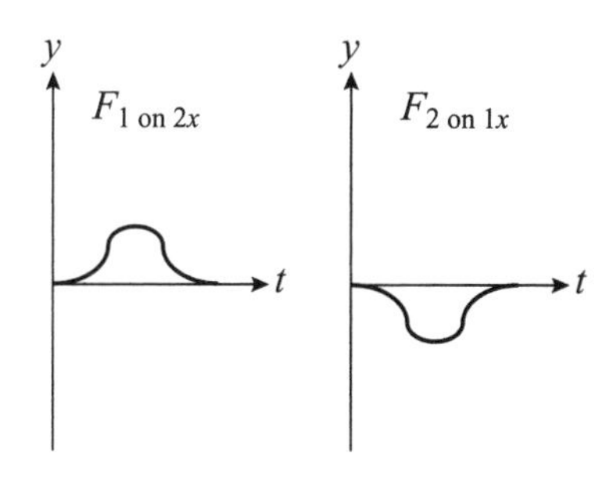

$m_1 = 2$ kg, $v_{1x} = -2$ m/s
$m_2 = 1$ kg, $v_{2x} = +1$ m/s

c.

a. Use the results provided in the illustration to devise a rule relating the force that cart 1 exerts on cart 2 and the force that cart 2 exerts on cart 1. Support your conclusion with evidence.

b. Does a heavier cart push more on a lighter cart than a lighter cart pushes on a heavier cart? Explain.

c. Does a faster cart push more on a slower cart than a slower cart pushes on a faster cart? Explain.

d. Are the results of this experiment consistent with the results of the experiment in 2.3.8? Explain.

2.3.12 Test your idea Design as many experiments as possible to disprove the rule that when two objects interact, they exert forces on each other that are the same in magnitude and opposite in direction. Use carts on a track and change their masses, speeds, and anything you can to disprove the rule. After you complete the experiments write the results.

2.4 Quantitative Reasoning

2.4.1 Represent and reason Complete the table.

Write a word description of the situation.	Sketch of the situation and circle the system object. Draw a motion diagram	Force diagram with perpendicular axes. Label the forces if needed.	Draw direction of acceleration and of net force. Are they consistent?	Apply Newton's second law in component form using symbols only.
An elevator is slowing down on its way up.		y x		

2.4.2 Represent and reason A book rests on a table.

a. Draw a sketch of the situation and identify objects that interact with the book.

b. Draw forces representing these interactions (a force diagram for the book).

c. If the book is stationary, these forces are equal in magnitude and opposite in direction. Can we say that they represent a Newton's third-law pair of forces? If not, why not?

d. Draw the Newton's third-law force pair for each force shown in the force diagram in (b) and identify the cause of each of these forces and the objects on which each of these forces is exerted.

2.4.3 Represent and reason A person pushes crate 1 which in turn pushes crate 2. The crates move faster and faster toward the right across a horizontal smooth surface. Make a motion diagram and a separate force diagram for crate 1 and for crate 2. Be sure to circle the object of interest and look for other objects interacting with it. Make the arrows in the force diagrams the correct relative lengths. Is it easier to push the crates in the order shown in the figure or in the opposite order?

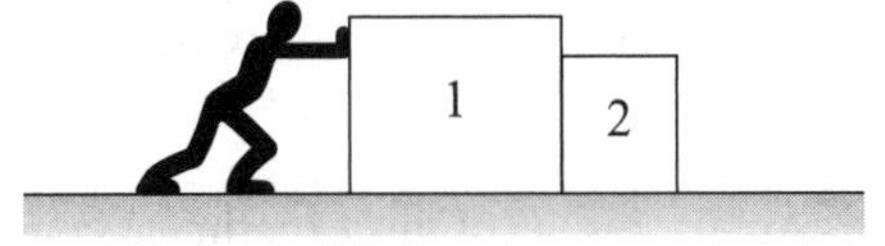

For the next problems use the problem-solving strategy for dynamics problems. Before you start solving any problem, make sure you read it at least three times and visualize the problem situation.

2.4.4 Regular problem The driver of a 1560-kg 4-door sedan traveling at 20 m/s on a level, paved road hits the brakes to stop for a red light. The car stops in 27 m. Determine the friction force that the road exerts on the car tires.

a. Construct an initial-final sketch of the situation. Choose a system object.

b. Model the system object as a point object and construct a motion diagram and a force diagram for it.

c. Use a kinematics equation to describe the process and the force diagram to help apply the horizontal component form of Newton's second law to construct another mathematical description of the process.

d. Use the mathematical descriptions to solve the problem. Evaluate the result to see if it is reasonable (unit, magnitude, and the value for limiting cases).

2.4.5 Evaluate the solution Identify any errors in the solution to the following problem and provide a corrected solution if there are errors.

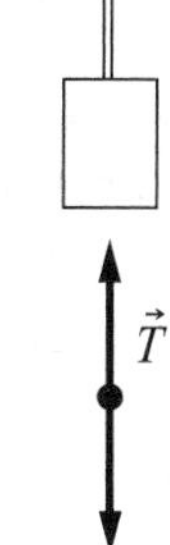

The problem: A 1000-kg elevator is moving down at 6.0 m/s. It slows to a stop in 3.0 m as it approaches the ground floor. Determine the force that the cable supporting the elevator exerts on the elevator as the elevator is stopping. Assume that $g = 10$ N/kg.

Proposed solution: The elevator at the right is the object of interest. It is considered a point object, and the forces that other objects exert on the elevator are shown in the force diagram. The acceleration of the elevator is:

$$a = v_0^2/2d = (6.0\text{ m/s})^2/2(3.0\text{ m}) = 6.0\text{ m/s}^2.$$

The force of the cable on the elevator while stopping is:

$$T = ma = (1000\text{ kg})(6.0\text{ m/s}^2) = 6000\text{ N}.$$

2.4.6 Evaluate the ideas Jim, Heather, and Minh disagree on what quantity a scale measures. Jim says it measures the force that Earth exerts on an object, Heather says it measures the force that the scale exerts on the object, and Minh says that scale measures the sum of the forces that Earth and the scale exert on the object. Describe an experiment that would test all three opinions and would have outcomes that might reject two of them.

2.4.7 Pose a problem Refer to the illustration at the right to pose a problem that can be solved using Newton's second law and kinematics. Make sure you provide enough information in your problem so that it can be solved.

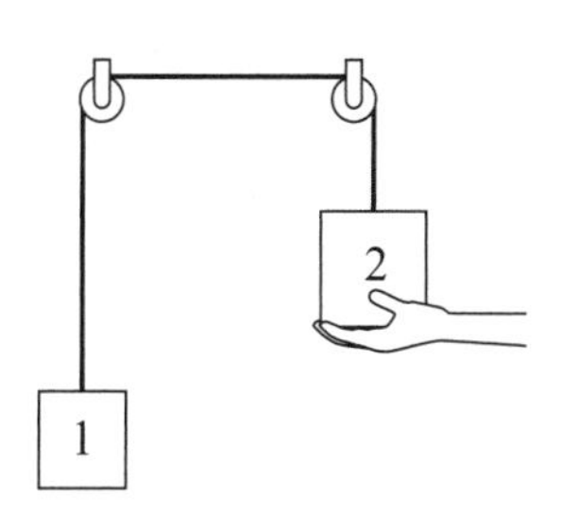

2.4.8 Reason The graph on the right shows the velocity of a book that jenny is pushing across a table. Draw five force diagrams for the book for selected clock readings (marked on the graph).

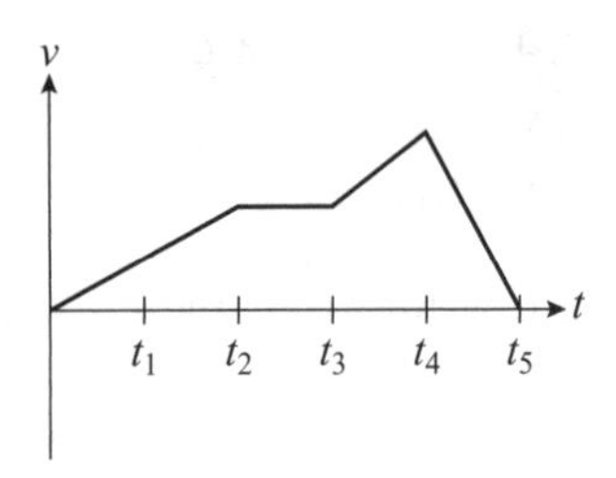

2.4.9 Represent and reason You throw a tennis ball upward.

a. Choose the ball as the system and draw a motion diagram and the force diagram for it for two instances: one when it is still in contact with your hand, the other when it is flying up.

b. Now choose your hand as the system and show the forces exerted on it by the ball for two instances: one when your hand is still in contact with it, the other when the ball is flying up.

c. Are the diagrams for the ball and the hand consistent? If not, what should you change? Did you apply Newton's third law answering these questions? Explain.

2.4.10 Test your reasoning Imagine you have the apparatus shown at the right. You hold the hanging object with your hand to prevent the cart from moving. To the left of the cart you place a motion detector that will record an acceleration-versus-time graph for the cart. First predict the shape of the graph if you remove your hand from the hanging object. Then predict the shape of the graph if you push the cart to the left and then remove your hand. Explain your prediction. Then perform the experiment and discuss whether it matched your prediction. What do you need to revise?

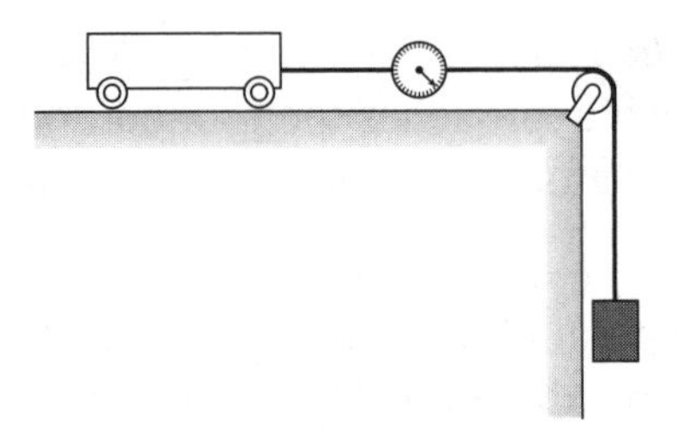

2.4.11 Reason Invent a story about motion represented on the acceleration-versus-time graph and draw force diagrams for the moving object for each of the accelerations.

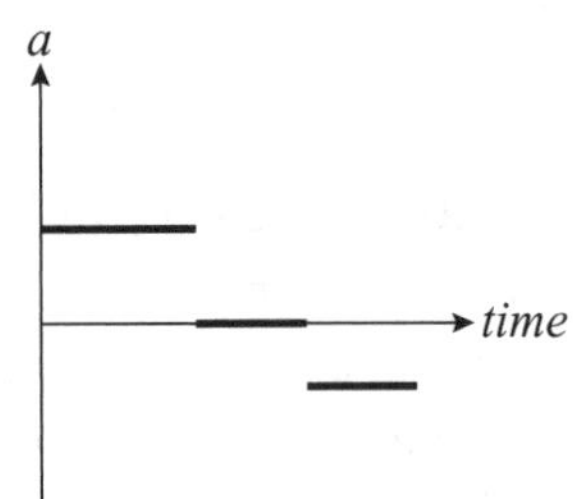

2.4.12 Design and analyze Otis Elevator Company is concerned that their elevators in an expensive hotel are running too slowly. They ask you to estimate the starting and stopping acceleration and the speed when moving at constant speed between stops. You tell them that you can determine the acceleration using only a bathroom scale. Fill in the table that follows to convince the elevator company that this method will work.

	Analysis when starting	Analysis when stopping
Write a description of your proposed starting and stopping experiments.		
Draw a labeled sketch of each process.		
Draw a force diagram (as needed) for each process.		
Write the physical quantities that you will measure and the quantities you will calculate.	To be measured: To be calculated:	To be measured: To be calculated:
Write the mathematical procedure you will use to determine the acceleration.		
List additional assumptions.		
List sources of experimental uncertainty and ways to minimize them.	Uncertainty: Ways to minimize:	Uncertainty: Ways to minimize:

3 Applying Newton's Laws

3.1 Qualitative Concept Building and Testing

3.1.1 Represent and reason A force diagram is a physical representation used by physicists while analyzing the forces exerted on objects by other objects. You learned to construct and use such diagrams in Chapter 2, primarily for motion along a single axis. Review the construction of such diagrams using a more complex situation. It involves forces in one plane and motion across a horizontal surface. In the example below a person is pulling a crate.

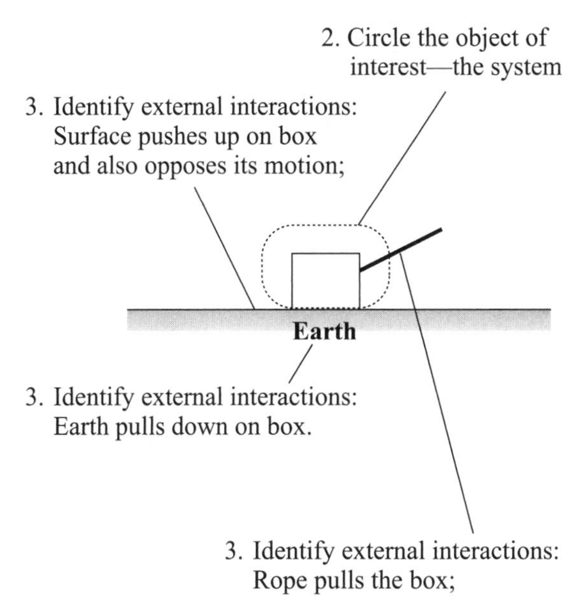

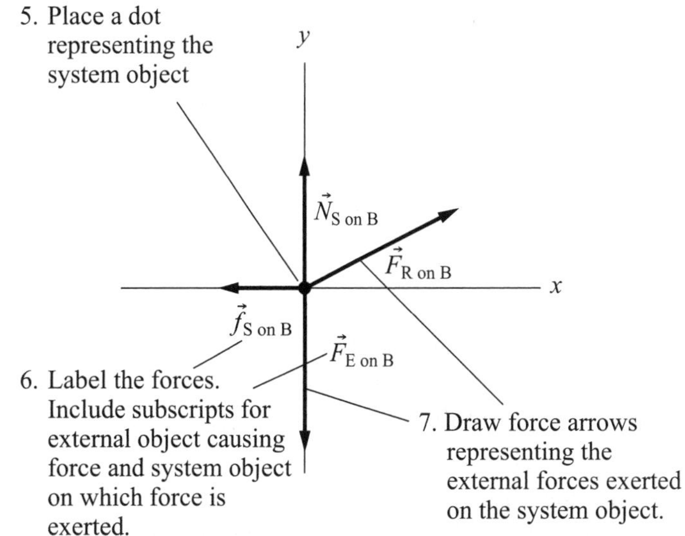

3.1.2 Reason

a. Use a piece of graph paper to redraw the force diagram from the skills box above. Measure the length units for each arrow in the vertical y direction and in the horizontal x direction. If an arrow points at an angle, measure its horizontal x length and its vertical y length.

b. Then decide whether all vertical units balance (add to zero) and all horizontal units balance (add to zero). If not, draw the direction of the unbalanced force. Discuss whether the result is reasonable.

3.2 Conceptual Reasoning

3.2.1 Reason The table below shows sketches of several situations. Construct a force diagram for the system object for each case. Be sure to label each force arrow in each diagram with two subscripts—the external object causing the force, and the system object. Make the arrow lengths perpendicular to direction of motion so they add to zero, because the object is not accelerating perpendicular to the direction of motion.

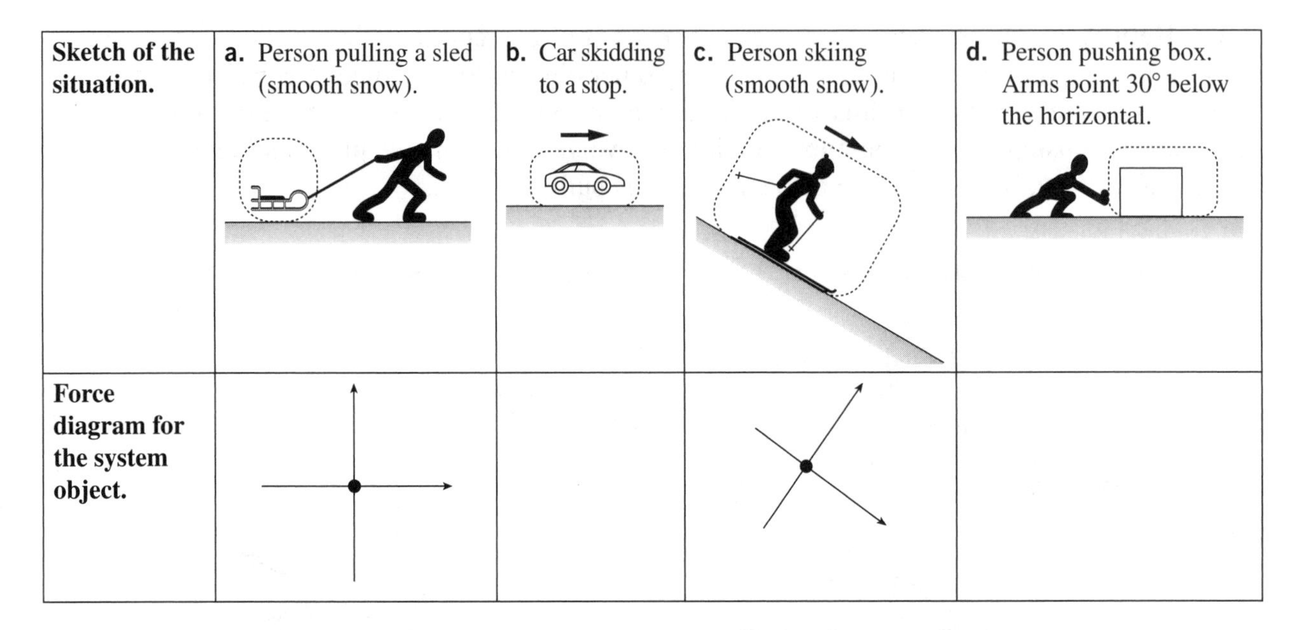

Sketch of the situation.	a. Person pulling a sled (smooth snow).	b. Car skidding to a stop.	c. Person skiing (smooth snow).	d. Person pushing box. Arms point 30° below the horizontal.
Force diagram for the system object.				

3.2.2 Diagram Jeopardy Unlabeled force diagrams for objects moving on a horizontal surface are shown below. For each instance, sketch and describe in words a process for which the diagrams might represent the forces that other objects exert on an object of interest. "S" stands for surface.

Force diagram (label the force arrows).	Sketch a situation consistent with the diagram.	Describe a process consistent with the diagram.
a.		
b.		

3.2.3 Diagram Jeopardy Unlabeled force diagrams for objects moving on an inclined surface are shown below. For each instance, sketch and describe in words a process for which the diagrams might represent the forces that other objects exert on an object of interest. "S" stands for surface.

Force diagram (label the force arrows).	Sketch a situation consistent with the diagram.	Describe a process consistent with the diagram.
a.		
b.		

3.2.4 Evaluate the solution A friend proposes that the force diagram at the right describes the forces exerted on a lawn mower during one instant while mowing the lawn; the mower is moving to the right. Could such a situation have occurred? If so, describe the situation and label the force arrows on the diagram. If not, explain why not.

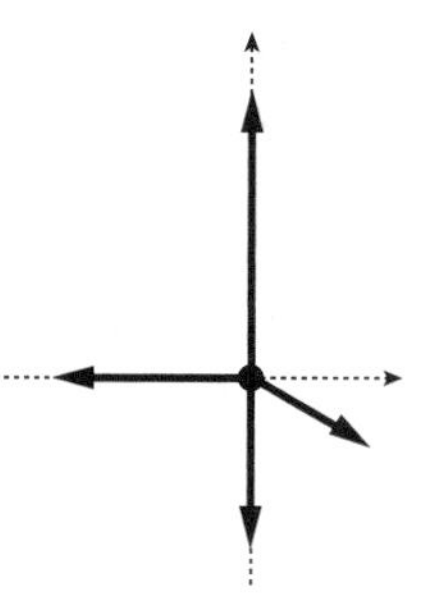

3.2.5 Represent and reason Two blocks are connected with a light rope. Another light rope is connected to block 2 and you pull it horizontally, exerting a force $\vec{F}_{\text{Y on R2}}$. The table and the blocks are smooth.

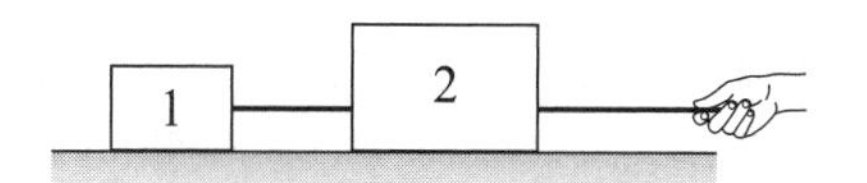

a. Fill in the table that follows.

Experiment	Draw a motion diagram.	List objects that interact with the system object.	Draw a force diagram and find the direction of the sum of the forces.
You pull the rope to the right exerting a force. Your system object is object 2.			
You pull the rope to the right. Your system object is object 1.			

b. Are your force diagrams consistent with the motion diagrams? How do you know?

c. Are your force diagrams consistent with Newton's third law? How do you know?

d. What assumptions did you make about the forces that the first rope exerts on both boxes?

3.2.6 Represent and reason Two blocks are connected with a light string that runs over two light pulleys. A hand supports the left block initially.

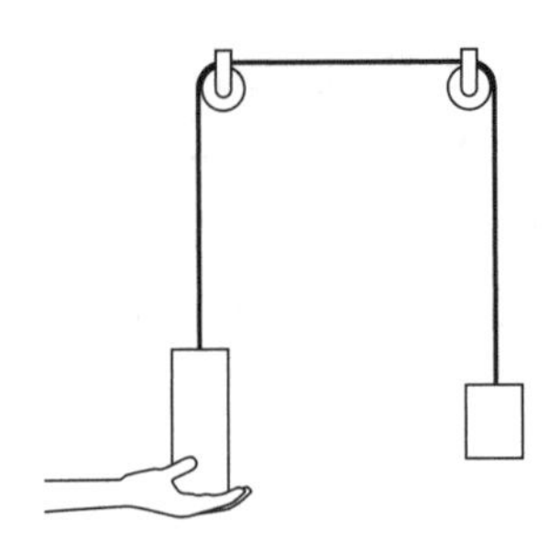

a. Fill in the table that follows.

Experiment	Draw a motion diagram.	List objects that interact with the system object.	Draw a force diagram and find the direction of the sum of the forces.
The left block is at rest while held by the hand. The left block is the system.			Direction of the sum of the forces:
The right block is at rest while the hand supports the left block. The right block is the system.			Direction of the sum of the forces:
The left block (the system) after the hand is removed.			Direction of the sum of the forces:
The right block (the system) after the hand is removed.			Direction of the sum of the forces:

b. List assumptions that you made about the magnitude of the force that the string exerts on both objects. How can you test whether those assumptions are valid?

3.3 Quantitative Concept Building and Testing

3.3.1 Components of force vectors The sketch below shows three strings pulling in different directions in a horizontal plane on a small ring (R) at the center. A force diagram for the ring is also shown on a grid. (a) Based on what you see in the diagram, explain why the ring does not accelerate in the positive or negative x-direction. Be explicit. (b) Repeat for the y-direction.

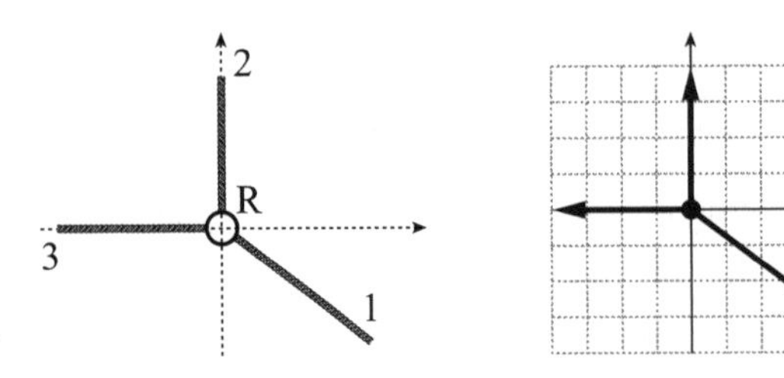

Comment Notice that string 1 exerts a 4-N force toward the right, which balances the 4-N force exerted by string 3 toward the left. Similarly, string 2 exerts a 3-N force upward, which is balanced by the 3-N downward pull exerted by string 1. If you don't see this, go back to the above diagram and try to visualize it. You should be able to realize that string 1 pulls in both the horizontal x direction and the vertical y direction. We say that $\vec{F}_{1\text{ on R}}$ has an x-component $F_{1\text{ on R}x} = +4$ N, and a y-component $F_{1\text{ on R}y} = -3$ N. Normally, we don't have force diagrams on grids that allow us to visualize the components so explicitly in this way. In the next activity, we will do the same analysis using trigonometry.

3.3.2 Components of force vectors The sketch on the right shows the same three strings pulling on the ring as in the previous activity. However, an angle is now shown for the pulling direction of string 1 relative to the x-axis. (a) How could you calculate the effect of string 1 pulling in the x-direction and (b) how could you calculate its effect pulling in the y-direction? That is, how could you calculate the x and y components of $\vec{F}_{1\text{ on R}}$ if you know only the magnitude of the force (5 N) and the direction of the force relative to the x-axis (37° below the positive x-axis)?

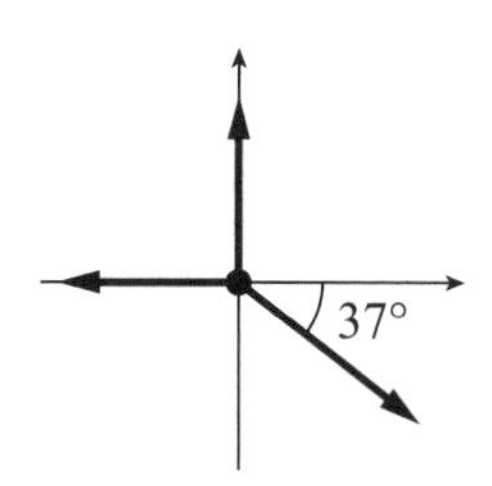

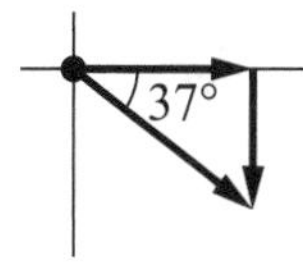

Comment Make a right triangle with force $\vec{F}_{1\text{ on R}}$ as the hypotenuse and the components as the sides of the triangle (see the figure on the left). Note that:

$$\cos 37° = F_{1\text{ on R}x}/F_{1\text{ on R}} = F_{1\text{ on R}x}/(5\text{ N})$$

$$\text{or } F_{1\text{ on R}x} = +F_{1\text{ on R}}\cos 37° = +(5\text{ N})\cos 37° = +4\text{ N}.$$

The x-component is positive because it points in the positive direction relative to the x-axis.

$$\sin 37° = F_{1\text{ on R}y}/F_{1\text{ on R}} = F_{1\text{ on R}y}/(5\text{ N}) \text{ or}$$

$$F_{1\text{ on R}y} = -F_{1\text{ on R}}\sin 37° = -(5\text{ N})\sin 37° = -3\text{ N}.$$

The y-component is negative because it points in the negative direction relative to the y-axis.

3.3.3 Represent and reason You are pulling a 20-kg sled up a smooth 20° snow slope. You are exerting a force of 170 N along the slope. What happens to the sled?

Draw a picture of the situation, label the known quantities.	**Draw a force diagram for the sled.**	**Orient the axes along and perpendicular to the slope and write Newton's second law in component form.**	**Write the expression for the acceleration of the sled. Does it make sense in the limiting case of 0° slope? 90° slope?**

3.3.4 Observe and find a pattern Put a wooden block of about 1 kg (can be replaced with a shoebox filled with sand) on a table. Attach a spring scale to the front of the wooden block (or the box). Pull the scale harder and harder. Observe carefully what happens to the scale reading. When does the block start moving? Record the scale reading just as the block begins to move, and then notice what happens to the scale reading just after it starts to move. Keep the block moving at constant speed and record the reading of the scale.

a. Fill in the table that follows by constructing a force diagram for the block (the system object) for these five situations.

The block sits on the table with no pulling.	The string pulls on the block, which does not start moving.	The string pulls harder but the block still does not move.	The block sits on the table at the moment just before it moves.	The string pulls the block at a slow constant velocity.

b. Describe in words how the magnitude of the horizontal force that the table surface exerts on the block varies in opposition to the force exerted on the block by the string pulling on it.

3.3.5 Observe and find pattern Instead of the block in Activity 3.3.4, you have rectangular blocks with different types of surfaces and different surface areas on which the block slides horizontally.

The force that the string exerts on the block (as measured by the spring scale reading) when the block just starts to slide is recorded in the table that follows. This force is equal in magnitude to the maximum static friction force (as we found in Activity 3.3.4). Examine the data in the table.

Mass of the block	Surface area	Quality of surfaces	Maximum static friction force
1.0 kg	0.1 m^2	Medium smooth	3.0 N
1.0 kg	0.2 m^2	Medium smooth	3.0 N
1.0 kg	0.3 m^2	Medium smooth	3.0 N
1.0 kg	0.1 m^2	A little rougher	4.0 N
1.0 kg	0.1 m^2	Even rougher	5.0 N
1.0 kg	0.1 m^2	Even more rough	7.0 N

Now decide how the maximum static friction force that the surface exerts on the block depends on the surface area of the block and on the roughness of the surfaces.

3.3.6 Observe and find a pattern

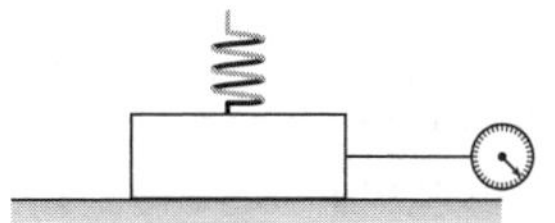

A spring scale pulls a 1-kg block over a medium smooth surface. The reading of the scale can be used to determine the magnitude of the maximum static friction force—in this instance, the force when the block starts to slide. In some experiments, a compressible spring also pushes vertically down on the block (see the second block).

a. Use the data in the table to draw a graph of the maximum static friction force versus the normal force exerted by the surface on the block.

Extra downward force exerted on the 1-kg block	Normal force exerted by the board on the block	Maximum static friction force exerted by the board on the block
0 N	10.0 N	3.0 N
5.0 N	15.0 N	4.5 N
10.0 N	20.0 N	6.0 N
20.0 N	30.0 N	9.0 N

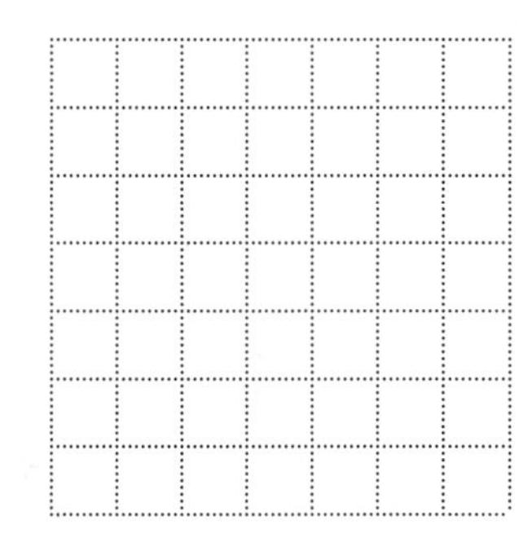

b. Express mathematically a relationship between the normal force and the maximum static friction force.

3.3.7 Test your idea Design two independent experiments to test whether the ratio of the maximum static friction force and the normal force exerted by the surface on an object is independent of the magnitude of the normal force. Describe the experimental set-up with words and pictures. Explain what data you will collect and how you will analyze them. Carefully describe how you will make the judgment about the relationship between the normal force and the maximum static friction force based on your experiments. (Do not forget the experimental uncertainties.)

3.4 Quantitative Reasoning

3.4.1 Components of force vectors A force diagram at the right shows the forces with which three strings pull on a ring. Determine all of the x components of the forces that the strings exert and all of the y components of the forces that the strings exert. Do the x and y components independently add to zero for the ring, which is not accelerating?

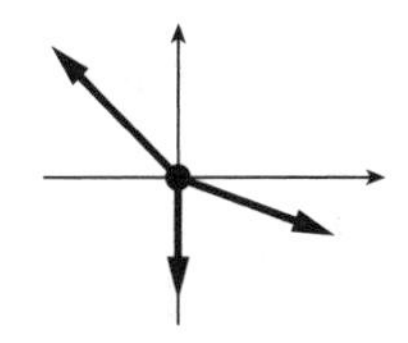

3.4.2 Components of force vectors A force diagram at the right shows the forces that other objects exert on a grocery cart being pushed on a smooth tile surface. Determine all of the x components of the forces that other objects exert on the cart, and all of the y components exerted on the cart. Add the forces in the vertical y direction and in the horizontal x direction. Is the cart accelerating? Explain.

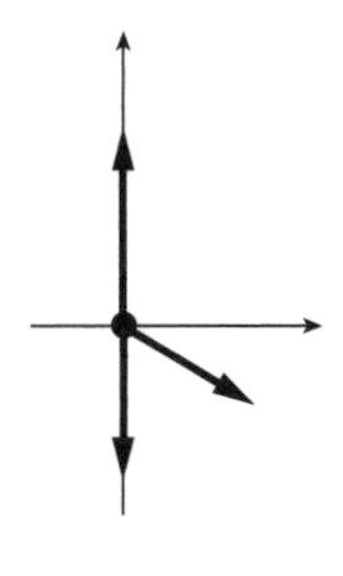

3.4.3 Represent and reason Complete the table that follows.

Write a word description of the situation.	Sketch of the situation and circle the system object.	Force diagram with perpendicular axes. Label the forces if needed.	Draw a motion diagram. Are the motion diagram and force diagram consistent?	Write Newton's second law in component form.
	30°	y, x, $\vec{N}_{\text{S on O}}$, $\vec{T}_{\text{R on O}}$, 30°, $\vec{F}_{\text{E on O}}$		
		y, x, $\vec{N}_{\text{S on O}}$, 60°, 30°, $\vec{F}_{\text{E on O}}$		
	1, 2	y, x, $\vec{T}_{\text{R on 1}}$, $\vec{F}_{\text{E on 1}}$; y, x, $\vec{T}_{\text{R on 2}}$, $\vec{F}_{\text{E on 2}}$		

3.4.4 Test your ideas In the preceding activity you analyzed the system that has two objects of different masses connected by a string going over two pulleys. Imagine you have a similar setup with the object on the left having a mass of 150 g and the object on the right, 200 g. You hold the left object on the table; the right one is about 1 meter above the table. Use your knowledge of Newton's laws and kinematics to predict how long it would take the object on the right to move through the distance of 80 cm if you let go of the object on the table.

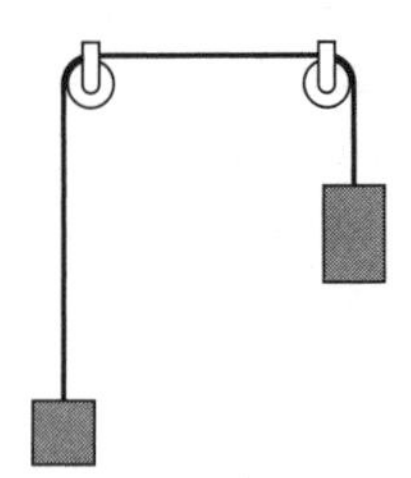

3.4.5 Regular problem Two of your neighbor's children (40 kg together) sit on a sled. You push on the back child exerting a 50-N force directed 37° below the horizontal. The sled slides forward at constant velocity. Determine the coefficient of kinetic friction between the snow and the sled. Use all the elements of the problem solving strategy.

3.4.6 Regular problem Determine the readings of all of the scales shown in the sketches in the table that follows. Each scale indicates the magnitude of the force that the rope exerts on the hanging block. That force is not changed as the mass-less rope passes around a pulley (the pulleys are frictionless and light; their purpose is to redirect the string). The force exerted by Earth on the blocks is 50 N.

Sketch the problem situation.	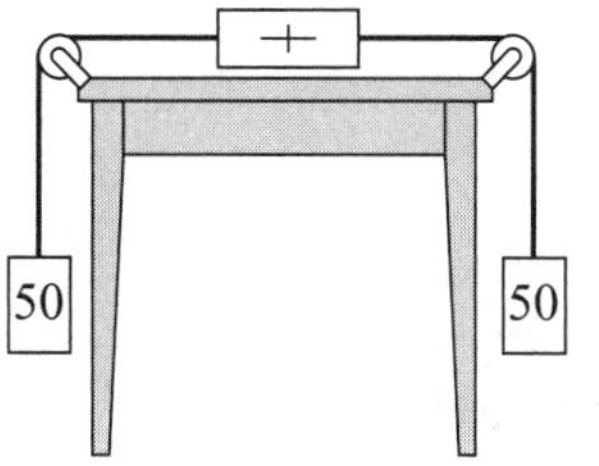 50 50	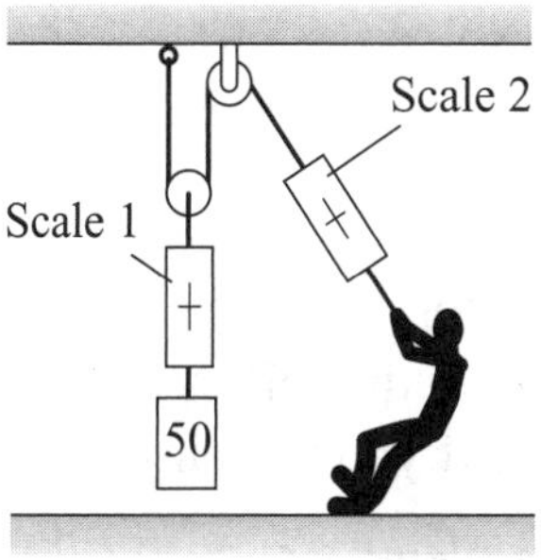 Scale 2 Scale 1 50
Draw a force diagram for a chosen system/systems. Remember the perpendicular *x* and *y* axes.		
Apply Newton's second law in component form (*x* and *y* axes) to the situation shown in the force diagram.		
Solve the equations for an unknown quantity and evaluate the results to see if they are reasonable.		

3.4.7 Equation Jeopardy Envision one process that the equations below might describe (there are many possibilities). Assume that $g = 10\text{ m/s}^2 = 10\text{ N/kg}$. (Note: Here N stands for the unit Newton and *N* stands for the normal force.) The object is on an incline.

x: $$0 + \left[-(100\text{ kg})(10\text{ N/kg})\cos 80°\right] = (100\text{ kg})a_x$$

y: $$N + \left[-(100\text{ kg})(10\text{ N/kg})\sin 80°\right] = (100\text{ kg})\,0$$

$$0 - 16\text{ m/s} = a_x t$$

Construct a sketch of a process the equations might describe and write in words a problem for which the equations might be used.

3.4.8 Equation Jeopardy Envision one process that the equations below might describe (there are many possibilities). Assume that $g = 10\text{ m/s}^2 = 10\text{ N/kg}$. (Note: Here N stands for the unit Newton and N stands for the normal force.) The object is on a level surface.

x: $$(100\text{ N})\cos 37° + 0 + (-0.40\,N) + 0 = (10\text{ kg})a_x$$

y: $$(100\text{ N})\sin 37° + N + 0 + [-(10\text{ kg})(10\text{ N/kg})] = (10\text{ kg})0$$

$$v_f - 0 = a_x(5.0\text{ s})$$

Construct a sketch of a process the equations might describe and write in words a problem for which the equations might be used.

3.4.9 Pose a problem Refer to the illustration at the right to pose a problem that can be solved using Newton's second law and kinematics. Make sure you provide enough information in your problem so that it can be solved.

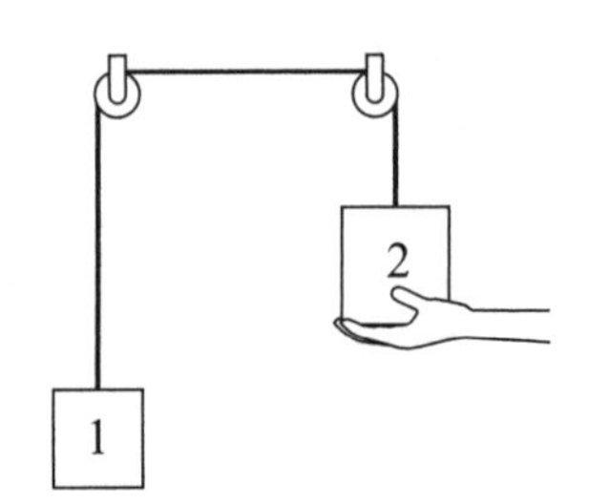

3.4.10 Evaluate the solution Identify any errors in the solution to the following problem and provide a corrected solution if you find any.

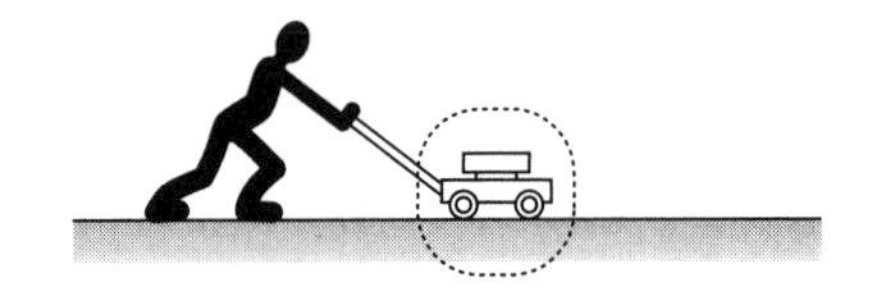

The problem: You push a 20-kg lawn mower, exerting a 100-N force on it. You push 37° below the horizontal. The effective coefficient of kinetic friction between the grass and mower is 0.60. Determine the acceleration of the lawn mower. Assume that $g = 10\text{ m/s}^2$.

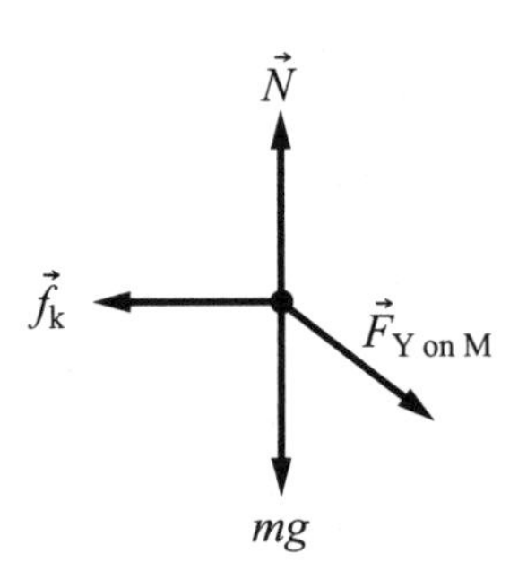

Proposed solution: The situation is pictured above at the right. The mower is the object of interest and is considered a particle. The forces that other objects exert on the mower are shown in the force diagram. The magnitude of the kinetic friction force is:

$$f_k = \mu kN = 0.60(20\text{ kg})(10\text{ m/s}^2) = 120\text{ N}.$$

The acceleration of the mower is:

$$a = (F - f_k)/m = (100\text{ N} - 120\text{ N})/(20\text{ kg}) = -1.0\text{ m/s}^2.$$

3.4.11 Design an experiment You are to determine the maximum coefficient of static friction between your show and floor tile in two different ways. You have the following equipment: the shoe, a spring scale, the floor tile, and a meter stick.

a. Devise a first method using the spring scale as your only measuring instrument. Include a sketch of your proposed method, a force diagram, and a detailed mathematical description that can be used to get a quantitative answer to the problem.

b. Devise a second method using the meter stick as your only measuring instrument. Include a sketch of your proposed method, a force diagram, and a detailed mathematical description that can be used to get a quantitative answer to the problem.

c. Compare the outcome of the two methods. Do they agree within expected uncertainties? Explain.

4 Circular Motion

Part A: Kinematics and Dynamics of Circular Motion

4.1 Qualitative Concept Building and Testing

4.1.1 Observe Perform or observe the following three experiments. For each experiment, fill in the blanks in the table that follows.

a. Roll a bowling ball along a smooth floor. As the ball moves, tap it with a rubber mallet, trying to make it move in a circle. The top-view diagram shows your successful attempt.

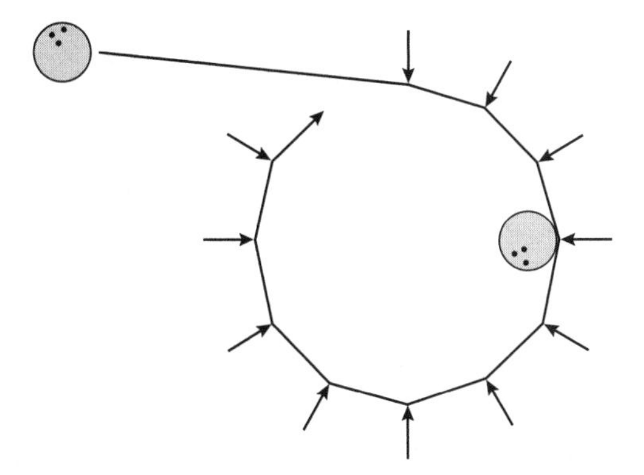

b. Swing a bucket attached to a rope and filled with water or sand at constant speed in a horizontal circle.

c. Imagine that Christine is wearing rollerblades; she hangs on to one end of a rope and a friend holds securely the other end. A third person pushes strongly on Christine and she moves in a straight line. When the rope becomes taut, Christine starts moving in a circle. As Christine moves at constant speed in the circle, her friend holding the rope turns, always pulling in on her.

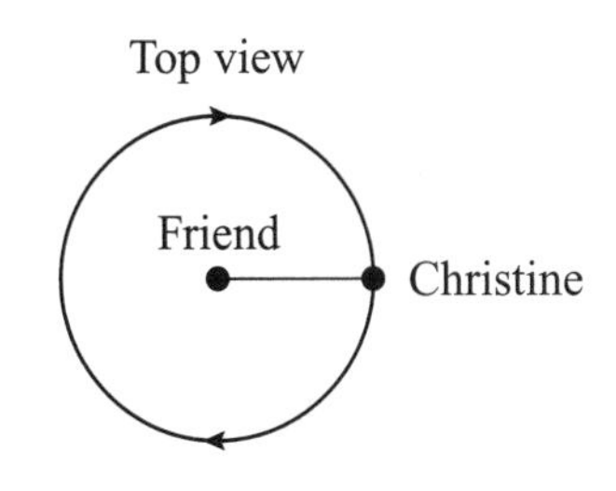

Fill the table below; assume that the friction forces exerted on all three objects are negligible.

Experiment; the circling object is in bold.	List objects that interact with the circling object.	Draw a top view force diagram for the circling object.	List forces or force components that balance.	Indicate the direction of the sum of the forces exerted on the object.
a. Tapping a **bowling ball.**				
b. Swinging a **bucket** in a horizontal circle.				
c. Pulling a rope attached to a moving **rollerblader** so she moves in a circle.				

4.1.2 Find a pattern Review your analysis recorded in the table for Activity 4.1.1. Based on your observations and on the analysis, find a pattern for the direction of the sum of the forces exerted on an object moving at constant speed in a circle.

Pattern:

4.1.3 Explain Devise one or more explanations for the pattern in the direction of the sum of the forces exerted on an object.

Explanation:

Top view

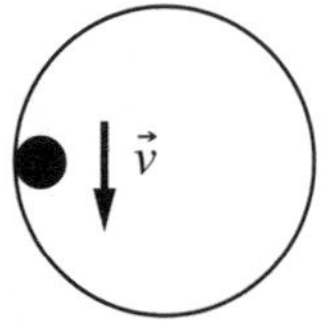

4.1.4 Test your explanation For the following two testing experiments, use the pattern that you formulated in Activity 4.1.2 and the explanation you formulated in 4.1.3 to predict the outcome of the experiment.

a. Inside a metal ring, roll a small ball or a marble on a smooth surface. Is the motion of the ball consistent with the pattern formulated in Activity 4.1.2? Explain.

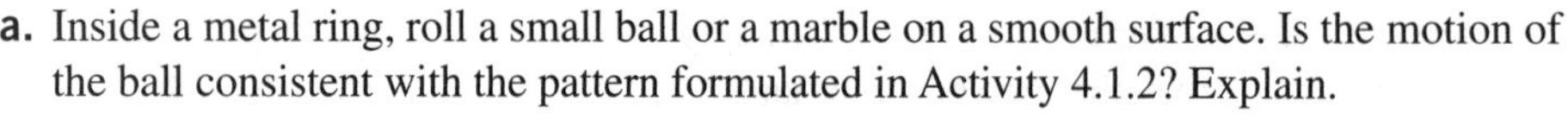

Later when loop is open

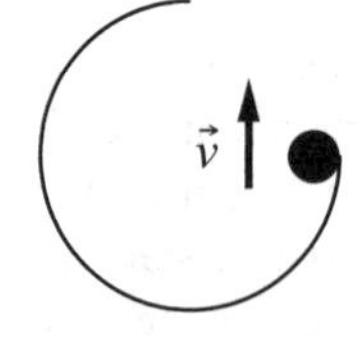

b. Predict what will happen to the ball if after the ball rolls for a couple of turns, you remove a quarter of the ring, as shown in the figure. Justify your prediction in words and with a force diagram.
Prediction:

c. After you make your prediction, perform the experiment to check the outcome.
Outcome and judgment:

4.2 Conceptual Reasoning

4.2.1 Represent and reason An object moves at constant speed in a circle.

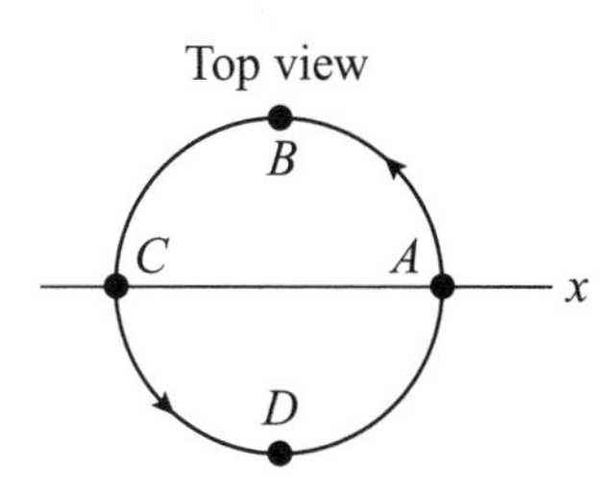

a. Determine its acceleration direction at each of the four positions shown in the illustration.

b. Is there a pattern in the acceleration directions? If so, what is it?

Pattern:

4.2.2 Explain Explain how the pattern you found in activity 4.2.1 accounts for the patterns you found in Activity 4.1.2. Does your explanation for why the sum of the forces exerted on an object moving in a circle at constant speed points toward the center of the circle match the one you constructed in Activity 4.1.3? If not, which one needs to be revised?

Explanation and revision:

4.2.3 Represent and reason Imagine that a golf cart moves on a level path around a curve shown in the top-view diagram to the right. For each situation below, use the graphical velocity subtraction method to estimate which arrow is closest to the direction of the cart's acceleration when passing the midpoint P.

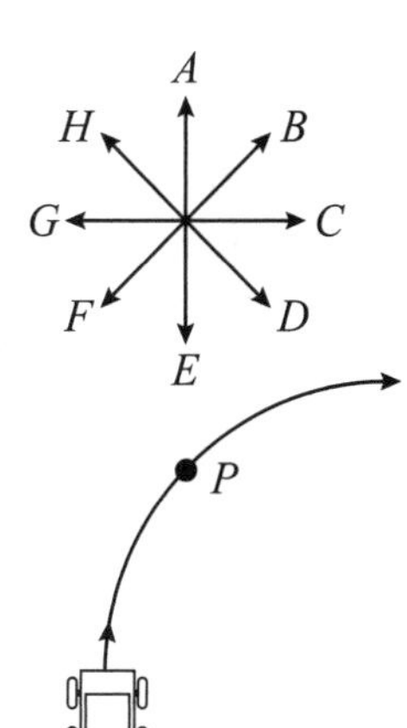

a. The cart moves at constant speed.

b. The cart's speed is decreasing.

c. The cart's speed is increasing rapidly.

4.2.4 Observe and find a pattern Observe a pendulum swinging in a vertical arc.

a. What can you say about its speed at different points; its velocity?

b. Use the velocity technique to determine the direction of the acceleration of the pendulum bob at four locations: at the top of the swing on the right; at some intermediate point when the pendulum is moving down; at the bottom of the swing; and at some point on the left before the pendulum reaches the highest point.

c. What patterns do you notice? Does the acceleration of the pendulum bob always point towards the center of circle? Can you explain why it does not?

d. Explain why the acceleration points straight up at the bottom of the swing.

4.2.5 Test your ideas Now that you know how to determine the direction of the acceleration of the pendulum bob at the bottom of the swing, use everything you learned about circular motion so far to make a prediction of the outcome of the following experiment:

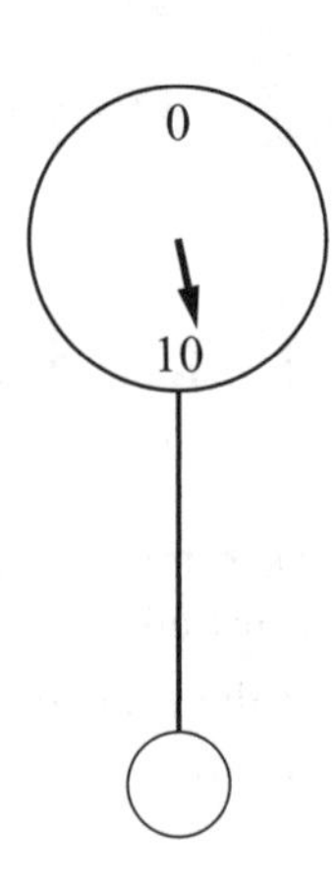

a. A 1.0-kg ball hangs from a 1.0-m long string. The other end of the string is attached to a Newton force measuring scale. The string pulls up on the ball exerting a 9.8-N force and the string and ball in turn pull down on the scale exerting a 9.8-N force—the scale reads 9.8 N. Imagine that you pull the ball to the side and release it so that the ball swings like a pendulum at the end of the string. Predict the scale reading as the ball passes directly under the scale (i.e., is it more, less or equal to 9.8 N?).

b. Perform the experiment; record the outcome and compare it to the prediction. Did the outcome support the pattern?

4.2.6 Represent and reason A battery-powered toy car moves at constant speed across the top of a circular hump, as shown in the illustration to the right. Fill in the table that follows.

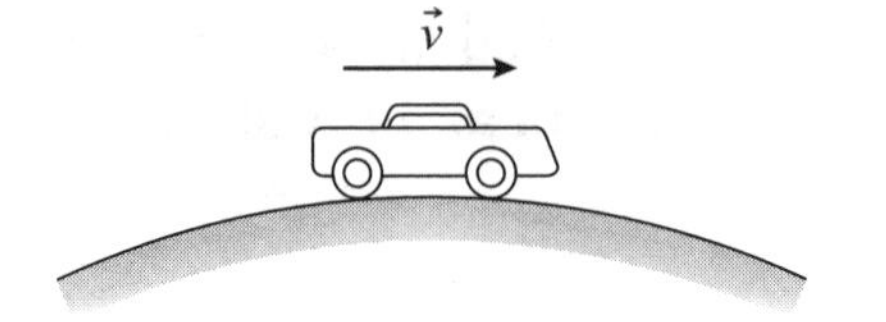

Indicate the direction of the acceleration of the car at the top of the hump.	Draw a force diagram for the car when passing across the top of the hump. Make the force arrows the correct relative lengths.	Explain in words whether the results of the first two cells of this table are consistent with Newton's second law.

4.3 Quantitative Concept Building and Testing

4.3.1 Observe and find a pattern Imagine three small toy cars travel at constant speed in identical-radii horizontal circular paths (top view is shown below). Car A moves at speed v, car B at speed $2v$, and car C at speed $3v$. Use the graphical method that you learned above to determine how the magnitude of the acceleration of the cars depends on their speeds. Remember that acceleration is $\Delta\vec{v}/\Delta t$ and that you need to compare the velocity change $\Delta\vec{v}$ vectors for the three speeds and also the time interval Δt needed for the velocity changes in each of the three cases.

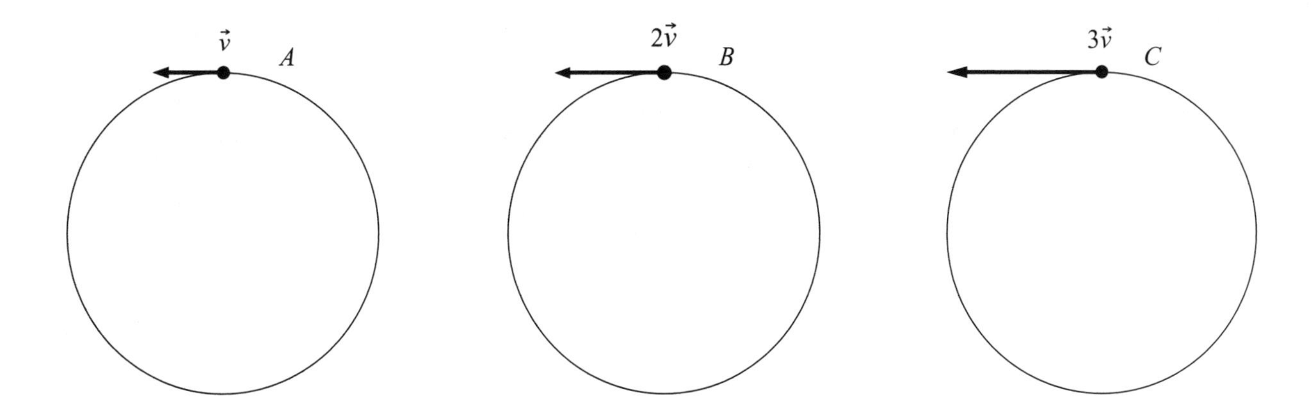

4.3.2 Observe and find a pattern Two small toy cars travel at the same constant speed in horizontal circular paths (top view is shown below). Car I moves in a circle of radius r and car II in a circle of radius $2r$.

a. Use the graphical velocity change method to determine how the magnitude of the acceleration of the cars depends on the radii of the circles. Do not forget to consider the time intervals needed for the velocity changes.

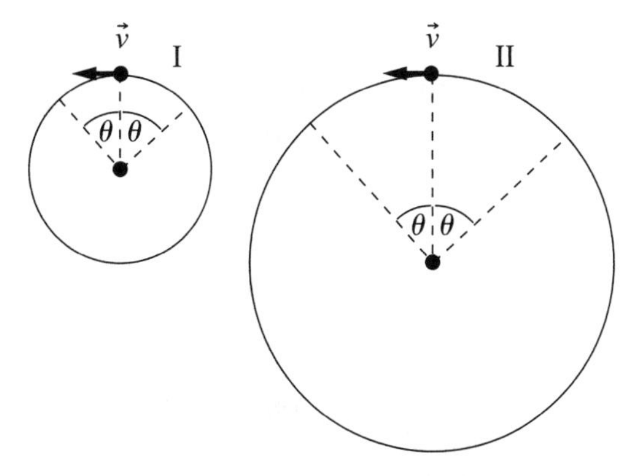

b. Combine the results from Activities 4.3.1 and 4.3.2a to write a general expression for the magnitude of the acceleration during constant speed circular motion.

4.3.3 Test the relation A 200-g ball traveling at speed 2.28 m/s swings at the end of a string that passes through a hollow tube down to another hanging block. The slanted part of the string that is moving in a circle is 0.50-m long and makes an angle of 37° with the horizontal.

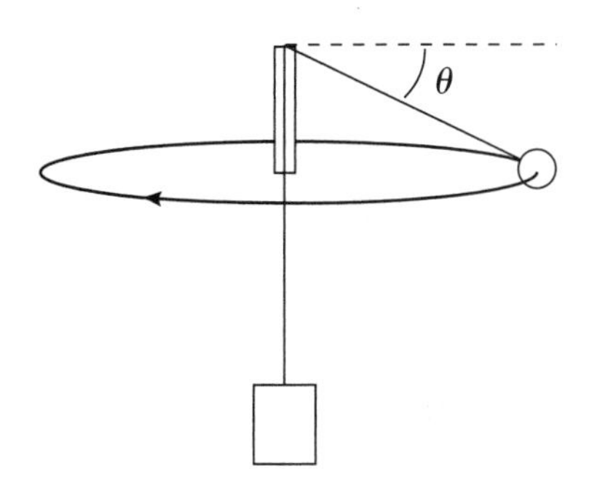

a. Fill in the table that follows to predict the mass of the hanging block that will prevent the string from moving up or down in the glass tube.

Draw a force diagram.	**Apply the vertical *y*-component form of Newton's second law.**	**Apply the radial *r*-component form of Newton's second law.**	**Use the results of your analysis to predict the mass of the hanging block.**
Ball:	Ball:	Ball:	Block mass:
Block:	Block:		

b. Now perform the experiment. Observe that the mass of the hanging block that prevents the string from going up or down is about 0.35 kg. Compare your prediction with this observation. What might have caused any discrepancy? (Hint: Think of the assumptions that you made.)

4.3.4 Test the relation Two objects of mass 500-g and 100-g with identical bottom surfaces are placed on a rotating platform that can turn with increasing speed. Fill in the table that follows.

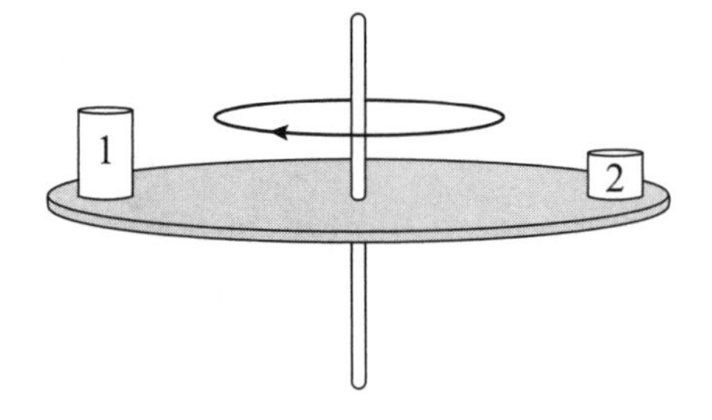

Draw force diagrams.	**Apply the vertical *y*-component form of Newton's second law.**	**Apply the radial *r*-component form of Newton's second law.**	**Use the results of your analysis to predict which object flies off first or if they fly off at the same time.**
Object 1:	Object 1:	Object 1:	
Object 2:	Object 2:	Object 2:	

Was your prediction consistent with the outcome of the experiment? Explain. What might have caused any discrepancy? (Hint: Think of the assumptions that you made.)

4.4 Quantitative Reasoning

4.4.1 Represent a process in multiple ways Some roller coasters have loops through which the cars travel. Car wheels and rails are constructed so that the cars cannot leave the track. For each situation described and sketched in the left column of the table below, fill in the rest of the table for the instant shown.

Roller coaster situation; circle the system.	Indicate the direction of $\vec{a}_c$.	Draw a force diagram.	Apply $ma_c = \Sigma F_{radial}$.
a. The roller coaster car glides at constant speed along a frictionless, level track.			
b. The roller coaster car moves past a frictionless circular dip in the track. $\vec{v}$			
c. The roller coaster car moves inverted past the top of a frictionless loop-the-loop. $\vec{v}$			

4.4.2 Regular problem Suppose the loop in a roller coaster has a 16-m diameter. How fast must the roller coaster car move across the top of the loop so that the force that the seats exert on its riders is half the force that Earth exerts on them? Be sure your solution includes the problem solving steps.

4.4.3 Car travels around a bend in the road Your 1600-kg mid-sized car moves around a level 50-m radius bend in the road. The coefficient of static friction between the car tires and the road is 0.80. Use the problem solving steps.

4.4.4 Equation Jeopardy 1 A situation involving circular motion is described mathematically below.

$$900\text{ N} - (50\text{ kg})(9.8\text{ m/s}^2) = (50\text{ kg})v^2/(12\text{ m})$$

a. Construct a sketch of a situation the equation might describe.

b. Write in words a problem for which the equation could be a solution.

4.4.5 Equation Jeopardy 2 A situation involving circular motion is described mathematically below.

$$200\text{ N} + (50\text{ kg})(9.8\text{ m/s}^2) = (50\text{ kg})v^2/(12\text{ m})$$

a. Construct a sketch of a situation the equation might describe.

b. Write in words a problem for which the equation could be a solution.

4.4.6 Evaluate the solution Identify any missing elements and errors in the solution to the following problem. Provide a corrected solution if you find missing elements or errors.

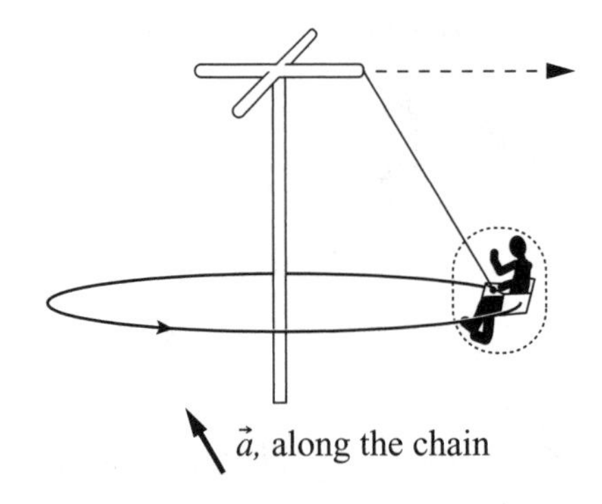

The problem: 80-kg Samuel rides at a constant 6.0-m/s speed in a horizontal 6.0-m radius circle in a seat at the end of a cable that makes a 50° angle with the horizontal. Determine the tension in the cable. Assume that $g = 10$ N/kg.

Proposed solution: The situation is pictured in the illustration.

We simplify by assuming that Samuel, the system, is a particle.

A force diagram for Samuel is shown at the right, along with the acceleration direction.

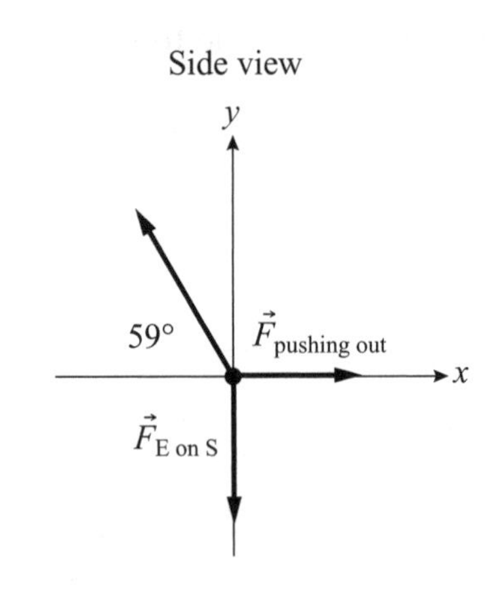

Represent mathematically and solve

$$F_{\text{C on S}} = m\frac{v^2}{r} = (80\text{ kg})\frac{(6.0\text{ m/s})^2}{(6.0\text{ m})} = 480\text{ N}$$

4.4.7 Design and analyze

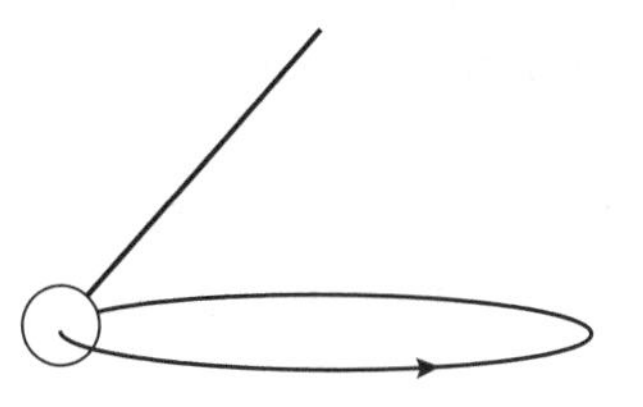

a. Design two independent methods to determine the sum of the forces ($\Sigma\vec{F}$) exerted on the bob of a conical pendulum by other objects as the bob moves at constant speed in a horizontal circle of a chosen radius. *Hint:* For more accurate measurements, use a circle with a large radius. Fill in the table that follows.

	Method 1	**Method 2**
Write a description in words of each of the methods you will use to determine the net force that other objects exert on the bob.		
Draw a labeled sketch.		
Draw a force diagram (as needed).		
Write the physical quantities that you will measure and quantities you will calculate.	To be measured: To be calculated:	To be measured: To be calculated:
Write the mathematical procedure you will use to determine the net force.		
List additional assumptions.		
List sources of experimental uncertainty and ways to minimize them.	Uncertainty: Ways to minimize:	Uncertainty: Ways to minimize:

b. Then perform the experiment and compare the two results. Discuss how assumptions and experimental uncertainties contribute to the discrepancy between the outcomes of the two experiments.

4.4.8 Pose a problem An Olympic throw for a 7-kg hammer (a ball connected to a cable) is a little less than 90 m. Invent two problems for this situation and use the principles of physics to solve these problems about the hammer-throwing situation depicted in the illustration.

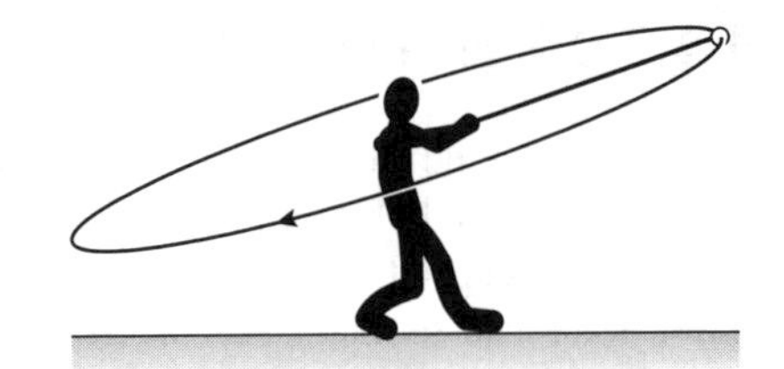

Part B: Gravitation

4.5 Quantitative Concept Building and Testing

4.5.1 Observe and explain

a. Newton found that the centripetal acceleration of the Moon when circling Earth was $2.69 \times 10^{-3}\ \text{m/s}^2$. This finding was based on a time for one orbit (27.3 days) and its orbital radius (about 3.8×10^8 m—the distance between the centers of Earth and Moon). Confirm the value of the Moon's acceleration.

This tiny number is exactly $1/3600$ times the free-fall acceleration on Earth's surface. Interestingly, the radius of the Moon's orbit about Earth is 60 times greater than the radius of Earth (6.4×10^6 m).

b. Newton assumed that the Moon moves in circular orbit due to the interaction with Earth. Draw a force diagram for the Moon. What relation between the force that Earth exerts on an object and the distance between the centers of Earth and the object could Newton propose based on this data? Explain your reasoning so another student can understand.

4.5.2 Explain Data on the acceleration of objects of different mass falling from small heights in vacuum tubes on Earth show that all of them fall with the same acceleration ($9.8\ \text{m/s}^2$). Newton assumed that in the absence of air, the only force causing this acceleration is the gravitational force Earth exerts on the falling object. What relation between the force of Earth on an object and the mass of the object could he propose based on this data? Explain your reasoning so another student can understand.

4.5.3 Explain According to Newton's third law, the force that a falling object exerts on Earth is equal in magnitude to the force that Earth exerts on the object. What relationship between the gravitational force that Earth exerts on a falling object and the mass of Earth could Newton propose based on his own third law? Explain your reasoning so another student can understand.

4.5.4 Test your explanation Using reasoning similar to that above, Newton decided that the gravitational force that Earth of mass M exerts on any object of mass m is directly proportional to the product of the two masses and inversely proportional to the square of the distance r between the centers of objects ($F_g = GMm/r^2$) where G is some constant number unknown to Newton—it was found later to be $G = 6.67 \times 10^{-11}\ \mathrm{N \cdot m^2/kg^2}$. Some scientists however, thought that the gravitational force was inversely proportional to the distance between the objects, not the distance squared. How could they decide whose ideas are more accurate?

Kepler, somewhat before Newton's time, devised several empirical relationships describing planetary motion using observational data collected by other astronomers. According to Kepler's third law, the period T squared of a planet about the Sun divided by the cube of the mean distance r of the planet from the Sun equals the same constant for all of the planets—that is, $T^2/r^3 = K$, a constant.

a. Use Newton's law of gravitation and his second law of motion as applied to circular motion to derive Kepler's third law. Then use the relationship for the gravitational force similar to Newton's but including the first power of distance to derive Kepler's third law.

b. Discuss which relationship led to a successful derivation and whether the gravitational force is inversely proportional to the distance between the objects or the distance squared.

4.5.5 Evaluate James thinks that the weight of a person on the Moon is less than on Earth because the Moon is farther from the center of Earth than is the surface of Earth where we normally weigh ourselves. Why would he have such opinion? Do you agree or disagree with him? If you disagree, what is your explanation?

4.5.6 Evaluate In the movie *Inception* the main heroes are in a bus that is falling from a bridge. You can see them floating freely inside the bus. One of them says that they are in zero gravity. Why would he say it? Do you agree or disagree? If you disagree how can you explain the floating?

5 Impulse and Linear Momentum

5.1 Qualitative Concept Building and Testing

5.1.1 Observe and find a pattern In the table that follows we describe a series of experiments. Fill in the table and think of a qualitative explanation that might account for all of the experimental outcomes.

Experiment	Draw initial and final sketches of the situation.	Discuss the direction and magnitude of velocities of the interacting objects before and after the interaction occurred.	Write a qualitative explanation that accounts for all three experiments. Focus on what was happening before and after the interaction occurred.
a. Pat, wearing rollerblades, is holding a medicine ball. She throws the ball forward and she in turn rolls backward. The initial speed of the ball is much larger than Pat's initial speed.	Before After	*Before:* Velocity of Pat Velocity of ball *After:* Velocity of Pat Velocity of ball	
b. Pat, still on rollerblades, stands still and catches a medicine ball thown at her. She rolls backward holding the ball. Her speed (and the speed of the ball after she catches it) is much smaller than the speed of the ball before she caught it.		*Before:* Velocity of Pat Velocity of ball *After:* Velocity of Pat Velocity of ball	
c. Pat is moving to the right and catches the ball thrown at her, which is moving left. She slows down after she catches the ball. Pat and the ball continue to move to the right slower than Pat was moving before she caught the ball.		*Before:* Velocity of Pat Velocity of ball *After:* Velocity of Pat Velocity of ball	

What patterns do you find in the change in velocity of the interacting objects and their masses?

5.1.2 Predict and test Use the explanation that you devised in Activity 5.1.1 to predict the results of the following experiments. Fill in the table that follows.

Experiment	Write your prediction.	Explain the prediction.	Perform the experiment; compare the results to your prediction.
a. Todd stands on a skateboard at rest; he jumps off it toward the rear of the board.			
b. A rolling railroad engine hits the coupler of a stationary train car and joins to it.			

Summarize what happens to the motion of the interacting objects. How do two objects affect each other?

5.2 Conceptual Reasoning

5.2.1 Explain Suppose you place a rifle on a platform capable of gliding with minimal resistance. When you pull the trigger, a bullet (mass 0.020 kg) shoots at high speed (300 m/s) out of the barrel, and the rifle (mass 2.0 kg) recoils back in the opposite direction at speed 3.0 m/s. Explain.

5.2.2 Explain At the National Transportation Safety test facility, videos are made of two identical cars initially moving at 80 km/h (45 mi/h) toward each other. Immediately after the collision, the cars are at rest and stuck to each other. Explain.

5.2.3 Explain You are wearing ice skates and standing on a frozen pond. How might you start moving without pushing off on the ice? Explain.

5.3 Quantitative Concept Building and Testing

5.3.1 Observe and find a pattern Imagine observing the following phenomena occurring with a two-object system. Two carts (the system) move on a dynamics track. Complete the table that follows for each experiment.

Experiment	Sketch the process before the collision and after the collision.	Determine if anything is the *same* before and after the collision. (*Hint:* Think of mass, speed, velocity, acceleration, or some combinations of these quantities.)
a. Cart *A* (200 g) moving left at constant 0.70 m/s speed hits identical cart *B* (200 g) that is stationary. Cart *A* stops and cart *B* starts moving at speed 0.70 m/s to the left.		
b. Cart *A* loaded with blocks (total mass of the cart with blocks is 400 g) moving left at 0.70 m/s hits stationary cart *B* (mass 200 g). After the collision, both carts move left, cart *B* at speed 0.86 m/s and cart *A* at speed 0.27 m/s.		
c. Cart *A* (200 g) with modeling clay attached to the front moves left at 0.70 m/s. Identical cart *B* (200 g) moves right at constant speed 0.70 m/s. The carts collide, stick together thanks to the clay, and stop.		
d. Repeat experiment (c) but this time cart *A* is loaded (total mass of the cart with blocks is 400 g). After the collision both carts stick together and travel left at speed 0.23 m/s.		

e. After you come up with a physical quantity that is the same before and after each collision, decide whether this quantity remains constant in all of the experiments.

5.3.2 Observe and explain The following table provides more data about the collisions of two dynamics carts, including the initial velocities of the carts before the collision (v_{ix}), the final velocities after the collision (v_{fx}), and the masses of the carts.

Before collision				**After collision**			
Cart 1	v_{ix1} (m/s)	Cart 2	v_{ix2} (m/s)	Cart 1	v_{fx1} (m/s)	Cart 2	v_{fx2} (m/s)
m	+2.0	m	0	m	+1.0	m	+1.0
m	+2.0	m	−2.0	m	−1.0	m	+1.0
$2m$	+2.0	m	−1.5	$2m$	+0.5	m	+1.5
$2m$	+2.0	m	−2.0	$2m$	0	m	+2.0

Determine whether the same quantity is constant in these experiments as was constant in 5.3.1.

5.3.3 Derive Your hand pushes horizontally on a cart of mass m exerting a force $\vec{F}_{\text{H on C}}$ for a time interval $(t_f - t_i)$. The forces exerted downward on the cart by Earth and upward on the cart by the track are balanced. The track is very smooth, so we assume that $\vec{F}_{\text{H on C}}$ is the only force exerted in the horizontal direction on the cart. The cart is initially moving at velocity $\vec{v}_{Ci}$ at the clock reading t_i, and after the pushing is moving at velocity $\vec{v}_{Cf}$ at the clock reading t_f.

a. Use Newton's second law and the definition of acceleration to show that:

$$\vec{F}_{\text{H on C}}(t_f - t_i) = m\vec{v}_f - m\vec{v}_i.$$

The term on the left, $\vec{F}_{\text{H on C}}(t_f - t_i)$, is called the *impulse* $\vec{J}$ due to external force $\vec{F}_{\text{H on C}}$ during that time interval. The term on the right, $m\vec{v}_f - m\vec{v}_i$, is called the *change in the momentum* of the cart.

b. Suppose that friction is not negligible. How would you modify the expression for the impulse on the cart to include both the effect of the hand's push and of friction?

c. Suppose the cart's forward velocity decreased due to some force. How would you write the expression for the impulse due to that force?

5.3.4 Predict Eugenia and David are both wearing rollerblades. Eugenia's mass is two-thirds of David's. Initially they are at rest, with one standing behind the other. Use your knowledge of momentum to complete the table that follows. Assume that friction is negligible.

Experiment	Eugenia is the system; what happens to her momentum? Consider movement toward the right to be positive.	David is the system; what happens to his momentum? Consider movement toward the right to be positive.	Eugenia and David are the system; what happens to their momentum? Consider movement toward the right to be positive.
a. Eugenia stands behind David and abruptly pushes him toward the right.			
b. David stands behind Eugenia and abruptly pushes her toward the right.			

c. Now use Newton's second and third laws and kinematics to make predictions about Eugenia's and David's final velocities in the experiment: What are the directions and relative speeds of Eugenia and David just after Eugenia stops pushing David toward the right?

5.4 Quantitative Reasoning

5.4.1 Reason The initial and final situations for four processes are shown below. Which ones are possible and which are not, according to your knowledge of momentum conservation? Explain your reasons. The numbers indicate the relative mass and the relative speeds of the blocks.

a.

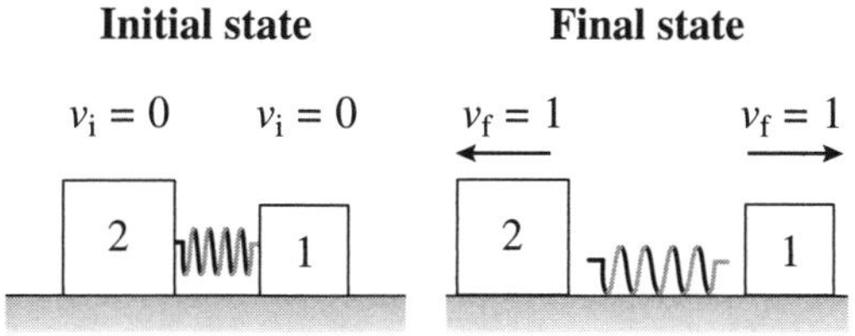

b.

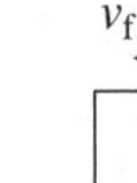

c.

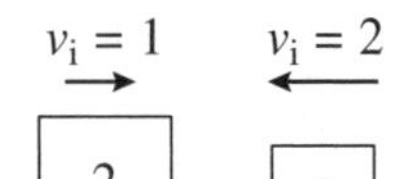

d.

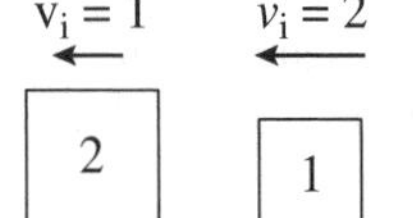

Use Reasoning Skill: Constructing a qualitative impulse-momentum bar chart.

5.4.2 Represent and reason Blocks 1 and 2 rest on a horizontal frictionless surface. A *compressed* spring of negligible mass separates the blocks. Block 1 has twice the mass of block 2. When the spring is released, the blocks are pushed apart. Answer the questions below concerning this process.

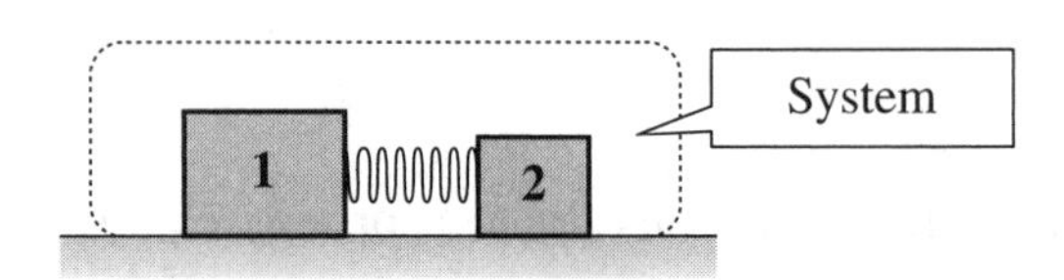

a. Compare the momentum *magnitudes* of block 1 and of block 2 after the spring is released.

b. Compare the combined momentum of both blocks before the spring is released and the combined momentum after the spring is released.

c. Compare the speed of block 1 to that of block 2 after the spring is released.

d. Represent the process using a qualitative impulse-momentum bar chart for a system consisting of the spring and two blocks. Do any other objects exert external forces on the system objects—especially in the horizontal direction?

5.4.3 Bar chart Jeopardy The bar chart on the right represents an impulse-momentum process. Answer the following questions concerning this process.

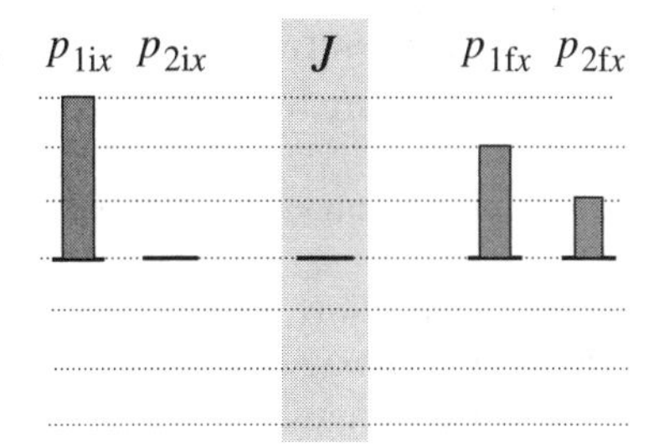

a. Describe in words and sketch a physical process that the bar chart might describe. Be specific.

b. Describe what would happen to the value of p_{1fx} if p_{2fx} had a value of positive three units instead of positive one unit.

c. Describe what would happen to the value of p_{1fx} if p_{2fx} had a value of positive four units instead of positive one unit.

5.4.4 Bar chart Jeopardy The bar chart on the right represents an impulse-momentum process. Answer the following questions concerning this process.

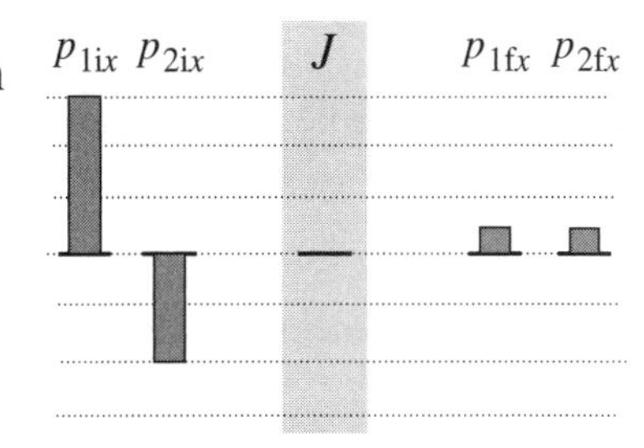

a. Describe in words and sketch a physical process that the bar chart might describe. Be specific.

b. Describe what would happen to the values of p_{1fx} and p_{2fx} if p_{2ix} had a value of negative three units instead of negative two units. Assume that the final momenta of the two objects are the same but different from before.

c. Describe what would happen to the values of p_{1fx} and p_{2fx} if p_{2ix} had a value of positive one unit instead of negative two units. Assume that the final momenta of the two objects are the same but different from before.

5.4.5 Reason Two equal-mass balls made of different material swing down at the end of strings and hit identical bricks. Ball 1 bounces back, whereas Ball 2 flattens and stops when it hits the brick. Ball 1 knocks the brick over and Ball 2 does not. Use your knowledge of impulse-momentum to explain why. Specify the system.

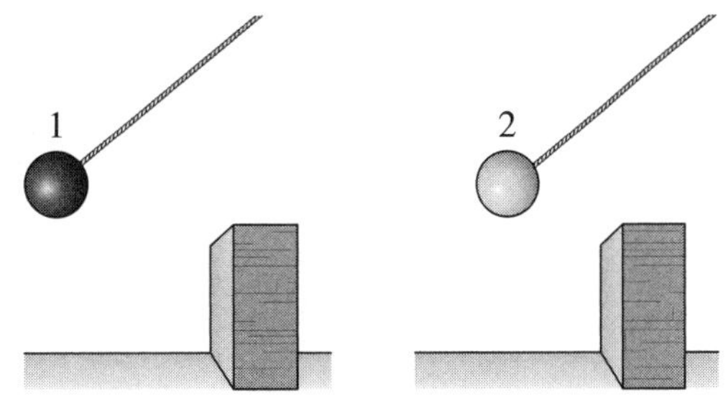

5.4.6 Represent and reason Imagine that you drop a ball from a window. After the ball falls 2.0 m, it acquires considerable speed.

a. Use your knowledge of impulse and momentum to fill in the table that follows.

Draw a bar chart for the process using the ball as the system.	Draw a bar chart using the ball and Earth as a system.	Explain why the ball speeds up as it falls, using your knowledge of impulse and momentum.	Explain why the ball speeds up as it falls, using your knowledge of Newton's laws.
p_{1iy} J p_{1fy}	p_{1iy} p_{2iy} J p_{1fy} p_{2fy}		

b. Discuss the differences in the bar charts because of the choice of system.

c. Discuss whether the explanations based on impulse-momentum and Newton's laws are consistent with each other.

5.4.7 Represent and reason You are wearing rollerblades and holding a heavy medicine ball. You push off the floor once and continue rolling at constant speed across the floor. Then you drop the medicine ball. Describe everything you can about momentum in this process using words, bar charts, and mathematics. Decide what your system is and what the initial and final states are.

5.4.8 Equation Jeopardy 1 Devise a problem that is consistent with the mathematical description of a process shown below. Multiple processes may be described by the equation.

$$(24\text{ kg})(-2.0\text{ m/s}) + (30\text{ kg})(+3.0\text{ m/s}) = (24\text{ kg} + 30\text{ kg})v$$

5.4.9 Equation Jeopardy 2 Devise a problem that is consistent with the mathematical descriptions of a process shown below. Multiple processes may be described by the equations.

$$F(t_f - t_i) = (0.010\,\text{kg})[(+100\,\text{m/s}) - (+300\,\text{m/s})]$$

where

$$(t_f - t_i) = \frac{(0.040\,\text{m} - 0)}{\dfrac{(+100\,\text{m/s}) + (+300\,\text{m/s})}{2}}$$

5.4.10 Evaluate the solution

The problem: A 0.40-kg bundle of unexploded fireworks moves horizontally to the right at speed 2.0 m/s. It creates a small explosion, which breaks the bundle into two pieces. The first piece with 10% of the mass moves right at 40 m/s. What is the velocity of the remaining piece?

Proposed solution:

$$(0.40\,\text{kg})(2.0\,\text{m/s}) = (0.04\,\text{kg})(40\,\text{m/s}) + (0.40\,\text{kg})v$$

or

$$v = \frac{(0.80\,\text{kg}\cdot\text{m/s} - 1.6\,\text{kg}\cdot\text{m/s})}{(0.40\,\text{kg})} = 2\,\text{m/s}$$

a. Identify any missing elements/errors in the solution.

b. Provide a corrected solution if you find missing elements or errors.

5.4.11 Evaluate the solution

The problem: A 2000-kg pickup truck traveling at 20 m/s collides with a stationary 1000-kg compact car. The vehicles lock together and skid on a level surface for 16 m until stopping. Determine the coefficient of kinetic friction between the tires and the road.

Proposed solution:

Part I: The collision

Momentum conservation: $(2000\text{ kg})(20\text{ m/s}) = (3000\text{ kg})v$

$$v = 13.3\text{ m/s}$$

Part II: Skidding to a stop

Stopping time: $$(t_f - t_i) = \frac{(x_f - x_i)}{v} = \frac{16\text{ m}}{13.3\text{ m/s}} = 1.2\text{ s}$$

Stopping force: $$f_k = \frac{(mv_f - mv_i)}{(t_f - t_i)} = \frac{(2000\text{ kg})(20\text{ m/s})}{(1.2\text{ s})} = 33{,}300\text{ N}$$

Coefficient of kinetic friction: $$\mu_k = \frac{f_k}{N} = \frac{f_k}{mg} = \frac{(33{,}300\text{ N})}{(3000\text{ kg})(9.8\text{ N/kg})} = 1.1$$

a. Identify any missing elements/errors in the solution.

b. Provide a corrected solution if you find missing elements or errors.

5.4.12 Pose a problem Design a problem based on a situation that you can observe during a U.S. Open Tennis tournament that requires understanding impulse and linear momentum.

5.4.13 Design an experiment Obtain a steel ball and a ball made of modeling clay, and assemble a force probe connected to a computer. Set up the software so that the computer screen displays a force-versus-time graph for a collision (the force that the probe exerts on a ball during a collision in which the probe stops the ball).

a. Design a series of experiments to investigate the differences between the collision of the steel ball with the probe and the clay ball with the probe.

b. Perform the experiments and record the results.

c. Explain the results using your knowledge of impulse and momentum.

6 Work and Energy

6.1 Qualitative Concept Building and Testing

6.1.1 Observe and find a pattern We describe three experiments involving a well-defined system and a process in which the system changes from an initial state to a final state. At the end of this process, we find that the system has the potential to do something it couldn't do before—to smash a piece of chalk into many pieces. In the sketches of the situations, draw arrows indicating the direction of the external force that you (outside the system) exert on a system object and the displacement of the object while you exert the force. Fill in the table that follows to help you complete this activity.

a. The system includes a 1-kg block with a flat bottom and a string attached to the top, Earth, and a piece of chalk. You (outside the system) pull up on the string so that the 1-kg block slowly rises 0.5 m above the piece of chalk. After this lifting process, you release the block. It falls and breaks the chalk.

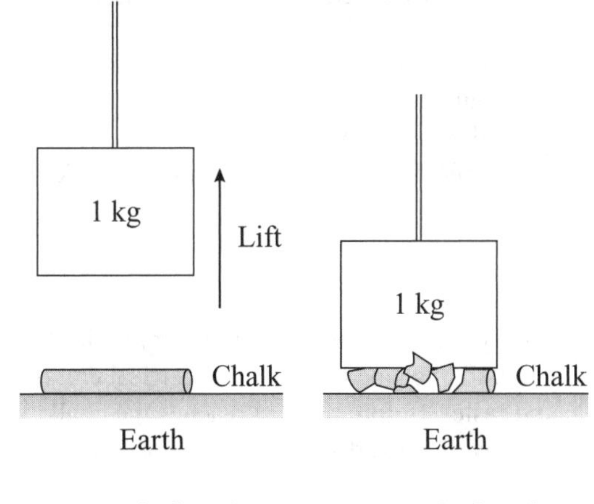

b. The system includes a 1-kg dynamics cart that can roll on a low-friction horizontal dynamics track and a piece of chalk that is taped to the fixed vertical end of the track. You (outside the system) push the cart so that it rolls faster and faster toward the chalk on the end of the dynamics track and breaks the chalk when it hits it.

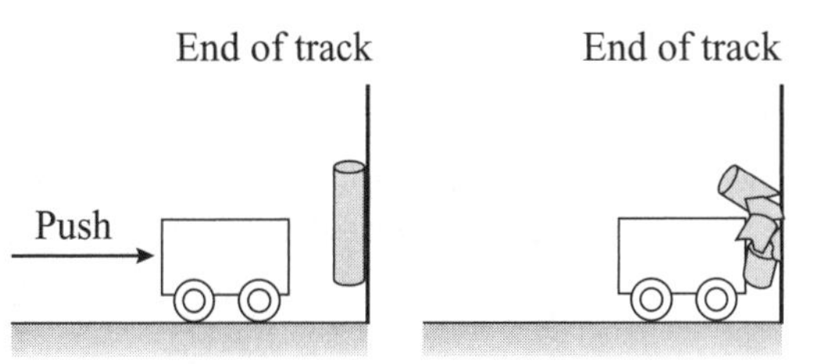

c. The system includes a slingshot that holds a piece of chalk. You (outside the system) slowly pull back on the sling. When you release the sling, the chalk shoots out at high speed and hits the wall, causing the chalk to break.

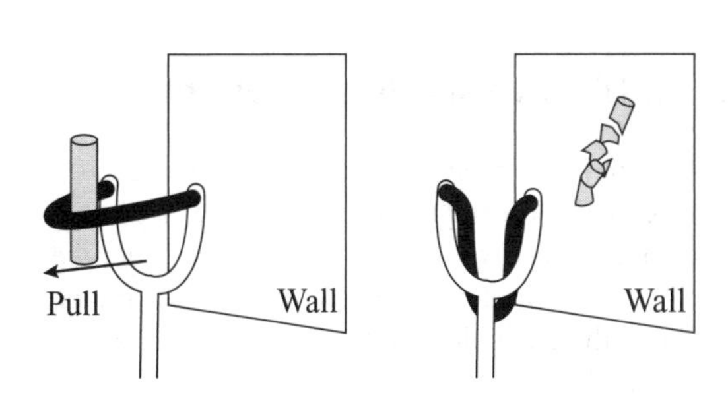

Experiment	**a.**	**b.**	**c.**
Draw arrows indicating the direction of the force you exerted on the system $(\vec{F}_{\text{Y on S}})$ and the displacement of the system object while you were exerting the force $(\vec{d})$.			

d. Look for a pattern of what was done to the system to give it the chalk-smashing potential. Then, devise a new physical quantity to describe this pattern. Be explicit.

6.1.2 Observe and find a pattern In Activity 6.1.1, you found that the external force you exerted on an object in a system gave the system the potential to smash a piece of chalk. The force you exerted on the object in the system was always in the direction of the displacement of that object. Suppose that a friend outside the system decides to save the chalk in the first two experiments by exerting with her hand an opposing force on the block or on the cart after they are released. In each case your friend pushes on the moving object opposite to the direction of its velocity. Describe for parts a and b of Activity 6.1.1 the direction of the force your friend exerts on the moving object relative to the displacement of the object as she stops it—in other words, when the system loses the potential to break the chalk.

a. The system includes a 1-kg block with a flat bottom and a string attached to the top, Earth, and a piece of chalk. You (outside the system) pull up on the string so that the 1-kg block slowly rises 0.5 m above the piece of chalk. After this lifting process, you release the block, and it starts falling. Your friend then starts pushing upward on the falling block, slows it down, and the block does not break the chalk.

b. The system includes a 1-kg dynamics cart that can roll on a low-friction horizontal dynamics track and a piece of chalk that is taped to the fixed vertical end of the track. You (outside the system) push the cart so that it rolls faster and faster. Before the cart reaches the chalk, your friend pushes on it opposite its displacement. This causes the cart to slow down and stop so that it does not break the chalk.

c. How could you modify the definition of the quantity you devised in Activity 6.1.1 to account for the system's loss of the chalk-breaking potential thanks to your friend's intervention?

6.1.3 Observe and find a pattern Consider a system that includes Earth and a 1-kg block.

a. You (outside the system) hold a string tied to the block so that it stays about 2 cm above a table. A piece of chalk is placed on the table under the block. If you release the block and it falls on the chalk, the chalk does not break (it's too close to the chalk).

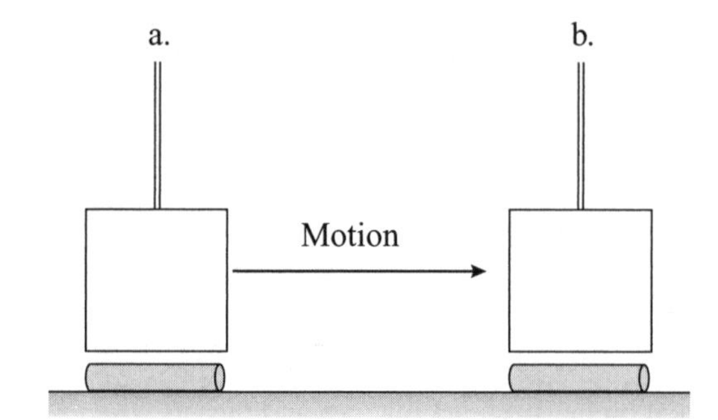

b. Next you slowly walk about 2 m beside the table, continually keeping the block 2 cm above the surface. After you have walked the 2 m, the block hangs over a second identical piece of chalk. Draw the force exerted by the string on the block and the displacement of the block as you walked the 2 m.

c. Discuss whether the *vertical* force the string exerted on the block while moving it *horizontally* above the tabletop caused the system to have a better chance of breaking the second piece of chalk than the first piece. Revise the quantity you devised in the last two activities to account for this result. Your revision will involve the angle between the external force exerted on the system and the system object's displacement. We call this quantity *work*.

6.1.4 Observe and find a pattern A system consists of a crate and a rough horizontal surface on which it sits. The surface is made of a special material that changes color when it changes temperature (these materials are called *thermoplastics*; they are made of regular plastic with added thermochromatic materials). You (outside the system) pull on a rope attached to the crate so that it moves slowly at constant velocity. You do positive work by pulling the crate for about 10 m. Describe how the system (block and surface) is different after you do the work than before the crate started moving.

6.1.5 Describe You do work on a system to change its potential to do something—for example, to break chalk or to make touching surfaces of objects in a system warm. In Activities 6.1.1 through 6.1.4, the work done on the system by the external force caused different types of changes in the system. For each situation below, describe each type of change in the system as a result of the work done on it and come up with a name for it.

a. The external force caused the block to move higher above Earth's surface.

b. The external force caused the cart to move faster and faster.

c. The external force caused the slingshot to stretch.

d. The external force caused the surfaces of the touching objects to warm.

6.1.6 Pose a problem Describe a real-life situation in which an external force does the following:

a. positive work on a system;

b. positive work on a system but with a value that is less than in part a;

c. negative work on a system;

d. zero work even though an object in the system moves.

6.1.7 Design an experiment For each item, describe one real-life experiment that is consistent with the work–energy process. (Do not mention the exact experiments described earlier.)

a. Positive work causes an increase in the gravitational potential energy of the system.

b. Positive work causes an increase in the kinetic energy of the system.

c. Positive work causes an increase in the elastic potential energy of the system.

d. Kinetic energy in the system is converted to gravitational potential energy.

e. Kinetic energy in the system is converted to elastic potential energy.

f. Gravitational potential energy in the system is converted to internal energy.

g. Gravitational potential energy in the system is converted to elastic potential energy.

6.2 Conceptual Reasoning

6.2.1 Represent and reason Fill in the table that follows.

Experiment: Description of system and process	Draw a sketch showing initial and final states.	Construct a qualitative work–energy bar chart.
A rope pulls a skier, initially at rest, up a hill. *Initial state:* A skier is at rest at the bottom of the hill. *Final state:* The skier is moving at moderate speed at the top of the hill. *System:* Includes the skier, rope, and Earth but excludes the motor that pulls the rope up the hill. Ignore friction.		K_i U_{gi} U_{si} W K_f U_{gf} U_{sf} ΔU_{int} + 0 –

6.2.2 Bar-chart Jeopardy In the table that follows, describe in words and then sketch a process (the system, its initial and final states, and any work done on the system) that is consistent with the qualitative work–energy bar chart shown below. There are many possible choices.

Bar chart for a process	Describe in words one possible consistent process.	Sketch the process just described.
K_i U_{gi} U_{si} W K_f U_{gf} U_{sf} ΔU_{int} (+, 0, −)		

6.2.3 Represent and reason Complete the table that follows for three processes to devise a graphical method to determine the work done by an external force on a system object. *Note:* P = person and O = object.

Word description of a process	Sketch the process.	Draw $F_{(\text{P on O})y}$ versus y (for vertical motion) or x (for horizontal motion) graphs.	Describe how to use the graph to find the work done by the specified force.
Rona lifts a backpack from the floor to the desk, exerting a constant upward force. The backpack and Earth (not Rona) are the system.			$\vec{F}_{\text{R on B}}$
Kruti catches a medicine ball in the gym. The ball and Earth are the system but not Kruti. Her hands move back toward her body while stopping the ball. Assume that she exerts a constant force on the ball.			$\vec{F}_{\text{K on B}}$
Carlos stretches a horizontal rubber cord (it behaves like a spring) with a spring constant k. The spring and Earth (not Carlos) are the system.			$\vec{F}_{\text{C on S}}$

6.3 Quantitative Concept Building and Testing

6.3.1 Find the relationships To develop mathematical expressions for gravitational potential energy and kinetic energy, we analyze the following situation: a cable lifts a block from vertical position y_i to vertical position y_f. When at position y_i, the block is moving up at speed v_{yi}, and when at position y_f, it is moving up at greater speed v_{yf}.

a. Sketch the situation. Include a labeled vertical axis.

b. Write an expression for the work the cable does on the block during its displacement $y_f - y_i$.

c. Draw a force diagram for the block. Use it to find an expression for the force that the cable exerts on the block in terms of its mass m, acceleration a, and the gravitational constant g. Substitute this expression into the expression in part b.

d. Use a kinematics equation to convert the acceleration a in the equation from part b into an expression involving the block's speeds v_i and v_f and its displacement $y_f - y_i$.

e. Substitute the expression from part d into the expression for the force that the cable exerts on the block found in part c.

f. Substitute the new expression in part e for the force that the cable exerts on the block into the expression for work in part b.

g. Examine the expressions that you derived in parts e and f. Do you see that the work that the cable did on the block equals the sum of the changes of two quantities: $mgy_f - mgy_i$ and $(1/2)mv_f^2 - (1/2)mv_i^2$? Discuss how these expressions can be used to write an expression for the gravitational potential energy of the block–Earth system and an expression for the kinetic energy of the block.

6.3.2 Find the relationship Determine an expression for the elastic potential energy of a stretched spring. The spring is the system, and you stretch it from its equilibrium position to some final position x. You are outside the system. The spring is very easy to stretch at first but gets more and more difficult as you stretch it farther. (Remember that $F_{\text{Y on S}x} = kx$; i.e., the force that "you" exert on the "spring.")

a. Examine the sketch of the situation, showing the system and the external object exerting a force on it. Note the origin of the x axis at the end of the unstretched spring that you start pulling.

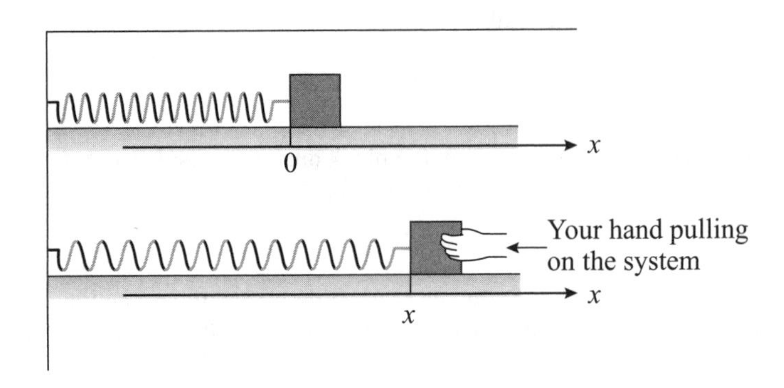

b. Draw a graph ($F_{\text{Y on S}x}$ versus x) and find the work that you did while stretching the spring from 0 to x as the area under the curve.

c. Examine the expression that you derived from part b; do you see that the work that you did on the spring equals $\frac{1}{2}kx^2$? How does this expression relate to the elastic potential energy of the spring that you have stretched?

6.3.3 Find the relationship Determine an expression for the change in internal energy due to friction in a system that consists of a crate and a rough horizontal surface on which it slides. You, outside the system, pull on a rope attached to the crate so that it moves slowly at constant velocity. At the end of the process, the bottom of the crate and the surface on which it was moving have become warmer.

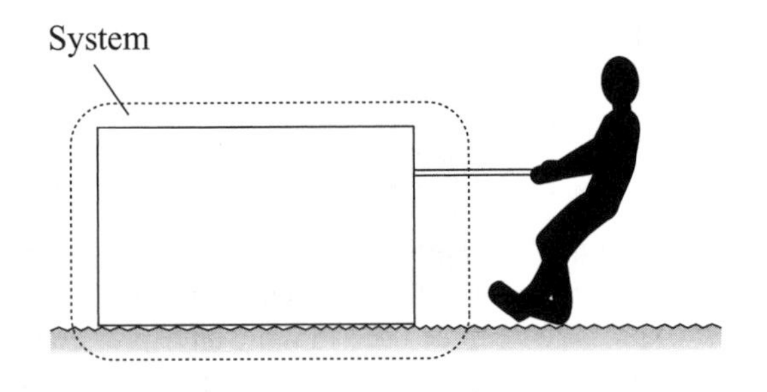

a. Write an expression for the work done on the system by the external force of the rope on the crate as the rope pulls the crate a distance s across the surface exerting a force $\vec{T}_{\text{R on C}}$

b. Choose the crate alone as the system (a different system than in the sketch) and draw a force diagram for the crate. Apply Newton's second law for the horizontal x axis. How are $\vec{T}_{\text{R on C}}$ (rope on crate) and $\vec{f}_{\text{S on C}}$ (surface friction on crate) related?

c. Now, combine parts a and b to write an expression for the work done by the kinetic friction force on the crate. Is it positive or negative? Write an expression for the work done by the rope on the crate. What is the total work done on the crate?

d. You found that when the crate is the system, the work done on it is zero. However, its bottom is warm. To explain where this internal energy came from, let's change the system. The system is the crate, the surfaces, and Earth. The rope is outside the system. In the initial state, the crate is moving but the surfaces are cool; in the final state, the crate is still moving with the same velocity but the surfaces are warmer. Now only $\vec{T}_{\text{R on C}}$ does work on the system. Represent the process with a bar chart.

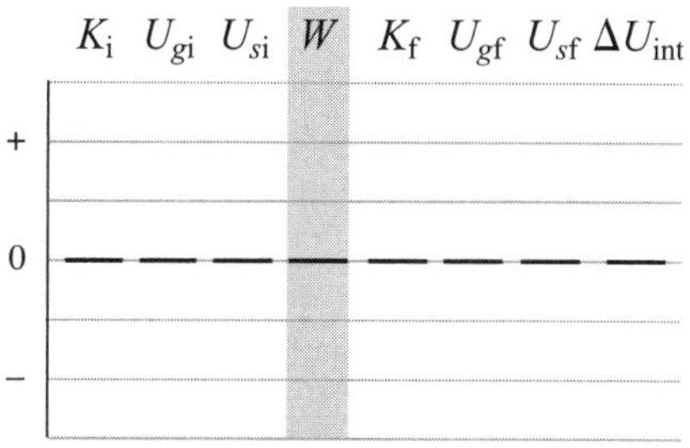

e. Examine the bar chart. Write an expression for the change in internal energy of the system and decide whether it increases or decreases.

6.3.4 Reason Summarize the results of Activities 6.3.1 through 6.3.3 to construct a generalized work–energy relationship—a relationship among the initial energy of a system, the external work done on the system, and the final energy of the system.

6.3.5 Test your idea You have a vertical metal rod, a light spring with one end closed and the other end open (so that it can slide down over the rod and then fly up when released), a spring scale, and a meterstick. Use this equipment to design an experiment whose outcome you can predict using the generalized work–energy relationship.

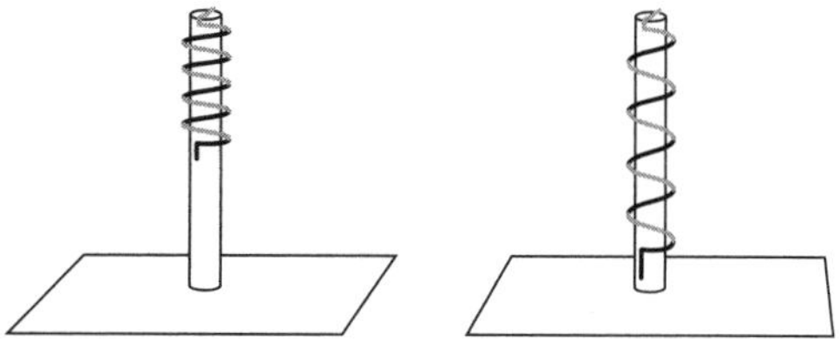

a. Fill in the table that follows to help you.

Draw an initial–final sketch of the experiment. List the quantities that you need to measure.	**Construct a bar chart and describe the process mathematically.**
	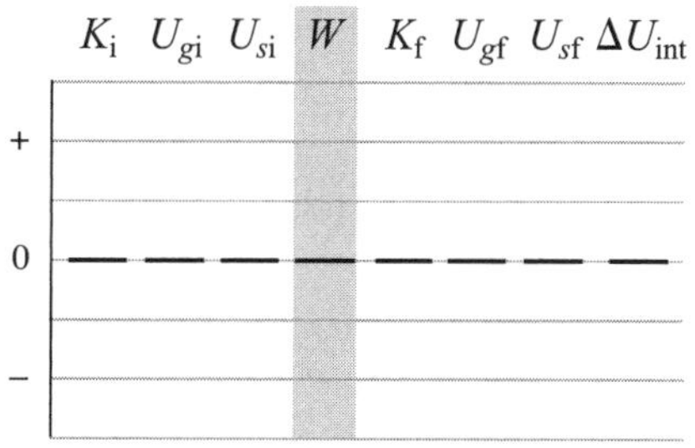
Use the equation you devised in 6.3.4 to make a prediction about the outcome of the experiment.	**List assumptions and how they should affect the predicted value.**

b. Perform the experiment several times and compare the results to the predicted value. How can you validate the assumptions?

c. Discuss whether the outcome supports the relationship you devised in Activity 6.3.4. How can you improve the experiment?

6.4 Quantitative Reasoning

6.4.1 Represent and reason Word descriptions and pictorial representations of two work–energy processes are provided in the table that follows. Complete the last column of the table. Do not solve for anything.

Word description	Picture description	Construct a bar chart for the process and apply in symbols the generalized work–energy principle.
A stunt car has an ejector seat that rests on a vertical spring compressed a distance x_i. When the spring is released, the seat with its passenger is launched out of the car and reaches a maximum height h_f above its starting position.	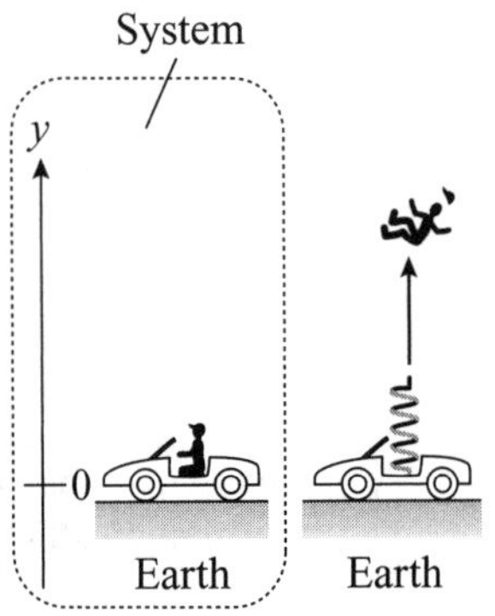 System y 0 Earth Earth t_i t_f $y_i = 0$ $y_f = h$ $v_i = 0$ $v_f = 0$ $x_i = 0$ $x_f = 0$	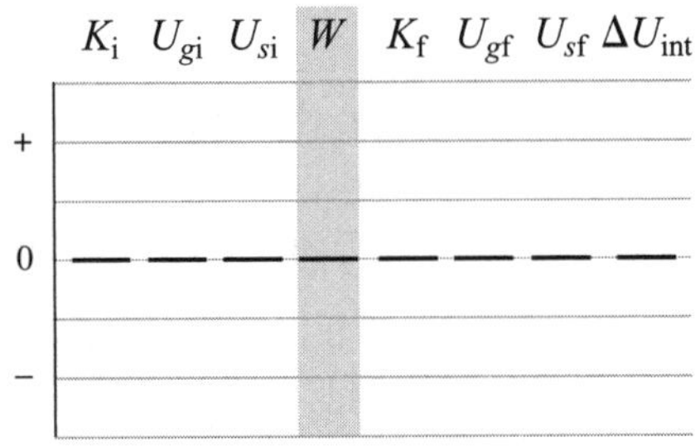 K_i U_{gi} U_{si} W K_f U_{gf} U_{sf} ΔU_{int} + 0 –

Word description	Picture description	Construct a bar chart for the process and apply in symbols the generalized work–energy principle.
An elevator, while moving down at speed v_i, approaches the ground floor and slows to a stop in a distance h.	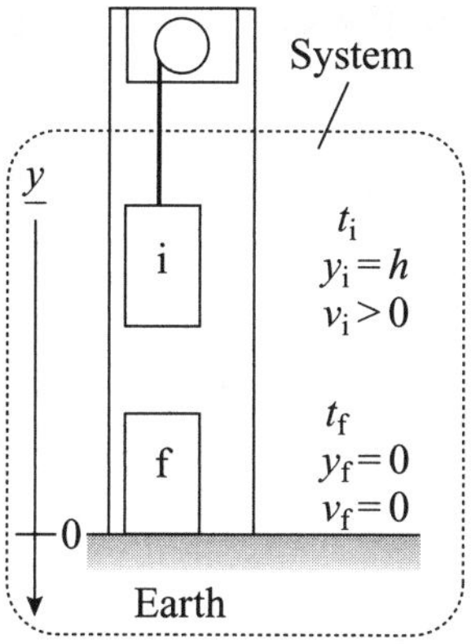 System; y; i: t_i, $y_i = h$, $v_i > 0$; f: t_f, $y_f = 0$, $v_f = 0$; 0; Earth	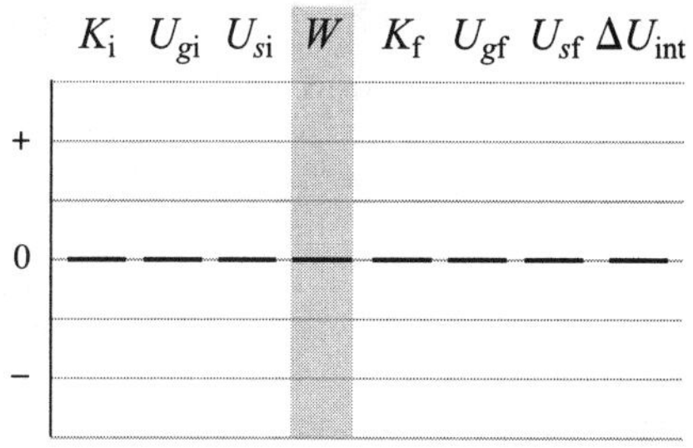 K_i U_{gi} U_{si} W K_f U_{gf} U_{sf} ΔU_{int}; +, 0, –

6.4.2 Bar-chart Jeopardy Each of the two bar charts below describes real processes. Fill in the table that follows to describe the possible processes. Do not solve for anything.

Bar chart	Draw an initial–final sketch of a process that might be described by the bar chart. Identify the system.	Describe the process in words.
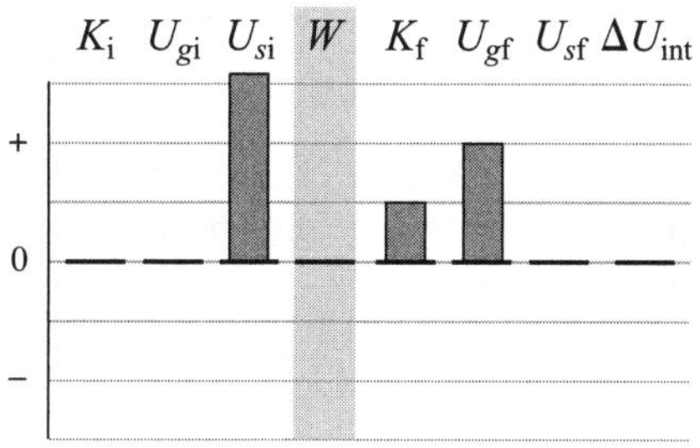 K_i U_{gi} U_{si} W K_f U_{gf} U_{sf} ΔU_{int}; +, 0, –		
	Use the bar chart to apply in symbols the generalized work–energy equation to the process.	

Bar chart	Draw an initial–final sketch of a process that might be described by the bar chart. Identify the system.	Describe the process in words.
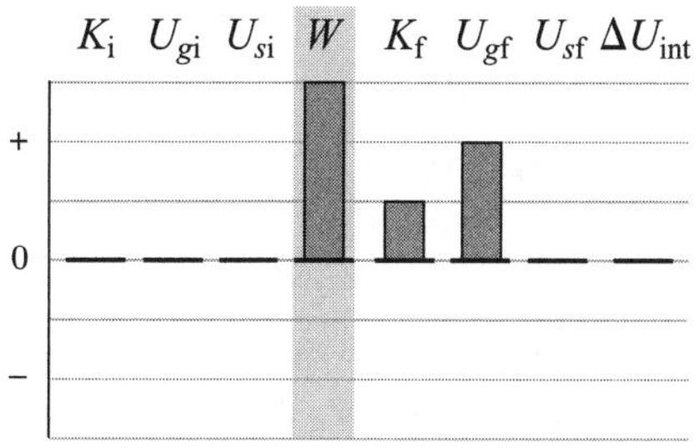 K_i U_{gi} U_{si} W K_f U_{gf} U_{sf} ΔU_{int}; +, 0, –		
	Use the bar chart to apply in symbols the generalized work–energy equation to the process.	

6.4.3 Regular problem Erin is interviewing for a new job—a car test driver. She is asked to determine whether a vehicle built to test the effects of high accelerations on people is safe. The car, moving at a speed of 20 m/s, runs into a piston that compresses the air in a cylinder and stops the car in 0.30 m. The 70-kg passenger stops in the same distance because of the force exerted by the shoulder straps and seat belts on her body.

a. Determine the average force that these restraints exert on the person during the stop.

b. What assumptions did you make?

6.4.4 Regular problem Erin wants to get in shape for her new job as a test driver. She sets her stationary bike on a high 100-N frictionlike resistive force. She cycles for 30 min at a speed of 8.0 m/s.

a. Determine the internal energy change of the bicycle.

b. If Erin's body is 10% efficient at converting her metabolic energy into work in pedaling the bicycle, determine Erin's internal energy change (equal to the metabolic energy used).

c. How long must Erin exercise to produce 3.0×10^5 J of internal energy while staying at this speed? This amount of energy equals the energy released by the body after eating a slice of bread.

6.4.5 Regular problem You are designing a new bungee-jumping system for beginners. An 80-kg cart (including its passenger) is to start at rest near the top of a 30° incline. The uphill side of the cart is attached to a spring. The other end of the spring is attached securely to a post farther up the hill. The spring is initially relaxed. After you are secure in the cart, it is released and you coast 40 m down the hill before stopping. What is the spring constant of the spring that you should buy for this invention? Follow the problem-solving strategy.

6.4.6 Regular problem An 80-kg skier comes off a slope traveling at speed 15 m/s on a level, snow-covered surface. The skier wearing a Velcro®-covered vest runs into a padded 20-kg cart, also covered with matching Velcro®. The skier and cart, now stuck together, compress a 1600-N/m spring on the other side of the cart (the spring's other end is mounted securely to a wall). The skier vibrates back and forth with the cart at the end of the spring. How far did the skier and cart compress the spring? Follow the problem-solving strategy.

6.4.7 Equation Jeopardy The first column in the table that follows applies the generalized work–energy equation to two different processes (in fact, there are many possible processes described by each equation). For each mathematical description, construct a sketch, word description, and bar chart that is consistent with the equation.

Generalized work–energy equation applied to a process	Sketch a process that might be described by the equation.	Construct a bar chart.
$(1/2)kx_i^2 = (1/2)mv^2 + mgy$	Identify a system and initial and final states.	K_i U_{gi} U_{si} W K_f U_{gf} U_{sf} ΔU_{int} + 0 –

Generalized work–energy equation applied to a process	Sketch a process that might be described by the equation.	Construct a bar chart.
$mgd \sin 20° = (1/2)kx^2$	Identify a system and initial and final states.	K_i U_{gi} U_{si} W K_f U_{gf} U_{sf} ΔU_{int} + 0 –

6.4.8 Evaluate the solution

The problem: A 40,000-N/m spring cart launcher, initially compressed 0.50 m, is released and launches you and your cart (100 kg total) up a 30° incline. What distance along the incline do you travel before stopping?

Proposed solution: $(1/2)(40{,}000\ \text{N/m})(0.50\ \text{m}) = (100\ \text{kg})(9.8\ \text{m/s}^2)y$ or $y = 10.2\ \text{m}$.

a. Identify any missing elements or errors in the solution.

b. Provide a corrected solution if you find any missing elements or errors.

6.4.9 Evaluate the solution

The problem: You are traveling in your 2000-kg pick-up truck at 20 m/s up a hill with a 6.0° incline when you see a goose crossing the road 24 m in front of you. You know from previous experience that when you hit the brakes, a 16,000-N friction force opposes the car's motion. Will you hit the goose?

Proposed solution: $(1/2)(2000\text{ kg})(20\text{ m/s})^2 = (16{,}000\text{ N})x$ or $x = 25\text{ m}$. Oops!

a. Identify any missing elements or errors in the solution.

b. Provide a corrected solution if you find any missing elements or errors.

6.4.10 Pose a problem Pose a problem that can be solved with the generalized work–energy principle using the situation depicted in the illustration. Make a list of necessary givens and assumptions to solve the problem.

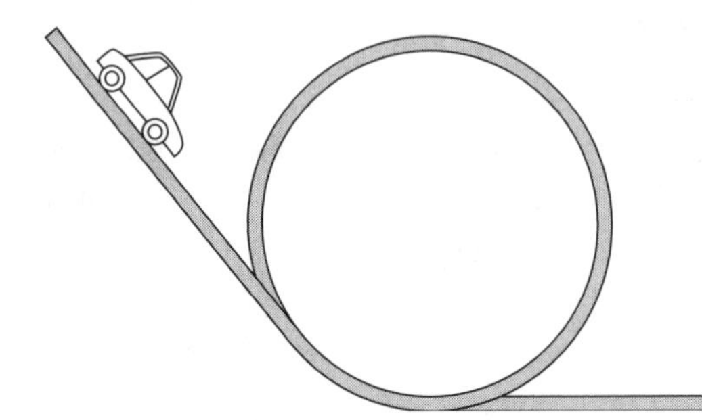

6.4.11 Design an experiment You have a Hot Wheels® car, Hot Wheels® track, a loop-the-loop piece for the track, a Hot Wheels® car launcher, a surface that can be inclined at different angles, masking tape, and a meterstick. Use any or all of this equipment to design an experiment to test whether the total energy of a Hot Wheels®–Earth system is constant if there are no external forces exerted on it by other objects. Then perform the experiment and decide whether its outcome supports the hypothesis under test.

a. Fill in the table that follows.

Describe the experiment in words.	Sketch the apparatus.	List quantities that you will measure.
Use the generalized work–energy principle and other principles (if needed) to make a prediction.	**List assumptions that you made.**	

b. Now perform the experiment several times and record the outcomes.

c. Compare the predicted value to the experiment outcome. Make a judgment about the hypothesis under test.

7 Static Equilibrium

7.1 Qualitative Concept Building and Testing

7.1.1 Observe and find a pattern Use a pencil eraser to push at several points on the edge of a thin, flat, irregularly shaped piece of plywood. The arrows in the illustration show the direction of the forces exerted by the eraser. The solid arrows indicate forces that cause the plywood to slide without turning. The dashed lines indicate the forces that cause the object to rotate while it is being pushed. Identify a pattern in the direction of forces that do not cause the object to rotate. *Hint:* Draw lines on the object in the direction of forces.

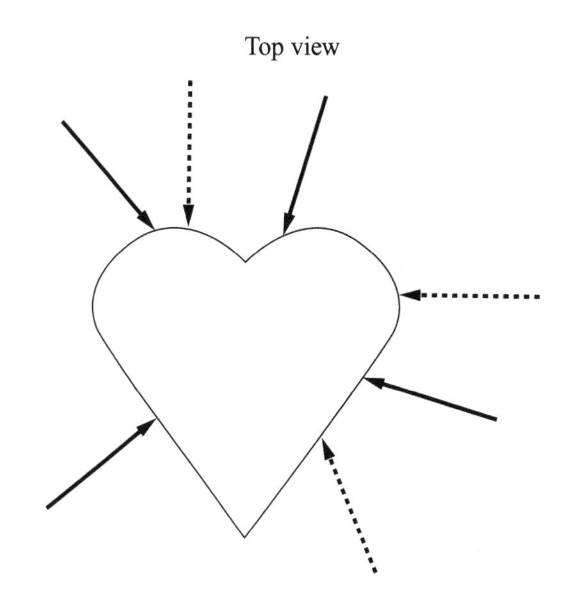

7.1.2 Observe and find a pattern Repeat Activity 7.1.1, only this time place a 500-g block on top and near one side of the plywood. Does the pattern change in the direction of the forces that do not cause the object to rotate?

7.1.3 Test your idea In previous chapters we were concerned with objects that could be considered particles—they did not rotate or we neglected their rotation (e.g., blocks sliding down hills). In Activities 7.1.1 and 7.1.2 you encountered situations in which the force exerted by another object on an object of interest could cause it to rotate. Use your knowledge of the gravitational force and the knowledge of the center of mass to predict the outcome of the following experiment. Do not perform it yet. You can use the same irregularly shaped board as in Activity 7.1.1 and hang it on a nail going through one of the holes drilled at its edges; the board should hang freely. Then attach the end of a string to the nail and hang a 100-g block at the string's other end. Next, suppose you hang the plywood and string from other holes. Fill in the table that follows to predict where the lines (along which the string is oriented) intersect when you hang the board from different positions.

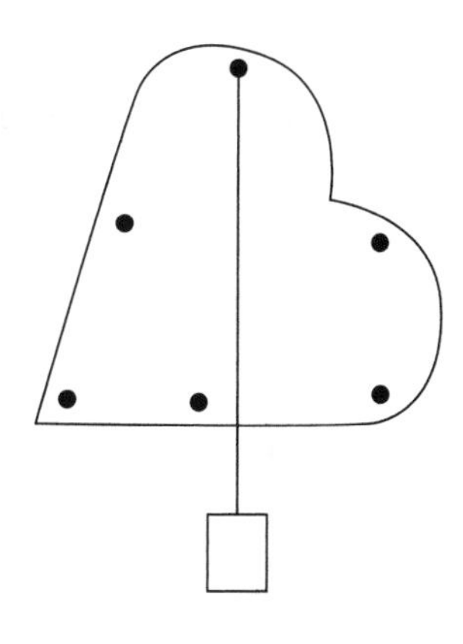

Write your prediction.	Explain your prediction.	Perform the experiment; record the outcome.	Reconcile the results with your prediction and decide whether you need to revise your idea.

7.2 Conceptual Reasoning

7.2.1 Represent and reason Imagine that you place a board on your desk and push it in different directions, as illustrated. Forces 1 and 3 cause the board to slide, and forces 2 and 4 cause it to slide and rotate. Find the center of mass of the board. What assumptions did you make?

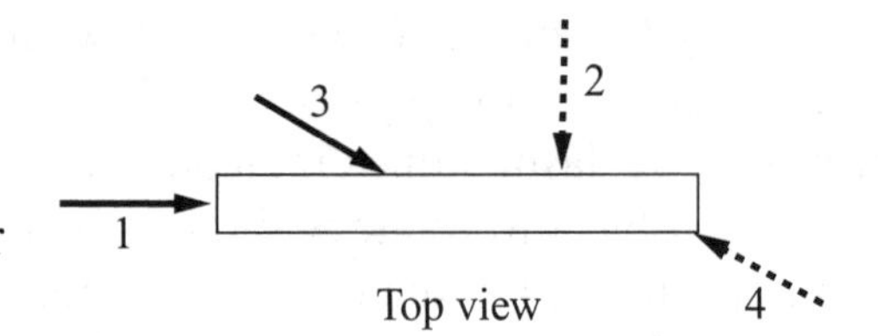

7.2.2 Reason You have a stand-alone ladder (irregularly shaped). Describe carefully how you can determine the location of the center of mass of the ladder.

7.2.3 Design an experiment Find three irregularly shaped objects and propose two experiments to determine the location of the center of mass of each object.

7.3 Quantitative Concept Building and Testing

7.3.1 Observe and find a pattern A meter stick is suspended by a string running through a hole at its center, as shown. You and your friend pull down on the stick at different positions with spring scales. When pulled as described in the table, the stick does not rotate. Find three other cases in which the meter stick does not rotate when you pull with two scales.

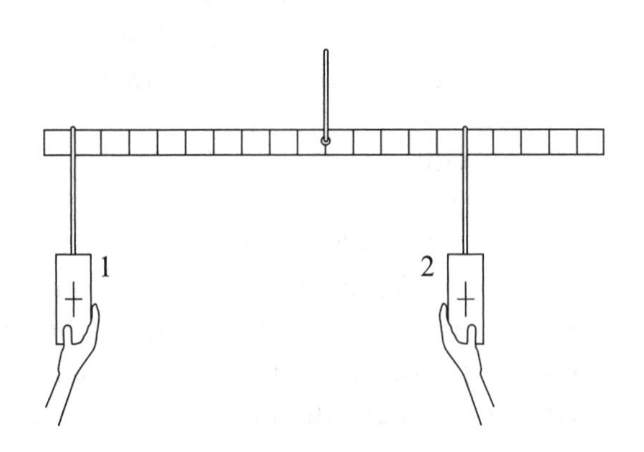

Experiment	Distance of spring scale 1 to left of center	Force 1	Distance of spring scale 2 to right of center	Force 2
a.	0.40 m	1.00 N	0.10 m	4.00 N
b.				
c.				
d.				

For each situation, draw a picture of the stick showing all of the forces exerted on it. Find a pattern that relates the distances of the spring scales from the point of suspension and the magnitudes of forces the spring scales exert on the meter stick. Record your work in the table that follows.

Experiment	Draw a picture of the stick showing all forces exerted on it.	Note the relationships between distances and forces.	Describe a pattern.
a.			
b.			
c.			
d.			

7.3.2 Observe and find a pattern Repeat Activity 7.3.1, but this time suspend the meter stick 20 cm to the left of the center of the stick—at the 30-cm mark. Collect the data and record it in the table. Find a pattern between the distances from the suspension point and magnitudes of the forces exerted on the meter stick by the scales and by Earth.

Experiment	Distance of spring scale 1 to left of suspension point	Force 1	Center of mass location to right of suspension point	Force exerted by Earth	Distance of spring scale 2 to right of suspension point	Force 2
a.			0.20 m			
b.			0.20 m			
c.			0.20 m			

Summarize below the observed pattern in words and symbols.

7.3.3 Explain Replace the string in Activity 7.3.2 with a third spring scale that pulls up on and supports the meter stick (see the figure). Record the readings of the third scale for the three experiments that you performed in Activity 7.3.2.

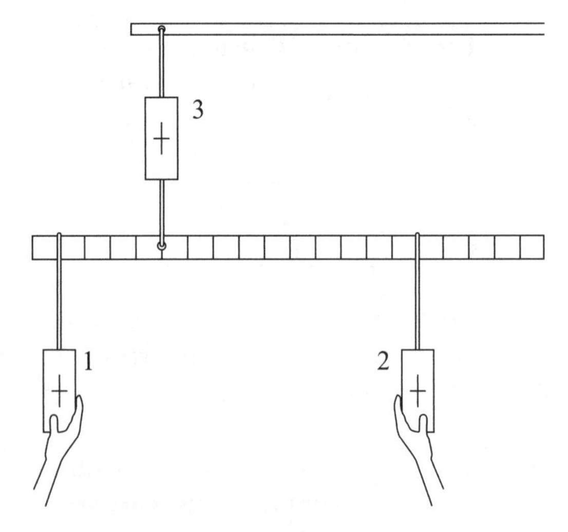

Experiment	Upward force exerted by scale 3	Distance of spring scale 1 to left of suspension point	Force 1 down	Center of mass location to right of suspension point	Force exerted down by Earth	Distance of spring scale 2 to right of suspension point	Force 2 down
a.				0.20 m			
b.				0.20 m			
c.				0.20 m			

Fill in the table that follows and explain these readings using your knowledge of Newton's laws.

Draw a picture of the stick for any of the three experiments showing all forces exerted on it.	Describe a new pattern.	Explain the third scale readings using Newton's laws.

7.3.4 Observe and find a pattern A meter stick is balanced at its center. When you hang different mass blocks from different positions on the stick, as shown in the illustrations that follow, the stick remains balanced. Draw *all* forces exerted on the stick by other objects. Remember the force exerted by Earth on the stick. Devise a rule to explain this behavior (or extend the rules developed in the previous activities). Be sure that the rule is compatible with all of the experiments.

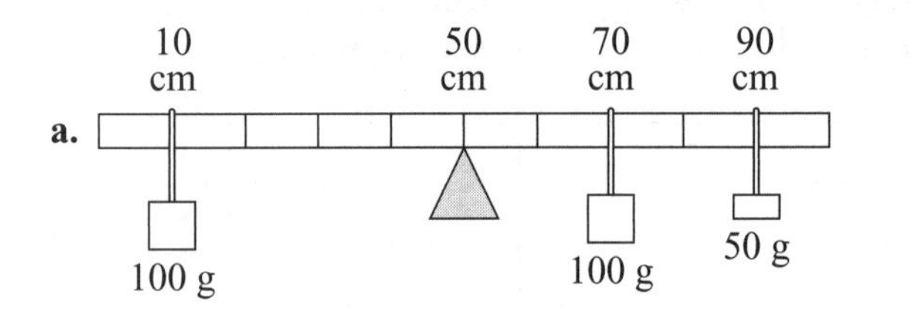

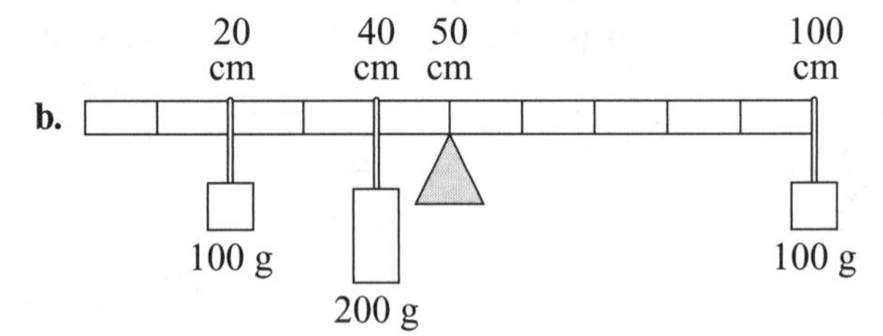

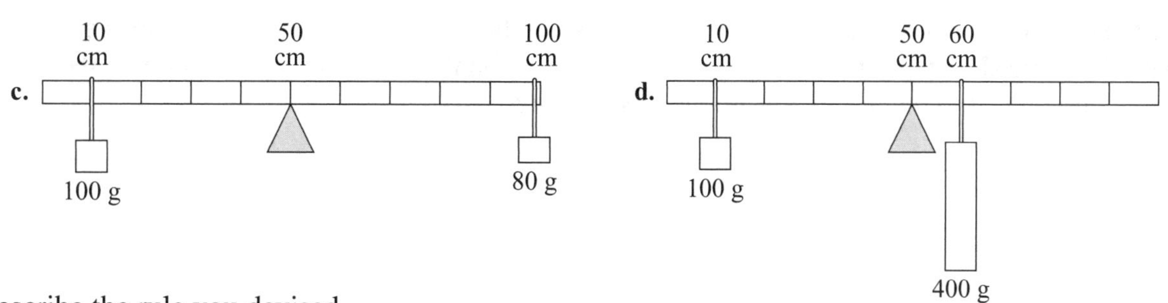

Describe the rule you devised.

7.3.5 Test your idea A fulcrum supports a uniform meter stick. Apply the rule or rules you developed earlier to fill in the table that follows. Use available objects to balance the meter stick as shown. Note the masses of objects you used on the figures.

Experiment	Apply the rules and predict the mass of the meter stick.	Measure the mass.	Reconcile the discrepancies.
0 cm, 20 cm			
0 cm, 20 cm, 65 cm			

7.3.6 Observe and find a pattern Measure and record the mass of the beam. Then, assemble the experiment as shown. Pull on the beam at the angles listed in the table and record the scale reading.

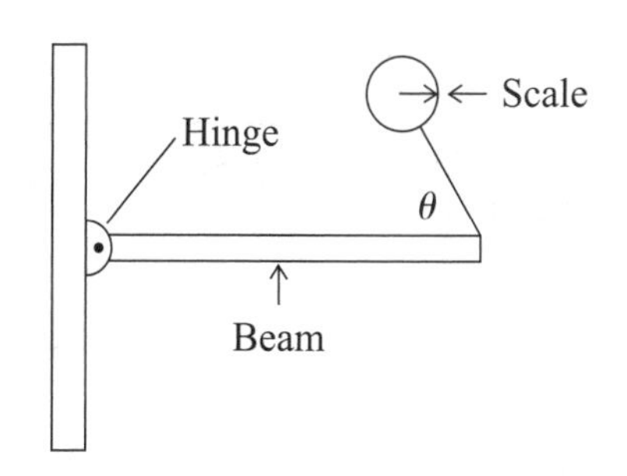

Angle (degrees)	90°	60°	45°	37°	30°
Force exerted by the string on the beam (N)					

Use the data that you collected to modify any rules you have made for the effect of a force on the rotation of a rigid body. In particular, how does the rotational effect of a force depend on its magnitude, direction, and the distance from the pivot point?

7.3.7 Explain Use the patterns and rules that you devised in the previous activities to summarize what you know about the sum of the forces exerted on an object that is in equilibrium and the sum of the torques caused by these forces. *Note:* Physicists give signs to torques. The torque caused by a force that tends to rotate an object counterclockwise about some pivot point is usually taken to be a positive torque, and the torque caused by a force that tends to rotate the object clockwise is taken to be negative.

7.3.8 Test an idea Your friend says that the mass of any object is distributed symmetrically around its center of mass. Design an experiment to test your friend's idea. You have a meter stick, a set of small objects of different masses, masking tape, and a mass measuring scale. Describe your experiment and predict the outcome based on your friend's idea.

7.3.9 Design an experiment Design an experiment to test the second condition of equilibrium (the torque condition of equilibrium). In the table, carefully show the equipment you will use and how it is arranged. Predict what you expect to observe.

List equipment.		**Sketch the apparatus.**	
Write quantities that you will measure.		**Use the conditions of equilibrium to make your prediction.**	
Record the outcome of the experiment.			

7.4 Quantitative Reasoning

7.4.1 Regular problem A 62-kg person stands at the end of a 4.0-m-long diving board, as shown in the illustration. Determine the magnitudes and directions of each of the forces that the two supports, separated by 1.0 m, exert on the board. What simplifications should you make to solve the problem?

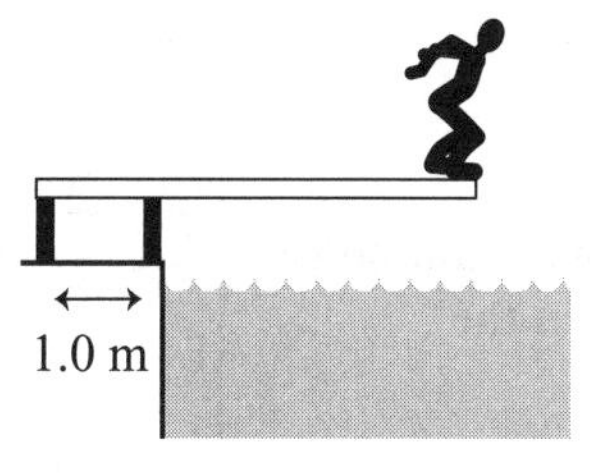

7.4.2 Regular problem Determine the force that the cable supporting a 3000-kg ski lift gondola exerts on it as the gondola moves across a canyon. The gondola hangs from a bar that is attached to a pulley that rolls across the cable. The cable tilts up at a 10° angle on each side of the pulley.

7.4.3 Equation Jeopardy The conditions of equilibrium applied to two processes are shown below. Fill in the table that follows using the information in the equations. Remember that *N* stands for normal force and N for Newton, the unit of force.

a. Equations	$+(2.0\text{ N})(0.25\text{ m}) + N(0) + (1.0\text{ N})(0) - T_2(0.40\text{ m}) = 0$ $-2.0\text{ N} + N - 1.0\text{ N} - T_2 = 0$		
Determine the unknown quantities in the equations.		**Draw an extended-body force diagram consistent with the equations.**	
Describe a possible situation in words.		**Sketch the situation.**	
b. Equations	$+F_{pin}(0) - (2\text{ N})(0.5\text{ m}) - (6\text{ N})(1.0\text{ m}) + T_1(1.0\text{ m})\sin 37° = 0$ *x* equation: $F_{pin}\cos\theta - T_1\cos 37° = 0$ *y* equation: $F_{pin}\sin\theta - 2\text{ N} - 6\text{ N} + T_1\sin 37° = 0$		
Determine the unknown quantities in the equations.		**Draw a force diagram consistent with the equations.**	
Describe a possible situation in words.		**Sketch the situation.**	

7.4.4 Evaluate the solution

The problem: An 800-N painter stands 3.0 m from the right side of a 10.0-m-long beam of a scaffold, which is connected to cables on each end. The gravitational force of Earth on the uniform beam is 200 N. Determine the forces that the cables exert on each end of the beam. Assume that $g = 10$ N/kg.

Proposed solution:

$$+T_{\text{left}}(10.0\text{ m}) + (200\text{ N})(5.0\text{ m}) + (800\text{ N})(3.0\text{ m}) + T_{\text{right}}(0) = 0$$

$$+T_{\text{left}} = -340\text{ N}$$

$$+T_{\text{left}} + 200\text{ N} + 800\text{ N} + T_{\text{right}} = 0$$

$$T_{\text{right}} = -660\text{ N}$$

a. Identify any missing elements or errors in the solution.

b. Provide a corrected solution or missing elements if there are errors.

7.4.5 Regular problem The illustration to the right shows a person lifting a 30-lb barbell. (*Warning:* DON'T TRY THIS YOURSELF. Each year more than half a million Americans sustain serious back injuries by lifting this way.) The illustration below is a mechanical model of the person's upper body. The beam is the backbone, and the cable is the back muscles (a complex set of muscles in the real back). The gravitational force that Earth exerts on the person's upper body at its center of mass is 30 lb. The gravitational force that Earth exerts on the person's head, arms, and the barbell is 50 lb at the end of the beam. The back muscles (cable) connect 0.20 m from the left end of the 0.60-m-long beam (the backbone) and make a 15° angle with the beam. Apply the conditions of equilibrium to the beam and use them to estimate the force that one primary back muscle (it is a complex system) exerts on the backbone and the compression of the beam at its joint on the left side.

7.4.6 Regular problem Suppose that your professor decides to lecture while sitting on a beam of length L, as shown in the illustration. The rope attached to the end of the beam passes over a pulley and is tied to the professor's waist. Apply the conditions of equilibrium separately to the professor and to the beam. Use the equations to determine the tension in the cable and the force that the professor exerts on the beam. The gravitational force that Earth exerts on the professor is 600 N. What assumptions did you have to make to solve the problem?

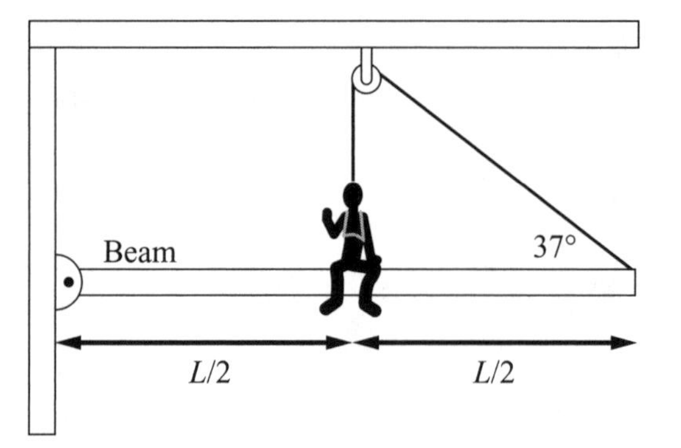

7.4.7 Design an experiment You have a meter stick of unknown mass and a small 100-g object. Fill in the table that follows to help you design an experiment to determine the mass of the meter stick using your knowledge of static equilibrium.

Draw a picture of the experimental setup.		**Describe the procedure in words.**	
Apply the concepts of equilibrium to develop equations that can be used to predict the mass of the meter stick. Then predict the mass.		**Use a scale to measure the mass and compare the result to the predicted value.**	

7.4.8 Pose a problem You walk by Tom Sawyer, who is painting his aunt's house while standing on a ladder that leans against the wall (he finished painting the fence already). Draw a picture of the situation. Identify forces that keep the ladder in equilibrium. Pose a problem related to the situation. Make sure you provide enough information to solve the problem.

8 Rotational Motion

8.1 Qualitative Concept Building and Testing

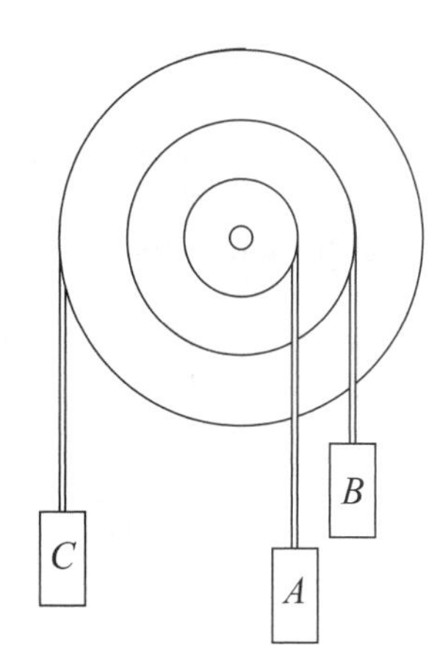

8.1.1 Observe Three blocks hang from strings wrapped around a multiradius pulley that is initially stationary. The radii are r, $2r$, and $3r$. By changing the masses of the hanging blocks, we can get the pulley to rotate clockwise (cw) faster and faster, to rotate counterclockwise (ccw) faster and faster, or to remain stationary. The table below indicates the results of five such experiments.

Experiment	Block A	Block B	Block C	Pulley behavior
Experiment 1	2 kg at r	1 kg at $2r$	2 kg at $3r$	Starts rotating counterclockwise faster and faster
Experiment 2	2 kg at r	2 kg at $2r$	2 kg at $3r$	Remains stationary
Experiment 3	2 kg at r	3 kg at $2r$	2 kg at $3r$	Starts rotating clockwise faster and faster
Experiment 4	2 kg at r	1 kg at $2r$	3 kg at $3r$	Starts rotating counterclockwise faster and faster, with the rate of rotational velocity change greater than in experiment 1
Experiment 5	3 kg at r	2 kg at $2r$	2 kg at $3r$	Starts rotating clockwise faster and faster, with a greater rate of rotational velocity change than in experiment 3

a. Find a pattern in the data from experiments 1 through 5 and formulate a general rule that will allow you to predict when the pulley does not start rotating and when it does start rotating.

b. Analyze the outcomes of experiments 1 and 4 and of experiments 3 and 5 to find a pattern in the rate of rotational velocity change.

8.1.2 Observe and explain Two objects of equal mass but different mass distribution that are attached to a pulley system are initially at rest, as shown in the illustration. Strings pull down on the objects, exerting a force of the same magnitude on the axles. The rate of rotation of object 2 increases more rapidly than that of object 1. Fill in the table that follows.

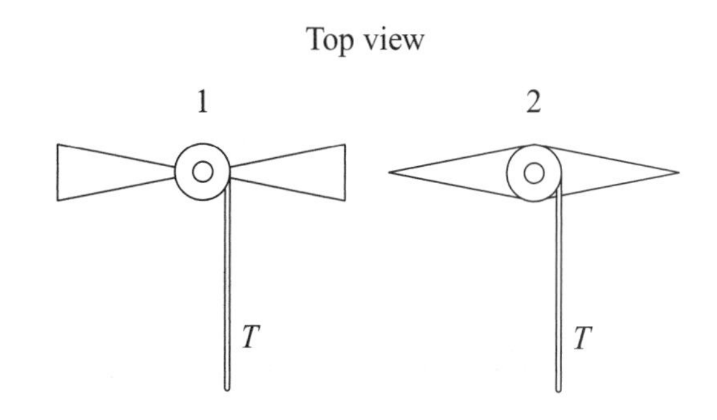

Write an explanation for this phenomenon.	Describe other phenomena from real life where the mass distribution makes it easier or more difficult to cause the object to start rotating.

8.1.3 Predict

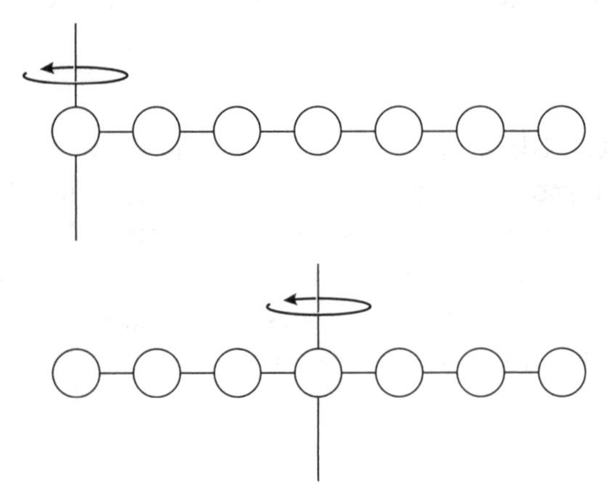

a. Use the explanation that you devised in Activity 8.1.2 to decide whether it would be more difficult, less difficult, or the same difficulty to change the turning motion of the set of balls shown in the illustration about the axis through the left ball or about the axis through the center ball. The balls are rigidly linked by a light rod going through them, and they spin in the plane perpendicular to the plane of the page. Be sure to justify your choice based on the explanation in Activity 8.1.2.

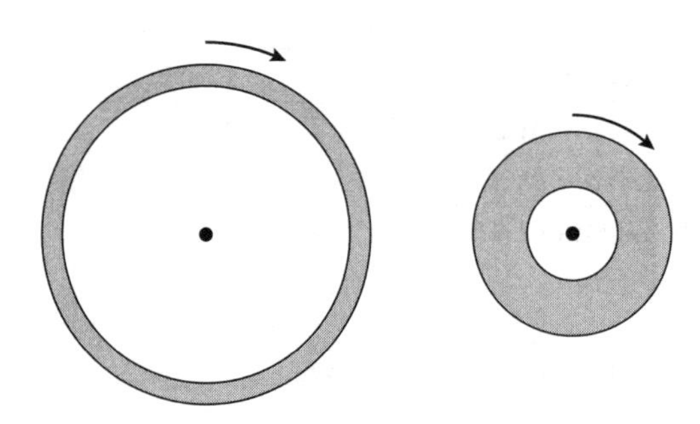

b. The two rings in the illustration have equal mass. Would it be more difficult, less difficult, or the same difficulty to start the larger radius ring spinning about an axis through its center compared to the difficulty of spinning the smaller radius ring about its center axis? Explain.

8.1.4 Observe and explain For each situation shown in the figure, determine the signs of the net torque and the initial velocity (see the curved arrows). The signs are + for counterclockwise, – for clockwise, and 0 for zero torque or for a stationary disk.

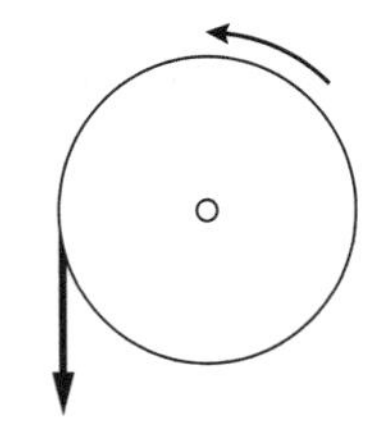

1. Rotating ccw faster and faster

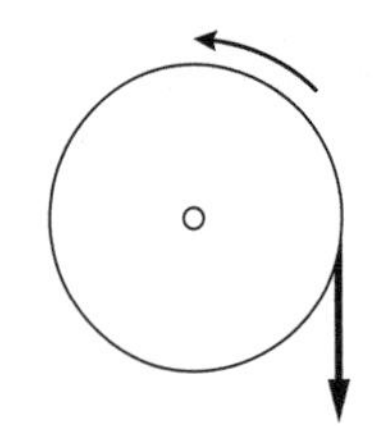

2. Rotating ccw slower and slower

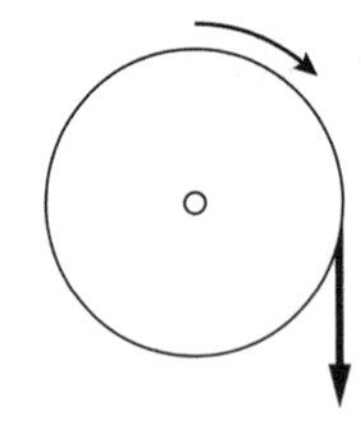

3. Rotating cw faster and faster

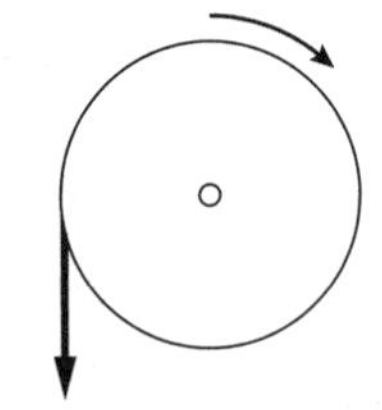

4. Rotating cw slower and slower

The sign of the rotational acceleration is the same as the sign of the rotational velocity if its magnitude is increasing, opposite the sign of the rotational velocity if its magnitude is decreasing, and zero if the rotational velocity magnitude is not changing. Note that the curved arrows on the illustrations represent the initial direction of rotation, and the straight arrows represent the direction of a rope pulling on the pulley.

a. Fill in the results in the table that follows. Note that $\Sigma\tau$ is the sum of the torques, ω is the initial rotational velocity, and α is the rotational acceleration.

Experiment	Sign of $\Sigma\tau$	Sign of ω	Sign of α
Situation 1			
Situation 2			
Situation 3			
Situation 4			

b. Is there a relationship between the sum of the torques and the rotational velocity? If so, what is the relationship? Explain your reasoning.

c. Is there a relationship between the sum of the torques and the rotational acceleration? If so, what is the relationship? Explain your reasoning.

8.1.5 Test your idea Use a bicycle for this experiment. Turn it upside down and place the saddle on the floor. Design an experiment to test your answers to questions 8.1.4 b and c.

8.2 Conceptual Reasoning

8.2.1 Represent and reason Answer the following questions for the situation shown in the figure. Note that the multiradius pulley is initially turning clockwise.

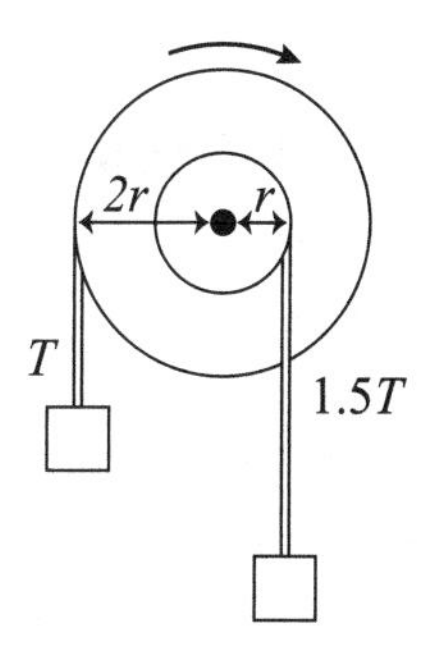

Write an expression for the net torque exerted on the pulley in terms of r and T.		Describe in words the behavior of the pulley as time progresses.	
Decide what force the left rope should exert on the pulley so that the rotational velocity of the pulley remains constant. Explain.		Decide what force the right rope should exert on the pulley so that the pulley turns faster and faster in the clockwise direction. Explain.	

8.2.2 Represent and reason Two pails hang from ropes suspended over a flywheel, as shown in the illustration. Ignore friction in the bearings; the rope does not slip on the pulley, and the flywheel has considerable mass. The pails are released from rest.

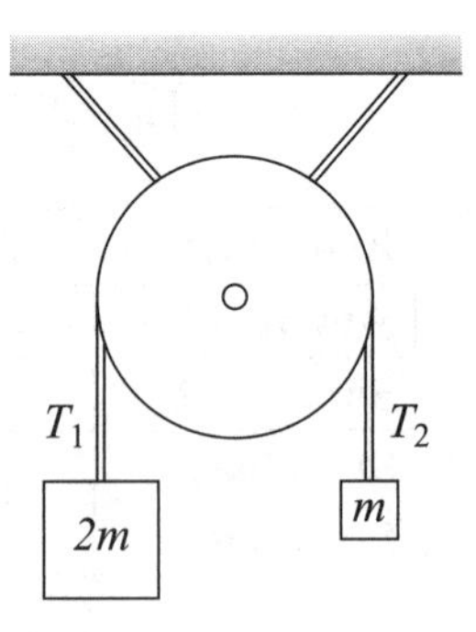

a. Fill in the table that follows. Draw the force arrows the correct relative lengths. Make sure different representations are consistent with each other.

Construct a motion diagram for each pail after its release.	Construct a force diagram for each pail after its release. Compare the magnitudes of T_1 and $2mg$ and of T_2 and mg.	Describe the rotational motion of the flywheel.	Construct a force diagram for the flywheel (include the rope wrapped around the wheel but not the pails).

b. (Optional) Conduct the experiment and record the outcome. Did it match the description that you provided in cell 3?

8.3 Quantitative Concept Building and Testing

8.3.1 Represent and reason A rod of negligible mass has a disk of mass m attached at one end. The other end has a pin through it, as shown in the illustration. The rod and disk start at rest and rotate in a horizontal circle about the pin on a frictionless air table. A person exerts a constant force $\vec{F}$ on the disk (see the arrow in the sketch), always pointing tangent to the circle. We extend Newton's second law to this rotational motion.

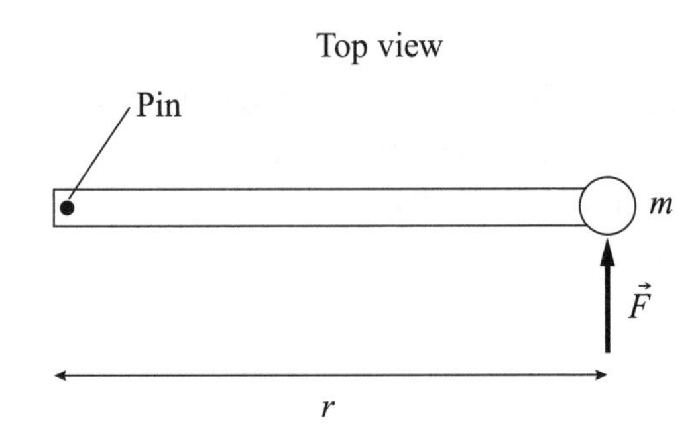

a. Apply Newton's second law to the disk in a direction tangent to the circle.

b. Write an expression for the torque τ that the force exerts about the pin.

c. Multiply both sides of the equation in part a by the same quantity so that the left force side of the equation becomes the torque determined in part b.

d. Multiply the right side of the equation by r/r and use one of these r's to convert the acceleration a to a rotational acceleration α. Discuss the result.

8.3.2 Test your idea You have a 50-g block that hangs from a string. The other end of the string wraps around a flywheel of radius 0.10 m and rotational inertia of $0.025\ \text{kg} \cdot \text{m}^2$. Use Newton's second law for rotational motion that you constructed in earlier activities to fill out the table that follows.

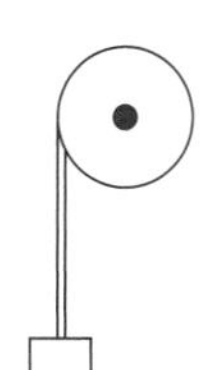

Predict the time interval needed for the block, initially at rest, to fall 1.0 m toward the floor.	List the assumptions you made.	Perform the experiment, record the outcome, and compare the actual time to the prediction.	Explain whether the experiment supported or disproved Newton's second law for rotational motion.
		Outcome: Comparison:	

8.3.3 Represent and reason In Chapter 3 we solved problems that involved two hanging objects (m_1 and m_2) connected by a string going over a pulley, as shown in the illustration. Because of the difference of the masses of the objects, the system accelerated:

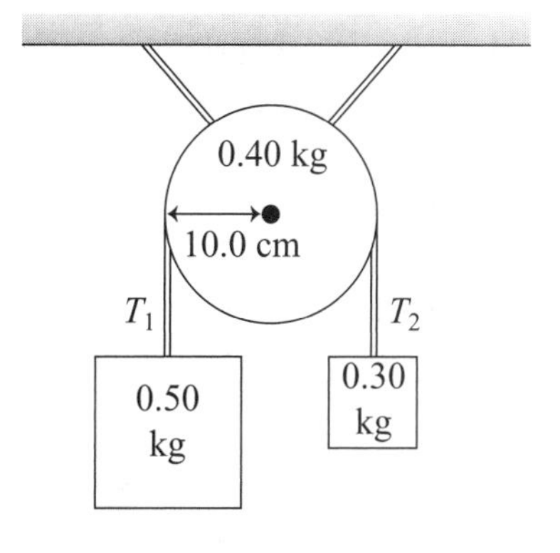

$$a = \frac{m_1 - m_2}{m_1 + m_2} g$$

The pulleys were light, and we neglected their masses. Use Newton's second law for rotational motion to decide whether the mass of the pulley can be neglected if the system consists of two objects (0.30 and 0.50 kg) connected with a string going over a pulley of mass 0.40 kg and radius 10.0 cm. List assumptions that you made.

8.4 Quantitative Reasoning

8.4.1 Represent and reason Fill in the table that follows. Remember to use symbols when writing the rotational form of Newton's second law for each situation (and the translational form for the pail). The rotational inertia of the multiradius pulley shown below is I, the inner radius is a, and the outer radius is $2a$.

Sketch and translate	T	$3T$ T	m
Simplify **Choose a system and diagram.**			
Draw a force diagram.			Pail: Pulley:
Represent mathematically **Write Newton's second law for rotational motion (or translational motion for the pail).**			Pail: Pulley:

8.4.2 Equation Jeopardy The equations in the table that follows describe many possible physical situations. Think of one situation that might be described by each set of equations.

Mathematical description	Sketch a situation the equation might describe.	Write in words a problem for which the equation is a solution.
$-(100\text{ N})(0.40\text{ m}) = (80\text{ kg}\cdot\text{m}^2)\alpha$ and $\omega = 0 + \alpha(3.0\text{ s})$		
$+(100\text{ N})(0.40\text{ m}) - (160\text{ N})(0.20\text{ m}) = (20\text{ kg}\cdot\text{m}^2)\alpha$ and $\theta = 0 - (4.0\text{ s}^{-1})t + (1/2)\alpha(2.0\text{ s})^2$		

8.4.3 Regular problem Suppose that some imaginary being (a Superperson) exerts a force of magnitude F at Earth's equator tangent to its surface for a period of 1 year. How large must the force be to stop Earth's rotation? List the assumptions that you made.

8.4.4 Regular problem As the tides slide across Earth, they exert a friction force that opposes Earth's rotation. Because of this, the time required for one Earth rotation increases by 0.0016 s every 100 years.

a. Determine the rotational deceleration of Earth.

b. Assuming that Earth is a uniform solid sphere, what torque do the tides exert on Earth?

c. If the friction force caused by the tides were all concentrated on the equator, how large would this force be?

8.4.5 Evaluate the solution

The problem: A 2.0-kg pail hangs from a rope that wraps around a 0.20-m radius disk of rotational inertia $0.80\ \text{kg} \cdot \text{m}^2$. Determine the rotational speed of the disk 2.0 s after the pail is released from rest. Assume that $g = 10\ \text{N/kg}$.

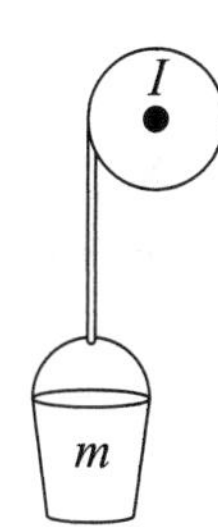

Proposed solution: The situation is pictured at the right. We choose the disk as the object of interest. The rope holding the pail exerts a downward force on the disk of magnitude $mg = (2.0\ \text{kg})(10\ \text{N/kg}) = 20\ \text{N}$. The rotational acceleration of the disk is:

$$\alpha = \Sigma\tau/I = (20\ \text{N})(0.20\ \text{m})/(0.80\ \text{kg} \cdot \text{m}^2) = 5.0\ \text{s}^{-2}$$

The rotational velocity after 2.0 s will be:

$$\omega = 0 + \alpha t = (5.0\ \text{s}^{-2})(2.0\ \text{s}) = 10\ \text{rad/s}$$

a. Identify any errors in the solution.

b. Provide a corrected solution if there are errors.

8.4.6 Pose a problem The person at the right is standing on a sled. Use the illustration to pose a problem which can be solved using the rotational form and the standard form of Newton's second law, rotational kinematics, and translational kinematics. Make sure you provide enough information to solve the problem.

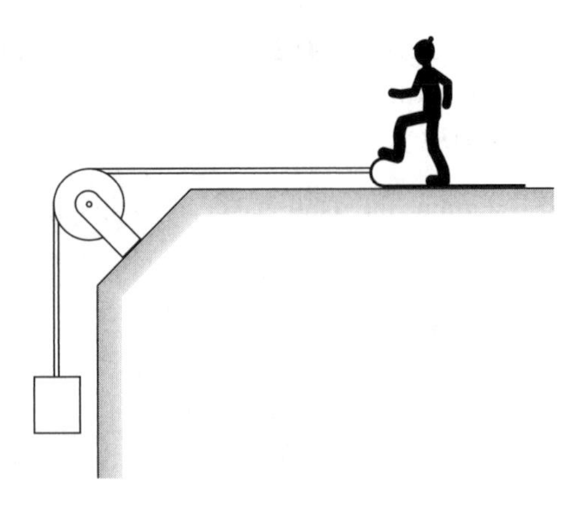

8.4.7 Design and analyze Design an experiment to determine the rotational inertia of the flywheel provided to you in lab. Your other available equipment includes a stand to hold the flywheel on a lab table, string, blocks with different mass, a stopwatch, and a meter stick.

Describe your experiment.	**Draw a labeled sketch of the experiment.**	**Draw one or more force diagrams (as needed).**	**Write the physical quantities that you will measure and quantities you will calculate.**
			To be measured: To be calculated:
Complete a mathematical solution that can be used to determine the rotational inertia.	**List additional assumptions you made.**	**List sources of experimental uncertainty and ways to minimize them.**	**Perform the experiment, record the data, and determine the rotational inertia.**
		Uncertainty: Ways to minimize:	

9 Gases

9.1 Qualitative Concept Building and Testing

9.1.1 Observe and explain Dip a piece of paper in rubbing alcohol (or rub the paper with alcohol) and place it on a table. The paper dries gradually.

a. What do you need to assume about the makeup of alcohol to explain the gradual disappearance of alcohol from the paper?

b. Think of possible mechanisms for the alcohol's disappearance. Suggest at least three different mechanisms. Fill in the table that follows.

Observation	List the possible mechanisms that explain the observation.
Alcohol disappears from the paper.	Mechanism 1:
	Mechanism 2:
	Mechanism 3:

9.1.2 Predict and test Mindy, Marc, Alex, and Nina are working on Activity 9.1.1. They agree that alcohol must be made of small parts to enable it to *gradually* "leave the paper" and dry. However, they disagree on the mechanism that allows these small parts to leave the paper. Fill in the table that follows to devise experiments whose outcomes you can predict based on each of the mechanisms given in the table that follows.

Mechanism	Describe a testing experiment.	Predict the outcomes of the testing experiments based on the mechanisms.
1. Mindy says that the parts did not *leave* the paper—that they are still there but that the students cannot see them.		If 1 is correct, then: If 2 is correct, then: If 3 is correct, then: If 4 is correct, then:

(continued)

Mechanism	Describe a testing experiment.	Predict the outcomes of the testing experiments based on the mechanisms.
2. Marc says that the small parts of alcohol went into the table because they soaked through the paper and on to the top of the table.		If 1 is correct, then: If 2 is correct, then: If 3 is correct, then: If 4 is correct, then:
3. Alex says that the air gradually absorbed the alcohol parts.		If 1 is correct, then: If 2 is correct, then: If 3 is correct, then: If 4 is correct, then:
4. Nina says that the parts moved out on their own—in effect jumped out because they are not stationary inside the alcohol but are constantly moving.		If 1 is correct, then: If 2 is correct, then: If 3 is correct, then: If 4 is correct, then:

9.1.3 Predict The following are testing experiments that Mindy, Marc, Alex, and Nina decided to perform. Predict the outcome of each experiment described below based on each of the four mechanisms given in Activity 9.1.2. (For example, if the small parts soak into the table through the paper and we hold the paper in our fingers when drying, then the paper should not dry—the table is not there to absorb the alcohol.)

a. Hold the paper that has been dipped in alcohol in your fingers without putting it on the table while it is drying.

Mechanism 1:

Mechanism 2:

Mechanism 3:

Mechanism 4:

b. Weigh the paper before the experiment, when it is wet, and then again when it is dry.

Mechanism 1:

Mechanism 2:

Mechanism 3:

Mechanism 4:

c. Take two identical pieces of paper and put the same amount of alcohol on each. Then place one under a vacuum jar and the other one just outside the jar.

Mechanism 1:

Mechanism 2:

Mechanism 3:

Mechanism 4:

d. Place a small drop of colored alcohol in clear alcohol but do not stir it.

Mechanism 1:

Mechanism 2:

Mechanism 3:

Mechanism 4:

9.1.4 Test Following are the results of the experiments described in Activity 9.1.3.

a. The paper dried.

b. The scale reading decreased as the paper dried and went back to the original reading before the alcohol was placed on the paper.

c. The paper under the vacuum jar dried faster.

d. The color dispersed after about 20 min.

What can you say about the proposed explanations based on the outcomes of the experiments? That is, which explanations led to the predictions that were not supported by the outcomes of the testing experiments?

9.1.5 Explain The only explanation for drying alcohol that could not be rejected by testing experiments was the explanation that alcohol consists of tiny particles (called *molecules*) that move randomly. How do you need to modify this explanation to account for the fact that not all of the particles leave instantly?

9.1.6 Observe and explain Open a bottle of strong perfume on your desk. A person next to you smells the perfume in 2 s, and a person across the room smells it in 3 min. Explain this phenomenon using the ideas from the previous activities or any other ideas.

9.1.7 Observe and explain Blow up a balloon and carefully observe how its shape changes during the process. Use the idea of moving particles to explain why it expands when you blow air into it. Explain why the balloon does not expand any more when you stop blowing. Describe an experiment you can perform to test your explanation.

9.1.8 Observe and explain Explain the observations in the table that follows using the model of particles moving randomly at different speeds.

Observation	Write your explanation.	List the assumptions you made while formulating your explanation.	Describe your proposed testing experiment.
You inflate a balloon indoors on a winter day and then take it outside; the balloon shrinks.			
You inflate a balloon outdoors at sea level and take it to a mountaintop; the balloon expands.			

9.2 Conceptual Reasoning

9.2.1 Represent and reason Imagine that a particle moves horizontally until it hits a vertical wall. Assume that it is an elastic collision (after the collision, the speed of the molecule is the same as before the collision) and the motion of the particle obeys Newton's laws. Fill in the table below to represent the motion of the particle.

Draw an arrow representing the momentum of the particle before the collision.	Draw an arrow representing the momentum of the particle after the collision.	Draw a momentum change arrow.
Draw the force exerted by the wall on the particle.	**Draw the force exerted by the particle on the wall.**	

9.2.2 Represent and reason Imagine that you have eyes that can see the particles of air in the room. Draw a picture representing the behavior of several particles as they move through the room. Think of their possible collisions and how the collisions will affect the directions of their motion and the magnitudes of their speeds.

9.2.3 Reason One mole (6×10^{23} particles) of chicken feathers is spread uniformly over the surface of Earth. Estimate the thickness of this layer of feathers. Justify any assumptions you make in your calculations.

9.3 Quantitative Concept Building and Testing

9.3.1 Simplify Fill in the following table to explain several observed phenomena, using the physical quantity of pressure and the simplified model that gases are made of small particles with empty space between them that move randomly and collide like hard billiard balls with the walls of the container.

Observed phenomenon	Write an explanation using the concept of pressure and the simplified gas model.
A balloon keeps a rounded shape when filled with air.	
If you put a sealed, deflated balloon under a vacuum jar, it will expand when you pump the air out of the jar.	
When you pump air into a rubber raft, it becomes bigger.	

9.3.2 Derive Imagine that the gas inside a container has such low density that its particles never collide with each other; they collide only with the walls of the container. Assume a model of the gas as tiny moving billiard balls obeying Newton's laws. We wish to calculate the pressure that the gas exerts on the walls of the container.

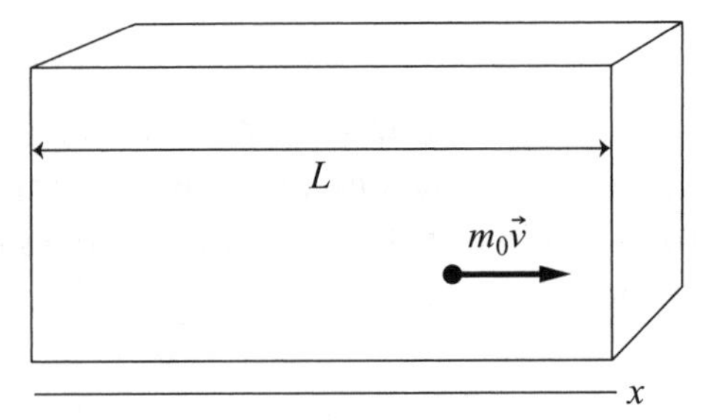

a. Start with one of the "balls" of mass m traveling at speed v parallel to the x axis. The ball bounces back and forth between two surfaces that are separated by distance L and that are perpendicular to the x axis. Use your knowledge of the impulse–momentum relationship to show that the impulse of the ball against one surface as a result of one collision has magnitude $2m_0v_x$.

b. Show that the time interval between hits for the one ball against that same wall is $2L/v_x$.

c. Use the results from parts a and b to show that the force of these impulsive collisions averaged over time F_{avg} is $m_0v_x^2/L$ and that the average pressure that N balls, or particles, will exert is $P = N(m_0v_x^2)/L^3$.

d. The v_x^2 in the average pressure equation in part c should more properly be designated as the average of the square of the x components of the velocities $\overline{v_x^2}$. The N particles inside the container move at different speeds in different directions. How is $\overline{v_x^2}$ related to the average of the square of the speeds $\overline{v^2}$? *Hint:* Assume that one-third of the particles move in each of the three directions (x, y, z).

e. Now consider the pressure that these particles exert on the walls of the container. Show that the pressure that they exert is equal to $P = \frac{1}{3}\frac{N}{V}(m_0\overline{v^2})$.

9.3.3 Reason Answer the following questions to decide if the expression for the pressure $P = \frac{1}{3}\frac{N}{V}(m_0\overline{v^2})$ in an ideal gas makes sense (where V = volume).

a. Why would you expect the pressure to depend on the total number of particles N in the container? Explain.

b. Why would you expect the pressure to depend on the average speed of the particles squared? Explain.

c. When we derived the expression $P = \frac{1}{3}\frac{N}{V}(m_0\overline{v^2})$, we neglected the size of the particles and the fact that they interact with each other. Discuss whether the pressure of a real gas for which these assumptions are not true should be larger or smaller than the pressure of the ideal gas calculated according to the derived expression. The gases are in containers of the same volume and have the same number of particles with the same average of the square of their speeds.

9.3.4 Observe and explain A sealed metal container with a low-density gas is alternately placed in baths of water at different temperatures. Then a different gas is placed in the container and the procedure is repeated. Data related to the pressure and the temperature of the gas inside the container are shown in the graph to the right.

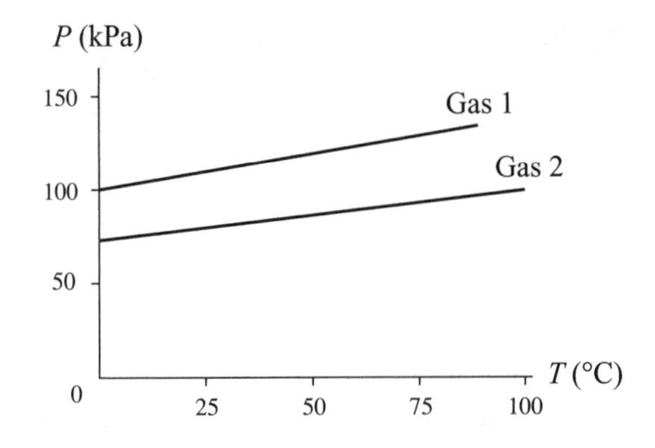

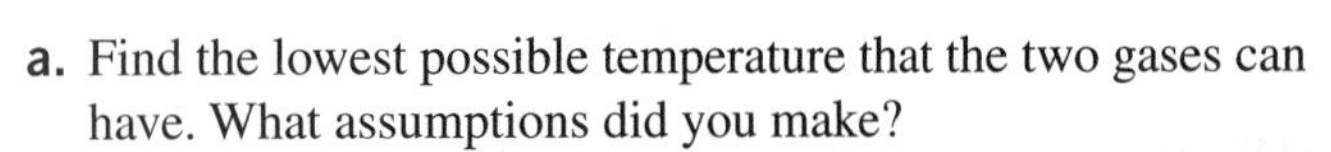

a. Find the lowest possible temperature that the two gases can have. What assumptions did you make?

b. Use the lowest temperature value you determined in part a to make a new temperature scale with the same temperature interval as in the Celsius scale but with the new zero point. This new scale is called the *absolute temperature scale*.

c. Use the expression in Activity 9.3.2e, $P = \frac{1}{3}\frac{N}{V}(m_0\overline{v^2})$, to discuss how the temperature on the absolute scale is related to the average kinetic energy of gas particles.

9.3.5 Reason In the late 19th century, scientist John Loschmidt found that there were 2.69×10^{19} molecules of any type of gas in 1 cm^3 when at 0°C and at atmospheric pressure. Use proportional reasoning to determine the volume of 1 mol of any gas under the same conditions.

9.3.6 Explain If we take different containers, each with 1 mol of different gases (for example, nitrogen, oxygen, and helium), and place each container in melting ice, we find that the ratio $\frac{PV}{N}$ is the same for all gases. If we place the same gases in a container with hot water, the ratio $\frac{PV}{N}$ is different, but again it is the same for all three gases. What is the meaning of the ratio? To answer this question, find the units of the ratio (the number does not have units) and think of when we encountered the same ratio before. After you decide on the units, think about why the ratio is the same when gas containers are placed in the same medium.

9.3.7 Reason The table below presents data collected when a constant-volume metal container with 1 mol of nitrogen ($N = N_A = 6.02 \times 10^{23}$ molecules) is placed in baths of very different temperatures. If we assume that the ratio $\frac{PV}{N}$ is proportional to the absolute temperature of the gas (i.e., $\frac{PV}{N} = kT$), we can find the coefficient of proportionality k. Solve for the proportionality constant k for the two sets of data.

Known quantities	$\frac{PV}{NT} = k$	**Known quantities**	$\frac{PV}{NT} = k$
$P = 1.01 \times 10^5\ \text{N/m}^2$	$k =$	$P = 1.38 \times 10^5\ \text{N/m}^2$	$k =$
$T = 273\ \text{K}$ (melting ice)		$T = 373\ \text{K}$ (boiling water)	
$V = 22.4 \times 10^{-3}\ \text{m}^3$		$V = 22.4 \times 10^{-3}\ \text{m}^3$	

a. Is the value independent of gas temperature? Explain.

b. Find a relationship between the absolute temperature of the gas and the average kinetic energy of its particles. *Hint:* Use the results of Activities 9.3.2 and 9.3.6 to help.

c. Find the total kinetic energy of all particles of nitrogen at the two different temperatures.

d. Does the absolute temperature of nitrogen depend on the total number of particles in the container? Explain.

e. Does the total thermal energy due to the motion of nitrogen particles depend on the number of particles in the container? Explain.

f. Compare and contrast two physical quantities: absolute temperature and thermal energy.

9.3.8 Reason The following two equations are fairly similar: $PV = \frac{1}{3}Nm\overline{v^2}$ and $PV = NKT$. How do these equations differ in terms of what they describe and the possibility of measuring the quantities that they relate?

9.4 Quantitative Reasoning

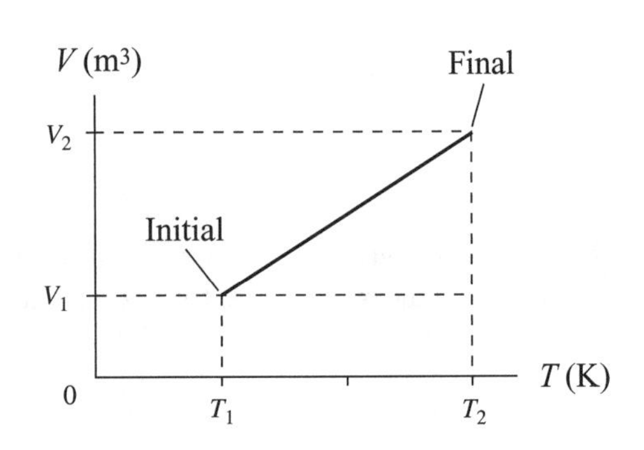

9.4.1 Represent and reason The graph to the right describes a process occurring to an ideal gas.

a. Fill in the table that follows.

Describe the process macroscopically.	Describe how you can carry out the process.
Represent the process mathematically.	**Explain the process microscopically.**

b. Represent the same process in P-versus-T and P-versus-V graphs.

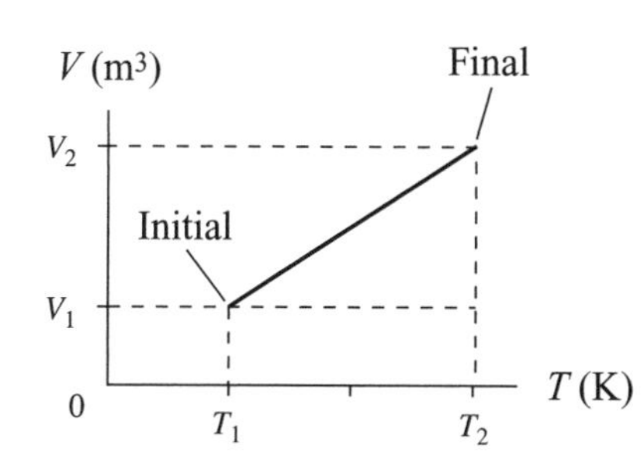

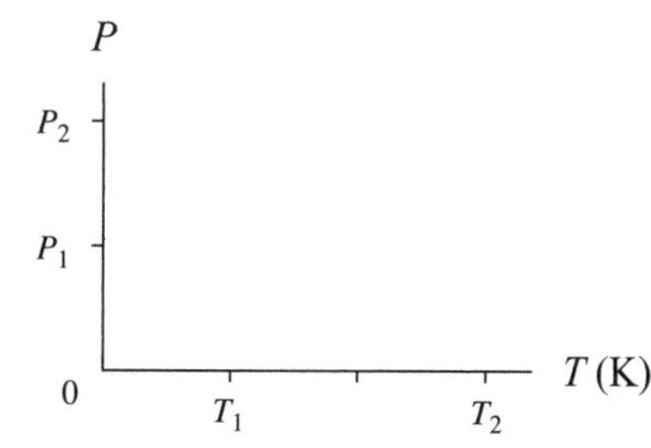

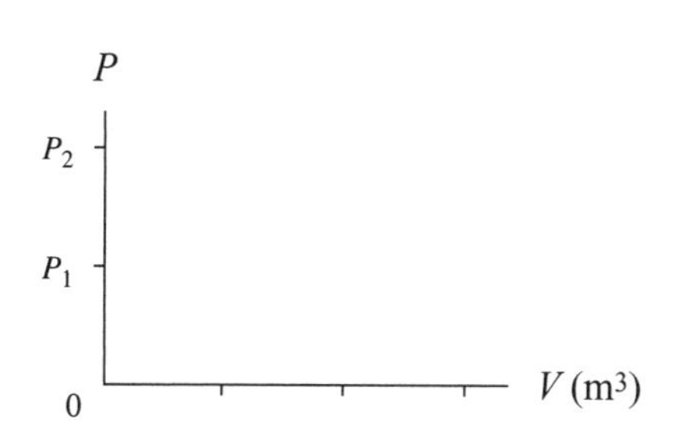

9.4.2 Represent and reason The P-versus-T graph to the right describes a real process.

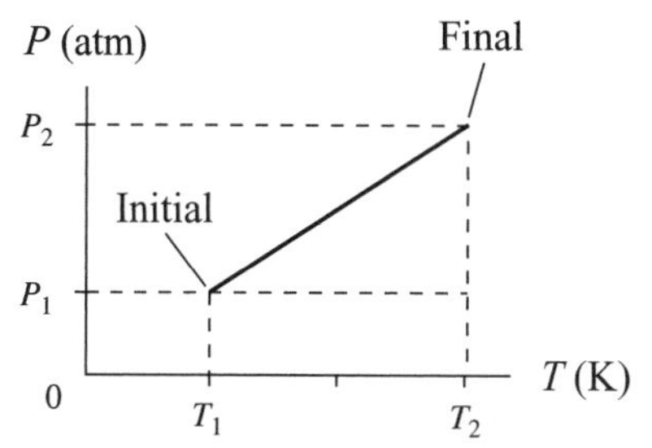

a. Fill in the table that follows.

Describe the process macroscopically.	Describe how you can carry out the process.
Represent the process mathematically.	**Explain the process microscopically.**

b. Now represent the same process in P-versus-V and V-versus-T graphs below.

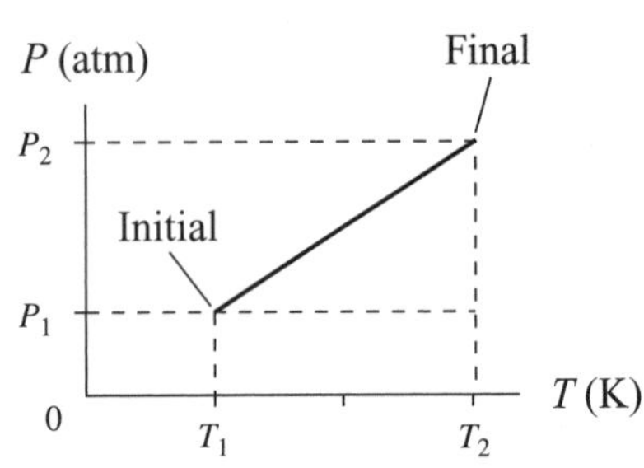

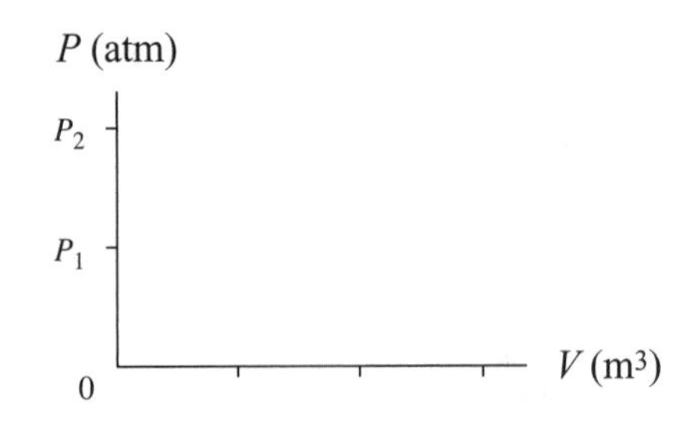

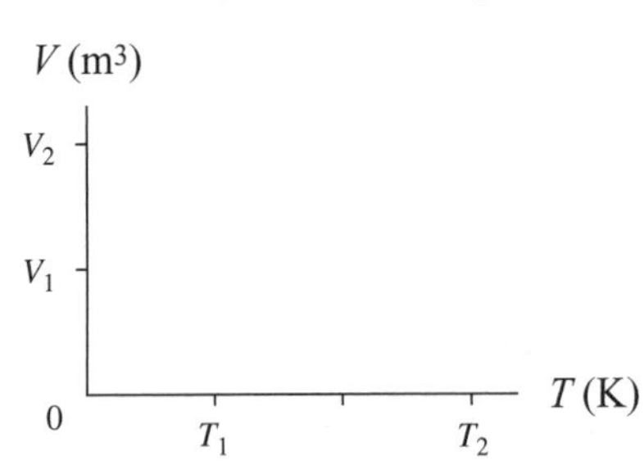

9.4.3 Represent and reason Write everything you can about the process represented in the graph below.

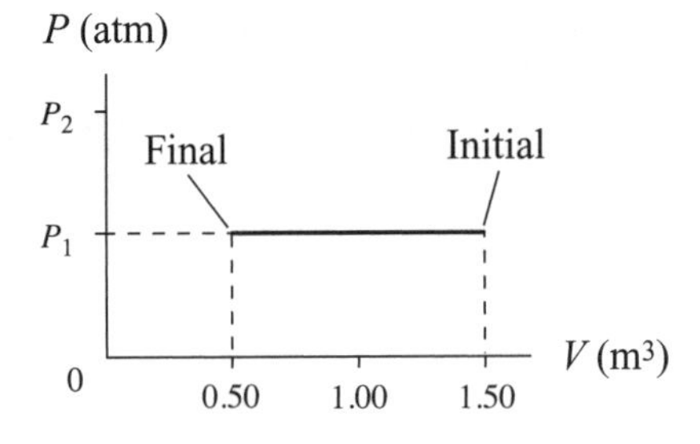

9.4.4 Represent and reason The P-versus-T graph in part b describes a cyclic process comprising three hypothetical processes. The mass of the gas is constant.

a. Describe the processes represented on the P-versus-T graph in part b by completing the table that follows.

Process	Write what happens to the pressure of the gas.	Write what happens to the temperature of the gas.	Write what happens to the volume of the gas.
$1 \rightarrow 2$	Remains constant		
$2 \rightarrow 3$		Increases	Remains constant (the line passes through the origin)
$3 \rightarrow 1$			

b. Use the information in the table to represent the processes in P-versus-V and V-versus-T graphs. Notice that we placed the P-versus-V graph to the right of the P-versus-T graph to keep the same scale for pressure and the V-versus-T graph under the P-versus-T graph to keep the same scale for temperature.

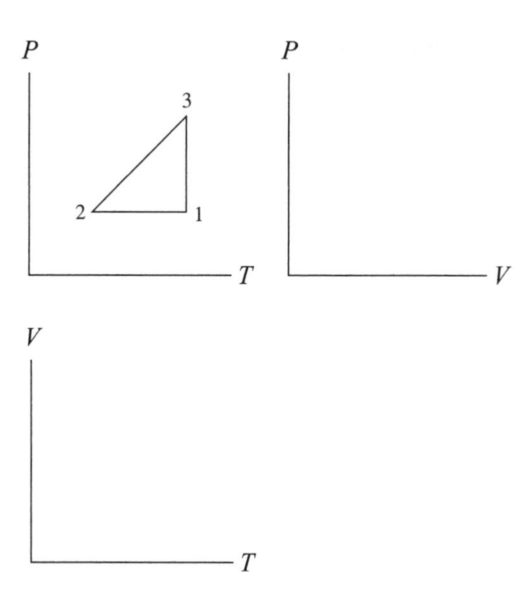

9.4.5 Represent and reason The V-versus-T graph in part b represents a cyclic process comprising four hypothetical processes. The mass of the gas remains constant.

a. Describe the processes represented on the graph in part b by completing the table.

Process	Write what happens to the pressure of the gas.	Write what happens to the temperature of the gas.	Write what happens to the volume of the gas.
$1 \rightarrow 2$			
$2 \rightarrow 3$			
$3 \rightarrow 4$			
$4 \rightarrow 1$			

b. Use the information in the table to represent the processes in P-versus-T and P-versus-V graphs.

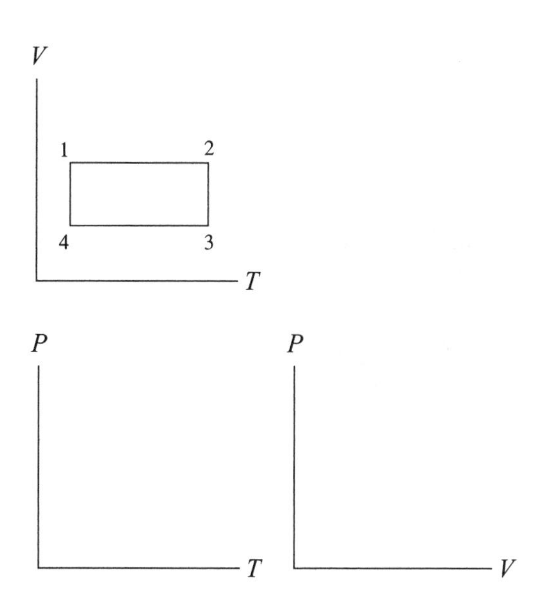

9.4.6 Regular problem The variation of the volume and pressure of a constant-mass ideal gas is provided in the table. The process starts at 0°C.

V (m^3)	P (N/m^2)
6.00×10^{-4}	1.00×10^5
5.45×10^{-4}	1.10×10^5
6.67×10^{-4}	0.90×10^5

a. Determine the number of gas particles.

b. Does the temperature change during the process? How do you know?

c. Explain the process microscopically.

d. If you carried out this process, what equipment would you need? What would you do?

9.4.7 Regular problem The constant volume (isochoric) variation of the temperature of 1 mol of an ideal gas is provided in the table below. Complete the table. (*Note:* Find the unknown pressures in the second column.)

T (K)	P (N/m^2)	**Describe the process. Explain the process microscopically and sketch a *P*-versus-*T* graph**
273	1.00×10^5	
336		
410		

9.4.8 Equation Jeopardy The equations in the following table describe a thermodynamic process (each equation can describe more than one process). Complete the table to describe one process for each equation.

Equations that describe one or more physical processes	Describe in words a process that is consistent with the equation.
$(0.012\ \text{m}^3 - 0.010\ \text{m}^3) = (0.25\ \text{mol})(8.3\ \text{J/K})/(1.0 \times 10^5\ \text{N/m}^2)(T - 300\ \text{K})$	
$(1.0 \times 10^5\ \text{N/m}^2)(0.018\ \text{m}^3)/(300\ \text{K}) = P(0.020\ \text{m}^3)/(280\ \text{K})$	

9.4.9 Regular problem Examine the air in the room in which you are solving this problem. Assume that the pressure is normal.

a. How many molecules of air are in 1 m^3 at normal conditions?

b. What is the average distance between molecules compared to the dimensions of the molecules? Consider the diameter of molecules to be about 10^{-10} m.

c. Can you consider air to be an ideal gas?

9.4.10 Regular problem Estimate the energy that the Sun possesses due to the kinetic energy of its particles. Assume that the Sun is made of atomic hydrogen. The mass of the Sun is 2×10^{30} kg, and the average temperature is about 100,000 K.

9.4.11 Evaluate the solution

The problem: A deflated balloon starts with 0.017 mol of air inside of it, an internal volume of $3.0\ \text{cm}^3$, and an inside pressure of 1.0 atm. You add air so that when filled, the balloon contains 0.33 mol of air and has a volume of $30\ \text{cm}^3$. The process occurs at constant temperature. Determine the final pressure in the balloon.

Proposed solution: For a constant temperature process, $P_0V_0 = PV$. Thus:

$$P = P_0V_0/V = (1.0\ \text{atm})(3.0\ \text{cm}^3)/(30\ \text{cm}^3) = 0.10\ \text{atm}$$

a. Identify any missing elements or errors in the solution.

b. Provide a corrected solution or missing elements if you find errors.

9.4.12 Design an experiment Observe the motion of your rib cage as you breathe. How does your body manage to inhale and exhale air from your lungs? After describing in words how this process works, use the suggested materials to construct an apparatus that can be used as a breathing model: an open plastic bottle with the bottom cut off, a surgical glove, 15-inch balloons, and rubber bands.

9.4.13 Estimate

a. Estimate the mass of the air in a typical home. Indicate any assumptions that you make about the size of the room, the air pressure, and the temperature.

b. Estimate the average speed of air particles in the room. Indicate any assumptions you make.

c. How many molecules are in 1 g of air at normal conditions? If these molecules lie uniformly on Earth's surface, *estimate* the number that would be under your feet right now. The radius of Earth is about 6400 km. Indicate any assumptions you make.

10 Fluids at Rest

10.1 Qualitative Concept Building and Testing

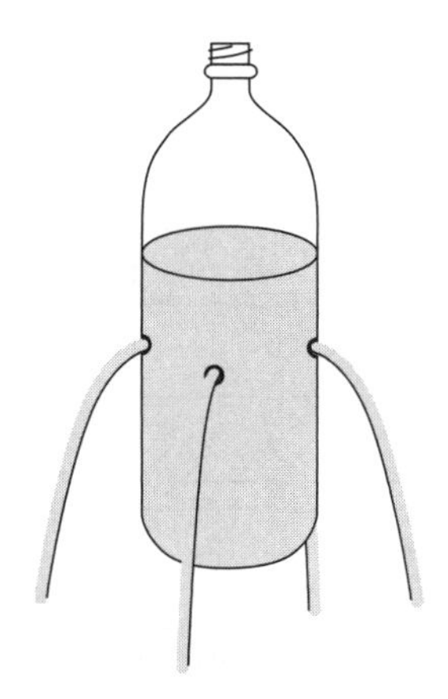

10.1.1 Observe and explain A 32-oz open plastic bottle full of water is punctured in four spots with thumbtacks. When the thumbtacks are removed from the bottle, narrow streams of water leave the bottle in identically shaped arcs perpendicular to the bottle's surface, as shown.

a. What do you need to assume about the pressure inside water at the level of the thumbtacks to explain the way water shoots out of the four holes?

b. Explain using your knowledge of the random motion of molecules why the water pressure is as described in part a.

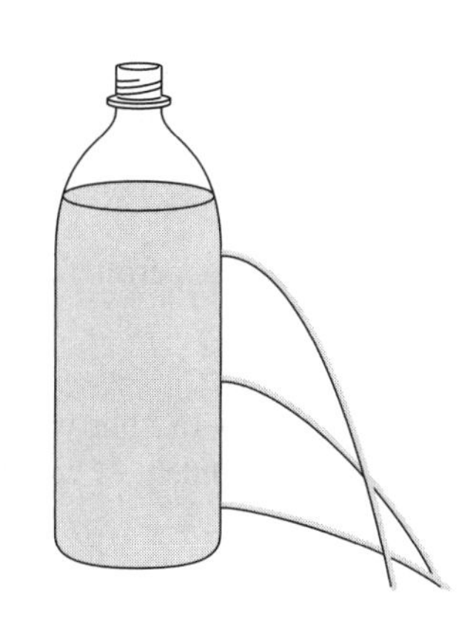

10.1.2 Observe and explain Three thumbtacks are poked into an open bottle full of water. When the thumbtacks are removed, the water streaming out of the holes takes the form that is shown in the illustration. To help explain this experiment, imagine dividing the liquid in the bottle into horizontal layers and examine the forces that layers exert on each other. Use Newton's third law for reasoning. Fill in the table that follows.

Draw force arrows to indicate the force that layer 1 exerts on layer 2, layer 2 exerts on layer 3, and layer 3 exerts on layer 4. Indicate the relative magnitudes of the forces by the lengths of the arrows.	Draw force arrows to indicate the force that layer 4 exerts on layer 3, layer 3 on layer 2, and layer 2 on layer 1. Indicate the relative magnitudes of the forces by the lengths of the arrows.	Draw arrows representing the pressure that the liquid exerts on very small surfaces inside the liquid shown below. Remember that liquids exert pressure in all directions.	Use the drawings in this table to help explain the experiment with the three thumbtacks.
1 2 3 4	1 2 3 4	1 2 3	

10.1.3 Test your idea Use the explanation that you constructed in Activity 10.1.2 to predict what happens if you take the same bottle, close the lid, and then remove only the top and the bottom tacks. After you complete your prediction, perform the experiment to test it.

Predict the outcome.	Write an explanation for your prediction.	Perform the experiment; record the outcome.	Make a judgment about the explanation that you used to make the prediction: Was it supported or not?

10.1.4 Test an idea Connect a pressure sensor to a computer, fill a beaker with water, and gather a ruler.

a. Describe an experiment that you can perform with this equipment to test the idea that fluids (liquids and gases) exert the same pressure on a surface at a particular depth in the fluid, independent of how that surface is oriented. What is your prediction?

b. Perform the experiment and record the outcome. Does it match your prediction?

c. Describe any reasons why fluids (liquids and gases) exert the same pressure on a surface at a particular depth in the fluid, independent of how that surface is oriented.

d. Describe everyday observations that are consistent with this idea.

10.1.5 Observe and explain You can lift objects immersed in water that are too heavy to lift when in the air (for example, you can easily lift your friend in a pool). Use the ideas developed in Activities 10.1.1 and 10.1.2 to explain this experience. Draw a force diagram to support your explanation.

10.1.6 Test an idea Hang a 2-liter empty plastic bottle from a spring; the spring stretches. Joshua says that the spring stretches because the air pushes down on the bottle. Taisha disagrees. She thinks that the air pushes up similar to the water pushing up on a person in the pool, and the spring stretches due to Earth pulling down on the bottle. How can you test whose idea makes more sense? You have the equipment shown below.

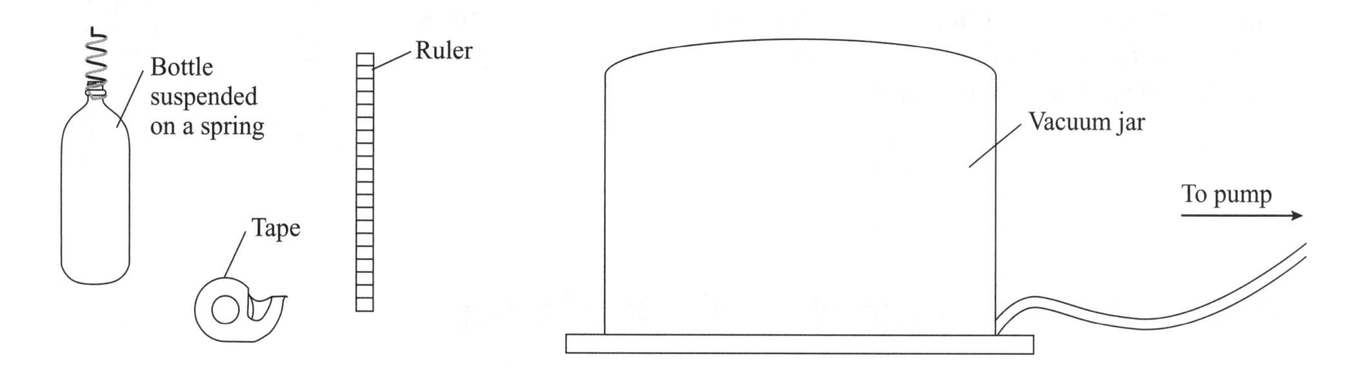

Fill in the table that follows.

Describe your experiment in words.	Predict the outcome based on Joshua's hypothesis.	Predict the outcome based on Taisha's hypothesis.	Perform the experiment, record the outcome of the experiment, and decide whose hypothesis needs to be rejected.
			Outcome: Hypothesis to be rejected:

10.2 Conceptual Reasoning

10.2.1 Reason You drink water from a plastic bottle while riding on an airplane. When the water is gone, you close the bottle and leave it under the seat. It looks like the bottle in part a of the figure. When the plane lands you pick up the bottle and to your surprise the bottle looks like the bottle in part b. Explain this observation by combining your knowledge of molecules, which you acquired in Chapter 9, with the new knowledge of the elevation dependence of fluid pressure that you acquired in Section 10.1.

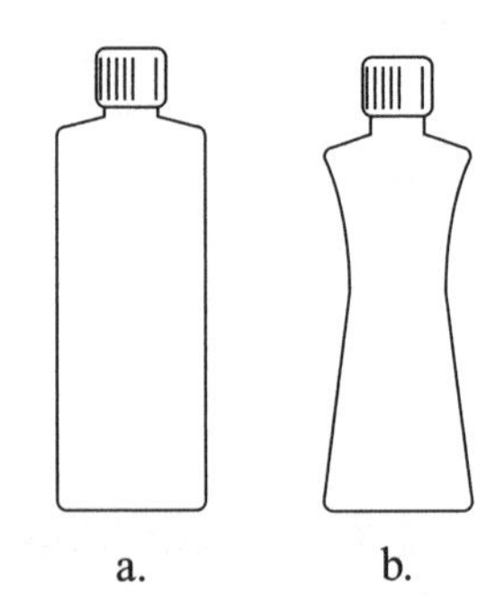

10.2.2 Reason A hydraulic lift has a small plunger on one side of a container of liquid and another large plunger on the other side. You can push down on the small plunger, exerting a relatively small force, and lift a heavy object sitting on the other plunger. Explain how this is possible. *Hint:* Think of the cross-sectional area under each plunger, the relationship between pressure and force and Pascal's Law.

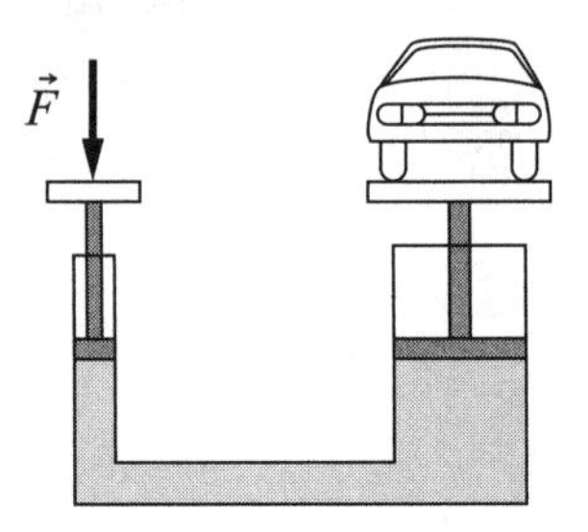

10.3 Quantitative Concept Building and Testing

10.3.1 Observe and find a pattern Chris uses a graduated cylinder and a scale to collect data of the masses and volumes of baby oil and water. He measures the mass of the empty cylinder, fills the cylinder with increasing volumes of oil, measures the mass of the cylinder with oil, and subtracts the mass of the empty cylinder. Then he repeats the same experiment with water. The volumes and masses of the liquids are provided in the table.

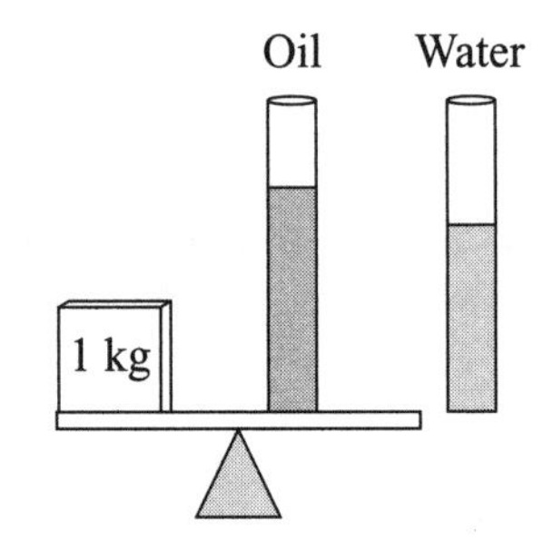

Volume (ml = cm^3)	Mass of water (g)	Mass of oil (g)
50	49	44
70	71	64
100	102	90
130	128	116
150	150	135
200	199	182
250	252	224
300	300	270
350	348	315

a. Graph mass versus volume for each liquid.

b. Examine the two graphs and discuss the similarities and differences. What is the meaning of the slope of the line for each liquid? Come up with a name that you can use for the ratio of the mass and the volume as represented by the slope of the graph.

c. Choose a particular value of the volume on the *V* axis and find the corresponding values of the mass of both liquids. Which mass is larger?

d. Choose a particular value of the mass on the *m* axis and find the corresponding values of the volume for both liquids. Which volume is larger?

10.3.2 Observe and explain You slowly lower a pressure gauge into a lake and measure the pressure at different depths in the water (see the table below).

Pressure (N/m^2)	Distance below surface (m)
95,000	0
203,000	10
298,000	20
405,000	30

Construct a graph of pressure versus depth.	Based on the graph line, write a relationship between pressure and the distance below the surface of the water.	Explain the relationship.

10.3.3 Observe and explain You repeat the previous experiment but this time in ocean water. The table shows the readings of the pressure gauge.

Pressure (N/m^2)	Distance below surface (m)
103,000	0
207,000	10
313,000	20
420,000	30

Explain why the readings are different from the readings in 10.3.2.

10.3.4 Test an idea Two of your friends disagree on how the pressure in a liquid depends on different physical quantities. Ari thinks that the pressure depends only on the depth—the deeper you go in the same liquid, the higher the pressure. Maria thinks that the mass of liquid above the level at which one measures the pressure matters.

a. Discuss supporting arguments for Ari's and Maria's hypotheses.

b. Fill in the table that follows to decide which friend is correct.

Describe an experiment you can perform to find out whose idea can be ruled out.	Make a list of equipment that you might need.	Predict the outcome of the experiment based on each hypothesis.	Perform the experiment; record the outcome and decide whose hypothesis was supported.
		If Ari is correct, then: If Maria is correct, then:	

10.3.5 Observe and explain A 2.0-kg block made of aluminum is attached to a Newton spring scale and is slowly lowered into a graduated cylinder filled with water. The experiment is repeated, only this time with a 2.0-kg copper block. The readings of the scales and the readings of the water level at the top surface of the water in the cylinder are shown in the table that follows.

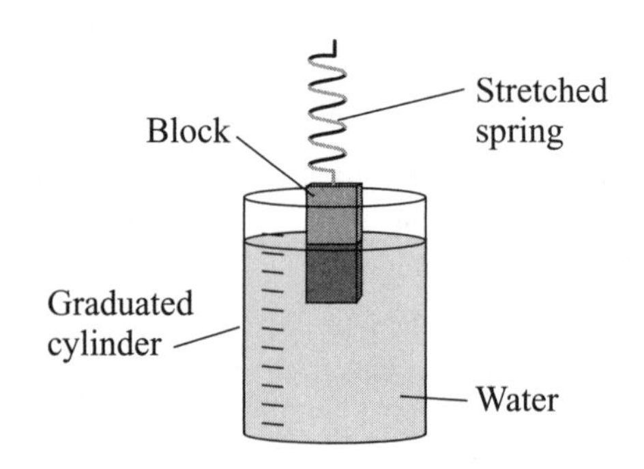

Scale reading for the aluminum block	Water level in the graduated cylinder	Volume of the aluminum block under the water	Scale reading for the copper block	Water level in the graduated cylinder	Volume of the copper block under the water
19.60 N	200 ml	0	19.60 N	200 ml	0
19.10 N	250 ml	50 ml	19.10 N	250 ml	50 ml
18.60 N	300 ml	100 ml	18.60 N	300 ml	100 ml
18.10 N	350 ml	150 ml	18.10 N	350 ml	150 ml
17.60 N	400 ml	200 ml	17.60 N	400 ml	200 ml
15.60 N	600 ml	400 ml	17.36 N	424 ml	224 ml (just submerged)
13.60 N	800 ml	600 ml	17.36 N	424 ml	224 ml (deeper in water)
12.20 N	940 ml	740 ml (just submerged)	17.36 N	424 ml	224 ml (even deeper)
12.20 N	940 ml	740 ml (deeper in water)	17.36 N	424 ml	224 ml (even deeper)

a. Use the data in the table to find a pattern in the amount of upward force that the water exerts on each block. *Hint:* Remember that the density of water is 1000 kg/m^3.

b. After the copper block is completely submerged, the scale reads 17.36 N. After the aluminum block is completely submerged, the scale reads 12.20 N. Both readings do not change when the blocks are lowered deeper under the water. Explain these results using the pattern that you found in part a.

10.3.6 Observe and explain The experiment described in the previous problem is repeated again, but this time the blocks are lowered into vegetable oil. Use the data recorded in the table to find a pattern in the amount of upward force that the oil exerts on each block. *Hint:* Remember that the density of oil is 900 kg/m^3.

Scale reading for the aluminum block	Oil level in the graduated cylinder	Volume of the aluminum block under the oil	Scale reading for the copper block	Oil level in the graduated cylinder	Volume of the copper block under the oil
19.60 N	200 ml	0	19.60 N	200 ml	0
19.16 N	250 ml	50 ml	19.16 N	250 ml	50 ml
18.70 N	300 ml	100 ml	18.70 N	300 ml	100 ml
18.26 N	350 ml	150 ml	18.26 N	350 ml	150 ml
17.80 N	400 ml	200 ml	17.80 N	400 ml	200 ml

10.3.7 Observe and explain

a. In Activities 10.3.5 and 10.3.6, what pattern emerges between the volume of the object under water and the effect of the fluid in supporting the object?

b. Use this pattern to devise an expression for the lifting force exerted by a liquid on an object partly or totally submerged in the liquid.

10.3.8 Test your idea Fill in the table that follows to predict the spring scale reading when you hang a block completely submerged in water of density 1000 kg/m^3.

Record the spring scale reading of the block in the air.	List the quantities that you need to measure to predict the reading of the scale when the block is submerged in water.	Write a procedure to predict the reading of the scale when the block is completely submerged.	Perform the experiment; record the actual value and then reconcile it with your prediction.

Will the reading change if you continue lowering the block deeper into the water?

10.3.9 Predict and test Use the values for the densities of water and oil to fill in the table that follows.

Predict what will happen if you pour water in a glass beaker partially filled with oil.	Explain your prediction.	Perform the experiment; record the results.

10.3.10 Observe and explain You have an unopened can of regular Coke® and an unopened can of Diet Coke®. You place the can of Diet Coke® in water and watch it float; then you place the can of regular Coke® in water and watch it sink. Fill in the table that follows to explain.

List the quantities that you will measure.	List the quantities that you will calculate.	Write a mathematical procedure that will help you explain the outcome.	Explain the outcome of the experiment.

10.3.11 Derive In Activities 10.3.5–10.3.10 you found that a fluid exerts an upward force on submerged objects.

a. Explain *why* the fluid pushes up on a submerged object.

b. Explain why the fluid would be expected to exert an upward force on a submerged object that is equal to the force that Earth exerts on the fluid displaced by the submerged object. *Hint:* Think about what the fluid supports when the object is not occupying that particular volume.

10.4 Quantitative Reasoning

10.4.1 Represent and reason The table below provides a word description for a situation involving the pressure of liquids. Complete the table.

Word description	Sketch the situation (identify two points in the water).	Apply the relationship $P = P_0 + \rho gh$ and any other principles needed to determine the net force that the air inside and water outside exert on the window.
A submarine is 100 m below the surface of the water, which has a density of 1000 kg/m^3. Compare the force of the water outside the submarine and the air inside (1.0 atm) on a 0.1-m × 0.2-m window.		

10.4.2 Regular problem The pressure above atmospheric pressure of blood in a person's head and in his or her feet when standing is shown at the right. Explain whether these numbers make sense.
Note: 780 mm Hg $= 1.0 \times 10^5$ N/m^2 $= 1.0$ atm.

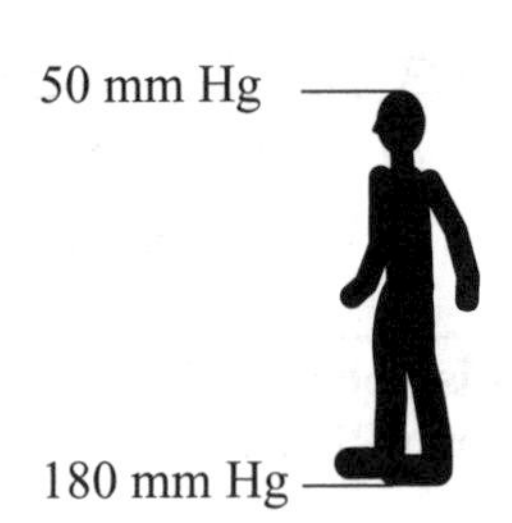

10.4.3 Represent and reason Word descriptions and pictorial representations are provided below for two situations involving the buoyant force. Complete the table that follows.

Word description	Picture description (system object circled)	Construct a force diagram for the system object.	Apply Newton's second law in component form for the system object.
A person of density 980 kg/m^3 floats partially submerged in salt water of density 1030 kg/m^3.			
You hold a 40-kg rock of density 2300 kg/m^3 that is completely submerged in water of density 1000 kg/m^3.			

10.4.4 Regular problem A 50-kg woman having a density of 980 kg/m^3 stands on a bathroom scale. Determine the reduction of the scale reading due to the air.

10.4.5 Equation Jeopardy Below we present principles of fluids applied to various situations. Complete the table that follows.

Principles of fluids applied to a situation	Sketch a situation that might be described by the equation.	Describe the situation in words.	Draw a force diagram for an object of interest (for parts b and c only).
a. $(1.6 \times 10^5\ \text{N/m}^2) - (1.0 \times 10^5\ \text{N/m}^2) = (1000\ \text{kg/m}^3) \times (9.8\ \text{N/kg})(0.0\ \text{m} - y_1)$			
b. $F + (1000\ \text{kg/m}^3) \times (9.8\ \text{N/kg})(0.010\ \text{m}^3) - (24\ \text{kg})(9.8\ \text{N/kg}) = 0$			
c. $(1000\ \text{kg/m}^3) \times (9.8\ \text{N/kg})V_{\text{displaced water}} - (24\ \text{kg})(9.8\ \text{N/kg}) = 0$			

10.4.6 Evaluate the solution

The problem: You slowly lift a 20-kg rock of density 2400 kg/m^3 from the bottom of a lake near the shore. Determine the force that you must exert on the rock while lifting it when it is under the water. Assume that $g = 10$ N/kg.

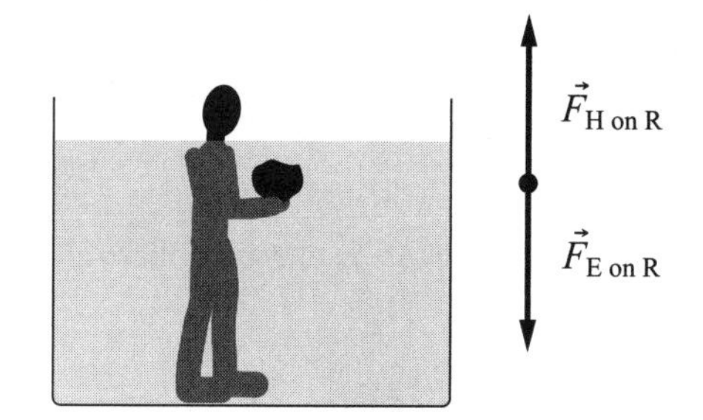

Proposed solution: Because you lift it slowly, the rock is not accelerating ($a = 0$). A sketch of the situation and a force diagram for the rock (the system) are shown. Applying the vertical component form of Newton's second law, we find:

$$+F_{\text{H on R}} - F_{\text{E on R}} = 0 \text{ or } F_{\text{hands}} = F_{\text{E on O}} = mg = (20\text{ kg})(10\text{ N/kg}) = 200\text{ N}$$

a. Identify any missing elements or errors in the solution.

b. Provide a corrected solution if there are errors.

10.4.7 Evaluate the solution

The problem: A 3.0-m $\times$ 1.0-m rectangular plastic container that is 1.0-m high has a mass of 1500 kg. The container floats in fresh water (density 1000 kg/m^3), partially submerged. Find the depth of the container submerged in water (marked as h in the figure). Assume that $g = 10$ N/kg.

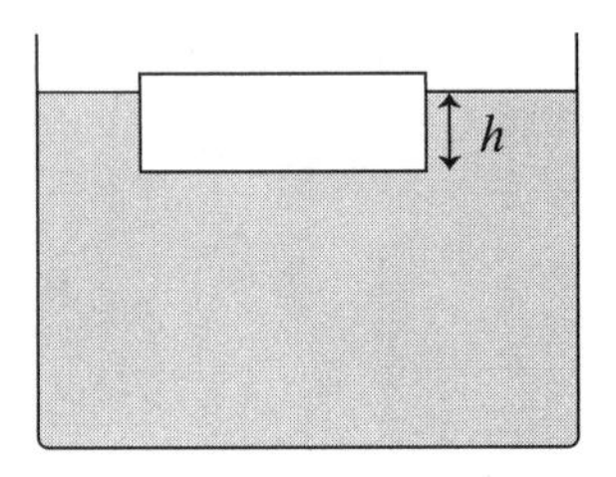

Proposed solution: The container is floating and has zero acceleration ($a = 0$). A force diagram for the (3.0 m)(1.0 m)(1.0 m) = 3.0-m^3 container (the system) is shown at the right. Applying the vertical component form of Newton's second law, we find:

$\vec{F}_{\text{W on C}}$

$\vec{F}_{\text{E on C}}$

$$\Sigma F_y = 0$$

$$+F_{\text{W on C}} - F_{\text{E on C}} = 0$$

$$\rho g V - mg = 0$$

$$[(1500\text{ kg})/(3.0\text{ m}^3)](10\text{ N/kg})(3.0\text{ m} \cdot 1.0\text{ m} \cdot h) - (1500\text{ kg})(10\text{ N/kg}) = 0$$

$$h = 3.0\text{ m}$$

a. Identify any missing elements or errors in the solution.

b. Provide a corrected solution if there are missing elements or errors.

10.4.8 Design an experiment You have an object of unknown material. Design and carry out two independent experiments that will allow you to determine its density. Reconcile any discrepancy in the two values you obtain.

10.4.9 Predict and test You have a rectangular block of wood, a large container with water—big enough for the block to fit into it—a ruler, and a scale. Predict how much of the block will be above water if you place the block into the container. Then perform the experiment and compare your prediction to the outcome.

a. Fill in the table that follows.

Describe the procedure that will help you make your prediction.	List quantities that you will need to measure and record the measurements.	Use the measured quantities and the procedure to make the prediction. Take into account experimental uncertainties.	Perform the experiment and record the experimental value of the height of the block above the surface.
Discuss whether the predicted and experimental values agree.			

11 Fluids in Motion

11.1 Qualitative Concept Building and Testing

11.1.1 Observe and explain For each experiment, observe what happens.

a. Then complete the table that follows.

Word description	**Sketch and identify a system.**	**Sketch a force diagram for the object of interest. Include the force that the fluid exerts on the system on both sides.**
Hold a piece of 8.5" × 11" paper at the corners and place your lips just above the paper between your fingers (the paper hangs down at the end away from your lips). Observe what happens if you blow hard across the top of the paper.	The hanging half of the paper farthest from your lips is the system.	Immediately after you start blowing, the end of the paper:
Place two empty soda cans on a smooth table separated by about 1 cm (or you can place cans on straws for a more dramatic effect). Blow hard between the cans.	One of the cans is the system.	Immediately after you start blowing, the cans start to move:
Fold a 5" × 8" index card in a U shape and place it on a level surface. Observe what happens to the card if you vigorously blow air through a straw under the card.	The horizontal surface of card is the system.	Immediately after you start blowing hard, the card:

b. Develop an explanation for the observations in these three experiments. Think of how the pressure of a fluid against a surface changes as the speed of the fluid moving across the surface increases.

11.1.2 Test your idea You have an industrial-style coffeemaker. An open narrow tube in the front of the pot indicates the amount of coffee that remains.

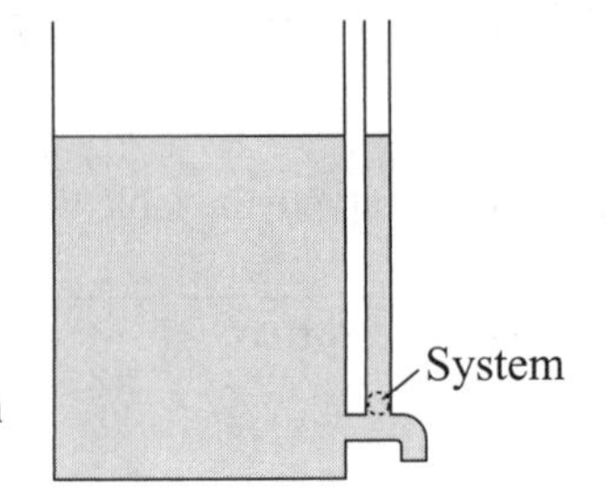

a. Apply the explanation devised in Activity 11.1.1b to compare the pressure of the coffee moving through the spigot (shown in the illustration) when the spigot is open and the pressure when the coffee is not flowing. Then, predict what happens to the level of the coffee in the tube when the coffee flows out of the spigot.

b. If you observe a similar coffeemaker, you will see that the level of coffee in the tube drops down when you open the spigot and stays down while the coffee is being poured. Then it goes up when the spigot is closed. Does this observation match your prediction? Explain.

11.1.3 Test your idea Fill in the table that follows to make a prediction about the experiment described in the table below.

Words	Sketch	Predict what will happen in this experiment; explain your prediction.	Perform the experiment and record your results. Reconcile any discrepancies between the results and your prediction.
Insert a straight straw into a soda pop bottle filled near the top with water; the top of the straw should be near the top of the water. What happens when you blow hard through a second straw placed horizontally so that the airstream moves across the top of the vertical straw in the water?	Blow		

11.1.4 Observe and explain

a. Observe how a syringe works and record your observations. Focus your attention on the speed with which you push the piston and the speed with which the water comes out of the syringe.

b. Explain the difference in the speeds.

11.1.5 Observe and explain You kayak down a narrow and swift mountain river. As you approach a meadow, you notice that the river becomes much wider and that the speed of the water decreases. Explain why.

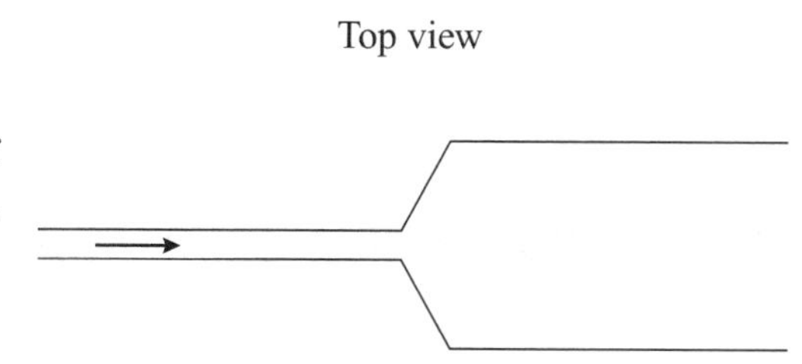

11.2 Conceptual Reasoning

11.2.1 Represent and reason Moving air can lift objects. Consider the example and fill in the table that follows.

Words	Sketch the situation.	Draw a force diagram for the canvas when it is being lifted.	Use the results from Activity 11.1.1b to explain why this happens.
The canvas cover of a moving trailer bows upward.			

11.2.2 Observe and explain Hold the sides of the handle of a spoon lightly between your thumb and a finger and bring the bowl of the spoon into a stream of water flowing at a high rate from a faucet. The spoon feels as though it is pulled into the stream. Complete the table that follows to explain this situation.

Sketch the situation.	Draw a force diagram for the bowl of the spoon; remember to include the forces that the air and water exert on that part of the spoon.	Explain the phenomenon.	Describe the effect when the bowl is held at the top of the stream compared to near the bottom. Is there a difference? If so, explain why.

11.2.3 Represent and reason Fill in the table that follows to explain the situation depicted in the art.

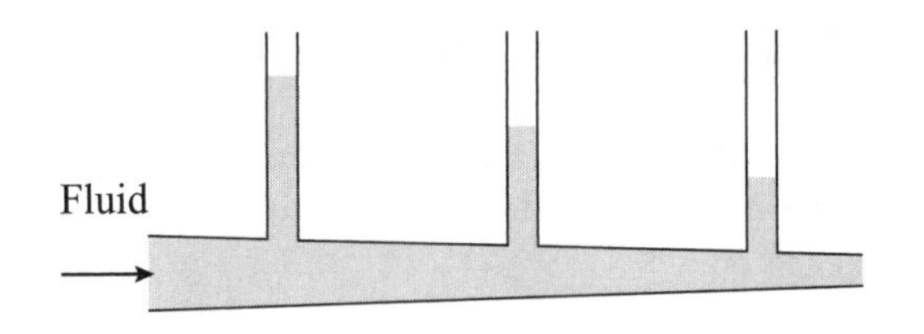

Describe in detail what you observe.	Explain why it is happening.	What is the difference between the description and the explanation?

11.3 Quantitative Concept Building and Testing

11.3.1 Derive The flow rate $Q = \Delta V / \Delta t$ of fluid through a vessel is defined as the ratio of the volume ΔV of fluid passing a cross section in the vessel and the time interval Δt needed for the fluid to pass.

a. Sketch a cylindrical vessel (filled with a liquid) as seen from the side and indicate a cross section in the vessel.

b. Suppose that all of the fluid a distance Δx from that cross section passes the cross section in time interval Δt. Show that $Q = \Delta V / \Delta t = vA$, where v is the average speed of the fluid through the vessel and A is the vessel's cross section.

c. Under what conditions is the flow rate the same in a subsequent part of the vessel where the radius and cross-sectional areas are different?

d. How does the speed compare for a narrower part of the vessel and for a part with a larger opening? Explain.

11.3.2 Derive Consider the vessel carrying fluid shown below.

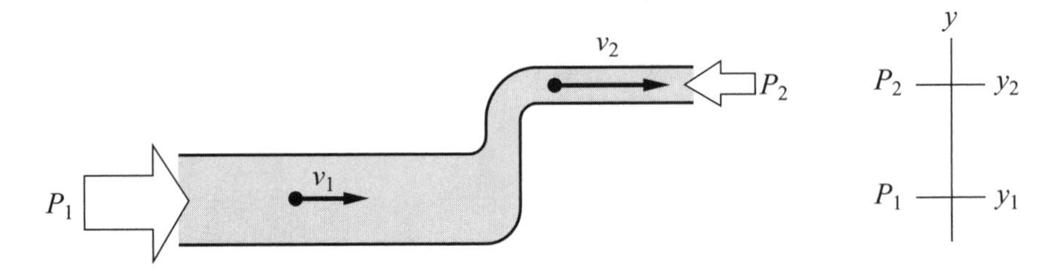

Fluid at the left end of the vessel pushes toward the right with pressure P_1 and exerts a force $F_1 = P_1 A_1$. Fluid at the right end of the vessel pushes back toward the left with pressure P_2 and exerts a force $F_2 = P_2A_2$. These two external forces do work on the system and cause its kinetic energy and the gravitational potential energy of the Water-Earth system to change.

a. Show that the fluid on the left pushing the system fluid forward a distance Δx_1 (the displacement on the left) does work $\Delta W_{left} = P_1A_1\Delta x_1 = P_1\Delta V$, where A_1 is the cross-sectional area of the vessel on the left and ΔV is the volume of fluid that has moved forward.

b. Show that the fluid on the right does work $\Delta W_{right} = -P_2A_2\Delta x_2 = -P_2\Delta V$. Note that Δx_2 is the distance the system fluid moves to the right.

c. Show that the fluid that has moved from the left side to the right side has mass $\rho\Delta V$ and has effectively changed elevation by $y_2 - y_1$—in other words, show that the change in gravitational potential energy is $\Delta U_g = \rho\Delta Vg(y_2 - y_1)$.

d. Show that the $\rho\Delta V$ mass of fluid that has moved forward has effectively changed kinetic energy by $\Delta K = (1/2)\rho\Delta V(v_2^2 - v_1^2)$.

e. Combine the results of parts a–d to show that:

$$P_1 - P_2 = (1/2)\rho(v_2^2 - v_1^2) + \rho g(y_2 - y_1)$$

Rearranging terms, we get:

$$(1/2)\rho v_1^2 + \rho g y_1 + P_1 = (1/2)\rho v_2^2 + \rho g y_2 + P_2$$

This is called *Bernoulli's equation.* Note that the sum of the kinetic energy density (kinetic energy of a unit volume of the fluid), the gravitational potential energy density (potential energy of a unit volume), and the pressure at position 1 equals the sum of these three terms at position 2.

f. What assumptions about the structure and the flow of the fluid did you make while doing this derivation?

g. Explain how Bernoulli's equation relates to energy conservation and how you can use it to explain experiments in Activity 11.1.1.

11.3.3 Test your idea Make a 4-mm-diameter hole on the side of a 2-liter open plastic bottle near the bottom. Cover the hole with tape. Fill the bottle with water and place it at the edge of a table, with the hole facing away from the table and toward the floor. Use Bernoulli's equation to predict quantitatively where water will reach the floor if you remove the tape and thus open the hole. (Make sure the top of the bottle is open.) Fill in the table that follows.

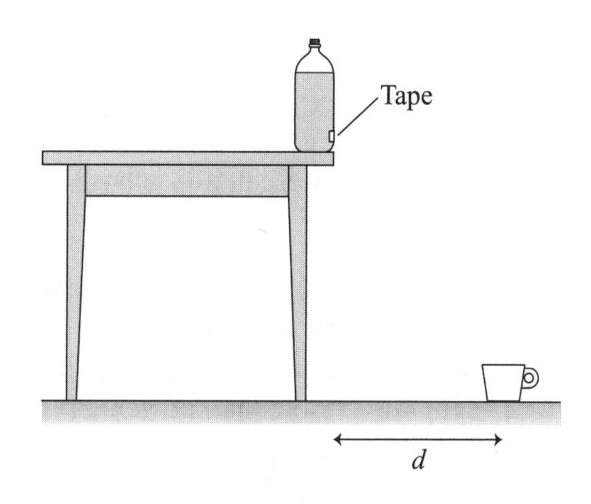

On the sketch, label quantities that you will measure and quantities that you will calculate. Measure the needed quantities and record them.	Outline a mathematical procedure to make the prediction. Then complete the procedure to predict where the water will reach the floor.	List the assumptions you made and describe how they will affect the result.	Perform the experiment. Record the results and make a judgment about whether Bernoulli's equation applies to the water flowing out of the bottle.
	Use Bernoulli's equation to find the speed of the water leaving the hole. Use the knowledge of kinematics to calculate how far the water jet will travel horizontally.		

11.4 Quantitative Reasoning

11.4.1 Represent and reason In the table that follows, examine the consistency among different representations for points 1 and 2 in the situation involving a moving fluid.

Words	Sketch	Bernoulli bar chart	Bernoulli's equation applied
A fire truck pumps water through a big hose up to a smaller hose on the ledge of a building. Water sprays out of the smaller hose onto a fire in the building.	y, 0, 1, 2	$K_1+U_{g1}+P_1=P_2+K_2+U_{g2}$; +, 0, −	$0.5\rho v_1^2 + P_1$ $= P_2 + 0.5\rho v_2^2 + \rho g y_2$

Are the different representations consistent with each other? Explain.

11.4.2 Represent and reason Sketches below represent two different processes. Fill in the table that follows to represent each process using a qualitative Bernoulli bar chart and the application of Bernoulli's equation (include only terms that are not zero). Compare the two dotted points in the fluid. Note that the velocity arrows in the sketches are not necessarily the correct relative lengths.

Sketch	Bernoulli bar chart	Apply Bernoulli's equation.
a. 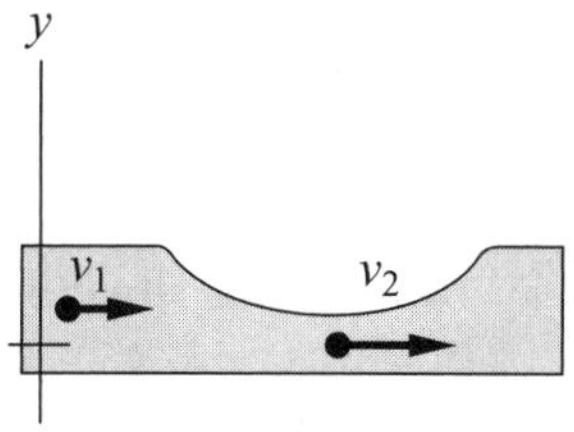 y, v_1, v_2	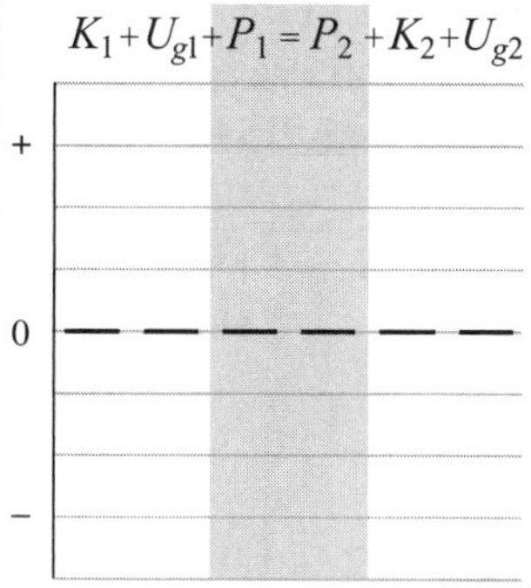 $K_1 + U_{g1} + P_1 = P_2 + K_2 + U_{g2}$ (+, 0, −)	
b. 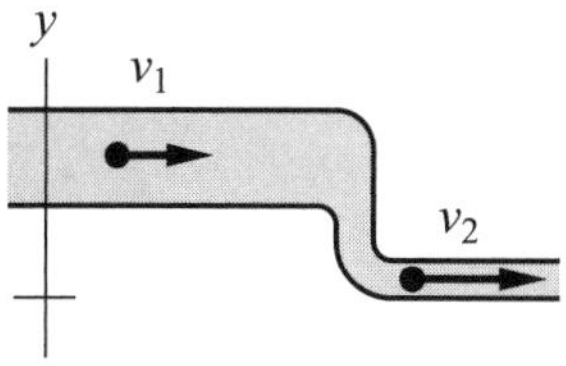 y, v_1, v_2	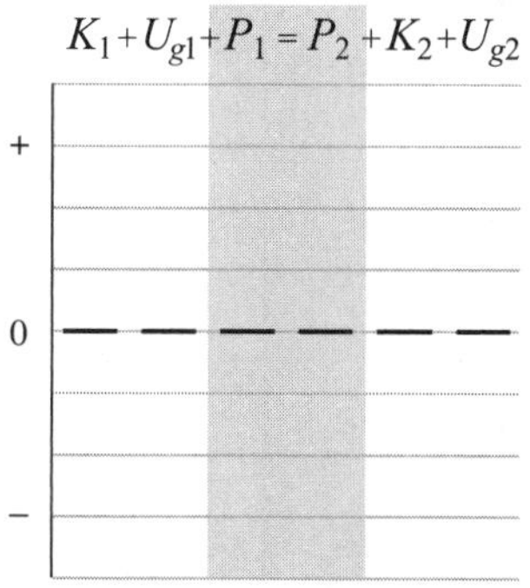 $K_1 + U_{g1} + P_1 = P_2 + K_2 + U_{g2}$ (+, 0, −)	

11.4.3 Bar-chart Jeopardy The qualitative Bernoulli bar charts in the left column of the table below each represent a different fluid-flow process. Fill in the table to devise a situation that can be described with this bar chart (there are many possibilities). Also make other representations of the processes.

Bernoulli bar chart	Describe the situation in words.	Sketch the situation.	Apply Bernoulli's equation.
a. 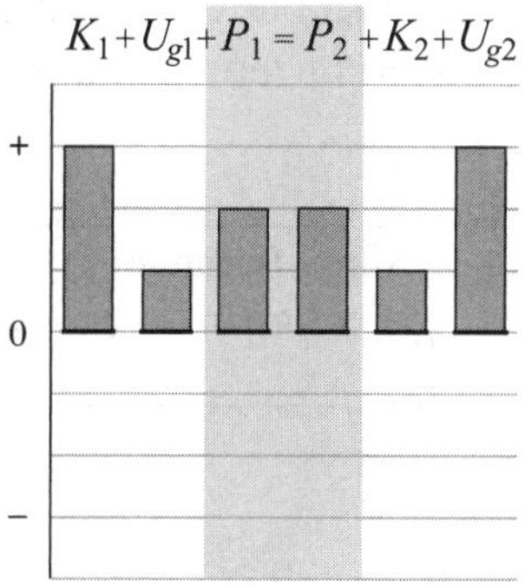 $K_1 + U_{g1} + P_1 = P_2 + K_2 + U_{g2}$ (+, 0, −)			
b. 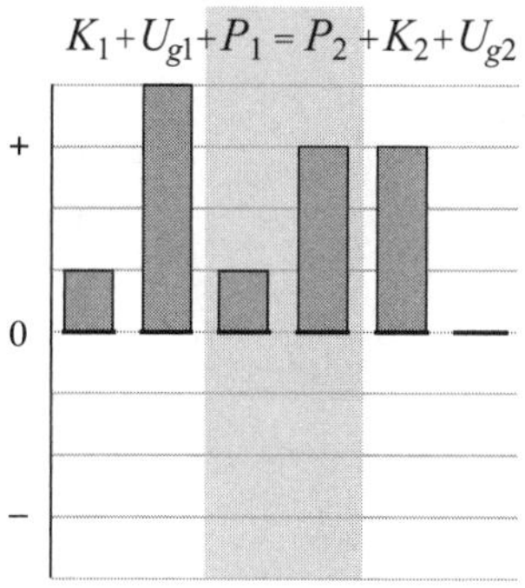 $K_1 + U_{g1} + P_1 = P_2 + K_2 + U_{g2}$ (+, 0, −)			

11.4.4 Equation Jeopardy The application of Bernoulli's equation (in symbols) for two processes is shown in the left column below. Fill in the table to construct a qualitative Bernoulli bar chart that is consistent with each equation and draw a sketch of a situation that might be represented by the equation (there are many possibilities).

Bernoulli's equation applied to a process	Construct a consistent Bernoulli bar chart.	Draw a consistent sketch of a situation.	Describe the situation in words.
a. $0.5\rho v_1^2 + P_1 = P_2 + 0.5\rho v_2^2 + \rho g y_2$	$K_1 + U_{g1} + P_1 = P_2 + K_2 + U_{g2}$ + 0 –		
b. $0.5\rho v_1^2 + P_1 = 0.5\rho v_2^2 + P_2$	$K_1 + U_{g1} + P_1 = P_2 + K_2 + U_{g2}$ + 0 –		

11.4.5 Represent and reason For the two situations described, fill in the table to represent the word description of the processes in other ways—sketches, qualitative Bernoulli bar charts, and the application of Bernoulli's equation to the processes. Be sure that the different representations are consistent with each other. Do not solve for anything.

Word description of the process	Sketch the process.	Construct a consistent Bernoulli bar chart for the process.	Apply Bernoulli's equation to the process.
The average speed of blood in the aorta is 72 cm/s, and the average pressure is 1.3×10^4 N/m^2 above atmospheric pressure. The blood splits into about 10 large arteries, where blood flows at average speed of 20 cm/s. Ignore elevation changes.		$K_1 + U_{g1} + P_1 = P_2 + K_2 + U_{g2}$ + 0 –	

The average speed of blood in the aorta is 1.0 m/s, and the average pressure is $1.3 \times 10^4\ \text{N/m}^2$ above atmospheric pressure. The blood passes plaque with a reduced cross-sectional area in which it travels at an average speed of 10 m/*s*.		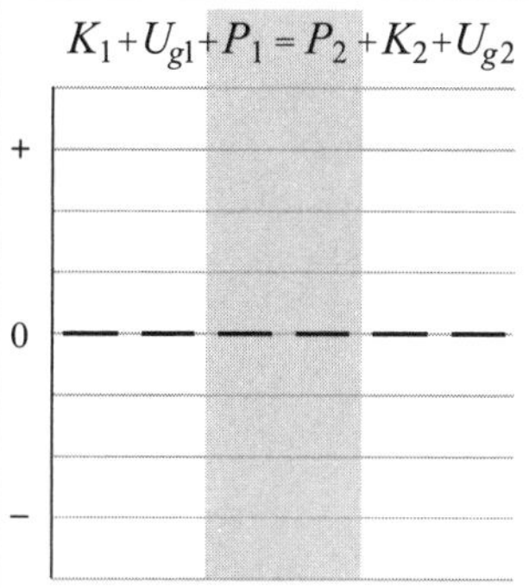	

11.4.6 Regular problem The large front yard (30 m $\times$ 50 m) of a farmhouse is watered from an irrigation canal. An 8-in-diameter pipe runs from the canal to the yard. The horizontal pipe is 0.60 m below the water surface in the canal. What time interval is needed to fill the yard with 0.10 m of water? Describe all assumptions that you make.

11.4.7 Evaluate the solution *The problem:* The surface of water in a community reservoir is 40 m above a 0.60-cm-radius hose connected to the faucet in your house. Assume that $g = 10$ N/kg. What is the flow rate of water from the nozzle of the hose? What assumptions did you make?

Proposed solution: Choose point 1 at the top surface of the reservoir and point 2 at the nozzle of the hose. Use Bernoulli's equation to determine the speed v_2 of water leaving the nozzle:

$$\rho g y_1 + P_{atm} = P_{atm} + (1/2)\rho {v_2}^2$$
$$v_2 = 2(1000\ \text{kg/m}^3)(10\ \text{m/s}^2)(40\ \text{m})/(1000\ \text{kg/m}^3) = 800\ \text{m/s}$$

The flow rate from the hose will be:

$$Q = vA = (800\ \text{m/s})[\pi(0.06\ \text{m})^2] = 9.04047808\ \text{m}^3/\text{s}$$

a. Identify any errors in the solution.

b. Provide a corrected solution if there are errors.

11.4.8 Pose a problem You are watering plants in your backyard. Pose a fluid dynamics problem that you can solve using a hose, a bucket, a watch with a second hand, and a mass-measuring scale.

12 First Law of Thermodynamics

12.1 Qualitative Concept Building and Testing

12.1.1 Observe and explain Close a test tube with a rubber stopper. Then place the bottom of the tube in an open flame. After about 20 seconds, the stopper will shoot out.

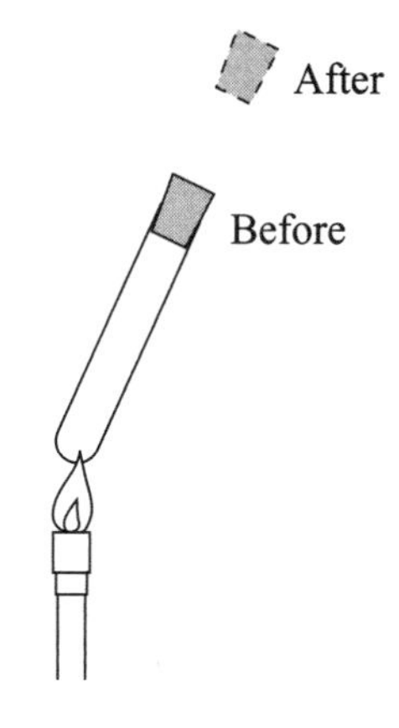

a. Construct a microscopic explanation for how the hot gas pushed out the stopper. Remember what you know about molecules of gas, their motion, and the pressure that they exert.

b. Choose the gas inside the test tube, the stopper, and the Earth (not the flame) as the system and use the concepts of work and energy to explain the experiment. If you need a new quantity or new quantities for your explanation, define them qualitatively.

12.1.2 Observe and explain Place a glass full of cold lemonade in a bowl of hot water. After about 10 min, the lemonade in the glass will warm to a new higher temperature and the water in the bowl will cool to that same temperature.

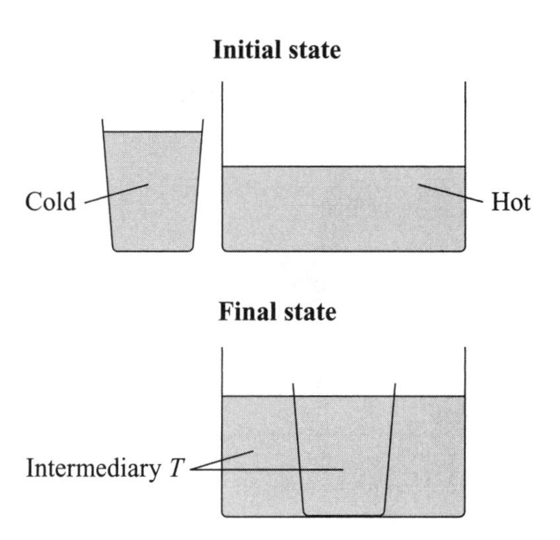

a. Consider the lemonade in the glass as the system and explain this observed process using your knowledge of molecules and their motion. Then use the generalized work–energy principle developed in Chapter 6 to explain what happened to the lemonade. If you cannot explain this process with this principle, try to modify the principle (for example, introduce a new physical quantity) to account for your observations.

b. Repeat part a, only this time consider the water in the bowl as the system.

c. Use your knowledge of molecules and their motion to explain why when two liquids of different temperatures mix together, the mixture will eventually reach some intermediate temperature (called the *equilibrium temperature*).

12.1.3 Observe and explain Vigorously rub two pieces of paper together, pressing the fingers of each hand firmly on the paper as you rub it. Consider one piece of paper as the system. Why did the thermal energy of that piece of paper increase?

12.1.4 Explain Water in a pan is at its boiling temperature (about 100°C). A gas flame continues to contact the pan holding the water, and after some time, all of the water has boiled away. The water vapor is at about the same temperature as the water at its boiling point (about 100°C).

a. Consider the water as the system and explain this process using your knowledge of work and energy. If you need any new quantity for your explanation, define it qualitatively.

b. After contact with the flame for some time interval, the pan, liquid water, and water vapor were at about the same temperature. On a molecular level, what happened to the energy transferred from the hot flame to the water?

12.2 Conceptual Reasoning

12.2.1 Explain Try to use the generalized work–energy principle (without heating), written in the form $(U_{gi} + K_i) + W = (U_{gf} + K_f + \Delta U_{int})$, to explain the following phenomena. If the principle accounts for the phenomenon, describe in symbols or words the work or energy changes. If it does not account for the phenomenon, describe the difficulty and see whether including heating $(U_{gi} + K_i) + W + Q = (U_{gf} + K_f + \Delta U_{int})$ in the principle helps to account for your observations, where Q stands for heating. Be sure to indicate the system used in your analysis of each experiment and any assumptions that you made about the process or the system.

Process described	Write a W/U word explanation (if possible).	Or write a $W/Q/U$ word explanation (if the left cell is not possible).
Drop a golf ball from a window. The process begins the instant the ball leaves your hand and ends just before it hits the grass below. The system is the golf ball and the Earth.		
Drop a golf ball from a window. The process begins at the instant the ball leaves your hand and ends just after the ball stops in the grass below. The system is the golf ball, the Earth, and the grass.		
Place a warm golf ball on top of a cube of ice. The process begins at the instant you place the ball on the ice cube and ends when the ice has melted and the ball has cooled. Choose the ice cube as the system.		
While walking on a golf course in the winter, you find a golf ball and slowly lift it from the frozen grass and place it in your pocket. The process starts with the ball at rest in the cold grass and ends 10 minutes later after it has warmed in your pocket. Choose the ball and Earth as the system (you are not in the system).		

12.2.2 Explain Use the generalized work–energy principle with heating, called the first law of thermodynamics $(U_{gi} + K_i) + W + Q = (U_{gf} + K_f + \Delta U_{int})$, to explain in words the processes described below. Decide whether the gravitational and kinetic energies of the system change.

Process	Write an explanation in words using the terms of energy, heating, and work.	Indicate the signs of the $W/Q/\Delta U_{int}$ terms in the equation.	Represent the process with a bar chart.
Gas originally at 20.0°C resides in a cylinder with a movable piston. Push on the piston, thus compressing and warming the gas. The cylinder is insulated so that there is no thermal energy transfer into or out of the gas. Choose the gas in the cylinder as the system.		*W*: *Q*: ΔU_{int}:	
Gas originally at 20.0°C resides in a cylinder with a movable piston. Push on the piston, thus compressing the gas, but this time the gas does not warm up. The cylinder has thin metal walls and slightly warms the air contacting the walls. Choose the gas in the cylinder as the system.		*W*: *Q*: ΔU_{int}:	
A burning match warms a paper cup that holds ice water. After 2 minutes of warming, the water is still at 0°C but now has less ice and more liquid water. Choose the ice water as the system.		*W*: *Q*: ΔU_{int}:	

12.3 Quantitative Concept Building and Testing

12.3.1 Observe and find a pattern You have a small electric heater and water in a calorimeter (an insulated container). The amount of energy provided to the system and the change in water temperature are shown below.

t (s)	ΔU	ΔT (°C) $= T_f - T_i$
0	0	0.0
10	1000	2.4
20	2000	4.8
30	3000	7.2
40	4000	9.6
50	5000	12.0
60	6000	14.4

a. Graph the data in the table to decide if there is a relationship between the amount of energy provided to the system (the water) and its temperature change ΔT. Think of what quantity is the independent variable and what quantity is the dependent variable.

b. Describe the relationship mathematically, if you find a relationship.

12.3.2 Observe and find a pattern In a second set of experiments, recorded in the table below, the same amount of energy (4000 J) is provided to the insulated water containers with different masses of water.

m (kg)	ΔT (°C)
0.10	9.6
0.20	4.8
0.30	3.2
0.40	2.4

a. Graph the data to decide if there is a relationship between the change in temperature and the mass of water (the system) when the amount of energy provided is the same.

b. Describe the relationship mathematically, if you find a relationship.

12.3.3 Observe and find a pattern The same amount of energy (4000 J) is provided to the identical-mass (1 kg) systems with different types of matter. Is there a pattern in the data listed in the table? If so, describe it in words and mathematically.

Substance	ΔT (°C)	
Freshwater	0.95	
Seawater	1.03	
Alcohol	1.65	
Mercury	28.47	

12.3.4 Reason You found in Activity 12.3.1 a relationship between the amount of energy provided to the system and the temperature change (for constant mass), and in Activity 12.3.2 a relationship between the mass and the temperature change. You also found that the same amount of energy provided to different types of materials caused somewhat random temperature changes. Combine these results to write a mathematical expression that shows how the change in the temperature ΔT of a system depends on the amount of energy provided, ΔU, the system's mass m, and the particular type of material of that system. You can account for the type of material using a new quantity c (called the specific heat of that particular type of material) measured in J/(kg · °C). Be sure that the units in your new equation are consistent.

12.3.5 Predict and test Describe two experiments, A and B, that you can perform to test the relationship devised in Activity 12.3.4. Show how the relationship allows you to make predictions concerning the outcomes of these experiments. You have: insulated containers, heaters, thermometers, scales, and objects made of known materials.

a. Perform experiment A and record the data and the value of the predicted quantity. Decide whether its deviation from the predicted value can be explained by the experimental uncertainties and assumptions.

b. Perform experiment B and record the data and the value of the predicted quantity. Decide whether its deviation from the predicted value can be explained by the experimental uncertainties and assumptions.

12.3.6 Derive You have a container with a gas at known conditions (its molar mass, the number of moles, and the temperature).

a. In Chapter 9, you learned how the average kinetic energy of a particle of ideal gas is related to the temperature of the gas. Use this knowledge to derive an expression for the internal energy of the gas, assuming the gas is ideal.

b. How will the internal energy of the same amount of gas change if you place it in a container with a bigger volume while keeping the gas temperature the same? Explain.

c. Imagine that you separate the gas into two equal containers half the size of the original container. Compare the internal energy of the gas in each container with the energy before the gas was split. Compare the temperatures of the gas. What is the difference between the internal energy of the ideal gas and its temperature?

12.4 Quantitative Reasoning

12.4.1 Represent and reason A graphical description of three processes for a gas in a container is provided in the table below. The gravitational and kinetic energies of the system do not change. Complete the table that follows. Note that the process in part C is cyclic—the system returns to its starting state. For part C you are to decide the changes in the three quantities for each part of the cycle (1 to 2 and 2 to 1) and for the complete cycle.

Part A

Describe the process in words.	Graphical description 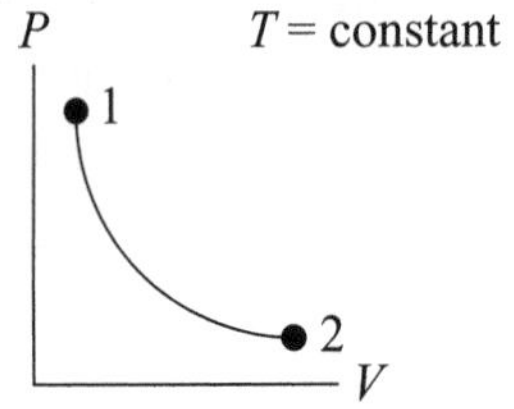 P T = constant 1, 2, V	Was ΔU_{int} +, −, 0?
Was W +, −, 0? (Also indicate this on the graph in the graphical description.)	Was Q +, −, 0?	Explain each process by using your knowledge of the motion of molecules in an ideal gas.

Part B

Describe the process in words.	Graphical description 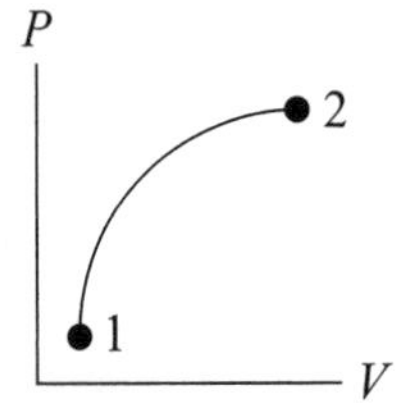 P 2, 1, V	Was ΔU_{int} +, −, 0?
Was W +, −, 0? (Also indicate this on the graph in the graphical description.)	Was Q +, −, 0?	Explain each process by using your knowledge of the motion of molecules in an ideal gas.

Part C

Describe the process in words.	Graphical description 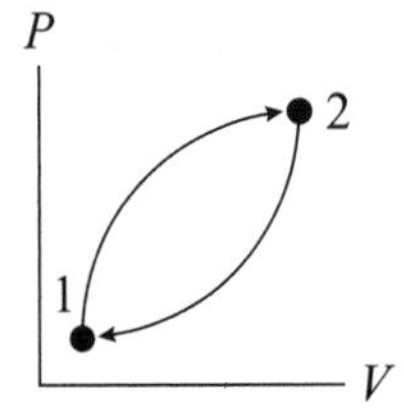 P 2, 1, V	Was ΔU_{int} +, −, 0? *For 1* $\rightarrow$ *2:* *For 2* $\rightarrow$ *1:* *For 1* $\rightarrow$ *2* $\rightarrow$ *1:*

Was $W +, -, 0$? (Also indicate this on the graph in the graphical description.) *For 1 → 2*: *For 2 → 1*: *For 1 → 2 → 1*:	Was $Q +, -, 0$? *For 1 → 2*: *For 2 → 1*: *For 1 → 2 → 1*:	Explain each process by using your knowledge of the motion of molecules in an ideal gas. *For 1 → 2*: *For 2 → 1*: *For 1 → 2 → 1*:

12.4.2 Represent and reason Carefully examine the following table, which represents a process in multiple ways—words, sketches, a graph, and an equation. Also see the graph at the right. The process starts with 0.10 kg of ice at −40°C and ends with 0.10 kg of steam at +150°C. *Note: c* stands for specific heat.

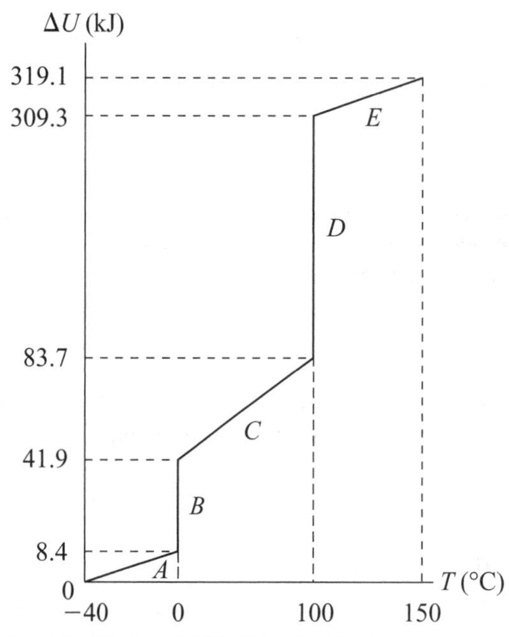

Words	Sketch the initial and final parts of the process.	Mathematical
A. Ice starts at –40°C and warms to its melting temperature at 0°C.	−40°C 0°C	$\Delta U_A = mc_{\text{solid}}\left[T_{\text{melt}} - (-40\,°\text{C})\right]$
B. Energy provided to ice causes the solid ice to convert slowly to liquid at 0 °C.		$\Delta U_B = +mL_{\text{f}}$
C. After the ice has melted completely, more energy causes the liquid water to warm from 0°C to 100°C, its boiling temperature.		$\Delta U_{\text{C}} = mc_{\text{liquid}}\left(T_{\text{boil}} - T_{\text{melt}}\right)$
D. Now, energy provided causes the water to convert slowly from the liquid to the gaseous state while still at 100°C.	100°C 100°C	$\Delta U_D = +mL_{\text{v}}$

(*continued*)

Words	Sketch the initial and final parts of the process.	Mathematical
E. Additional energy causes the gas to warm from 100°C to 150°C.	100°C 150°C	$\Delta U_E = mc_{\text{gas}}(150°\text{C} - T_{\text{boil}})$

a. Are the equations for parts B and D consistent with the corresponding parts of the graph? Explain.

b. Consider the so-called state changes that occur when ice is melting from a solid to a liquid in part B and evaporating from a liquid to a gas in part D. What happens to the temperature of the matter during these state-change processes? Explain.

c. Explain microscopically what is needed in terms of energy for parts B and D—that is, to convert the solid water to liquid water and then to convert the liquid water to gaseous water.

12.4.3 Reason Use the information in the graph in Activity 12.4.2 to determine the heat capacity for the solid, the liquid, and the gaseous phases of water and its latent heats of fusion and vaporization.

12.4.4 Represent and reason Energy provided through heating causes the internal energy of 1.0 kg of some substance to change from a solid at −174°C to a gas at +158°C, as shown in the graph at the right. Determine the following:

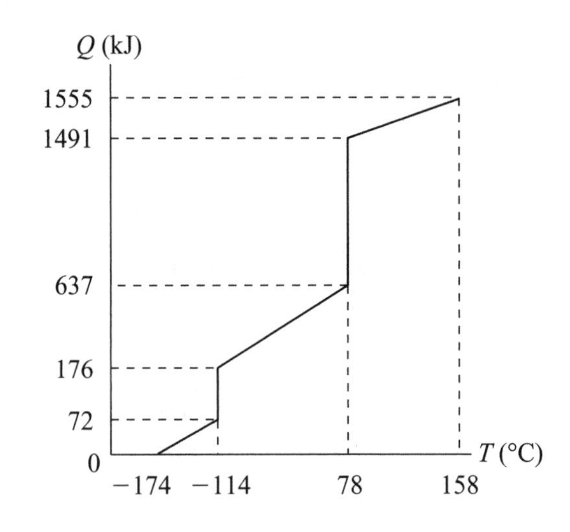

a. The freezing–melting temperature:

b. The boiling–condensing temperature:

c. The specific heat capacity of the solid:

d. The specific heat capacity of the liquid:

e. The specific heat capacity of the gas:

f. The latent heat of fusion:

g. The latent heat of vaporization:

h. Now write a mathematical expression that includes the total energy needed to convert the 1.0 kg of matter from −174°C to +158°C.

12.4.5 Represent and reason Complete the table that follows for the process described.

Word description	Sketch and translation	Write simplifying assumptions.	Represent mathematically in symbols (apply the temperature-change and state-change equations).
You add 20 g of ice at temperature 0°C to 200 g of coffee in an insulated cup at 80°C. The ice and coffee reach a final all-liquid temperature T_f.	**Initial** $T = 0°C$, $m = 0.02$ kg | $T_i = 80°C$, $m = 0.20$ kg **Final** $T_f = ?$ $m = 0.22$ kg Processes involved: • Melt ice. • Warm melted ice. • Cool hot coffee.		

12.4.6 Represent and reason Word descriptions for three processes are provided. Fill in the table that follows by describing each process with a picture description (initial and final states and what happens to each type of matter in going from the initial state to the final state) and the application of the first law of thermodynamics. Do not solve for anything.

Word description	Sketch and translate; include the changes in going from the initial to the final state.	Write simplifying assumptions.	Represent mathematically in symbols (apply the temperature-change and state-change equations).
A 50-g metal spoon at temperature 20°C is placed in an insulated cup with 200 g of coffee at 100°C. The spoon and coffee reach a final temperature T_f.			
25 g of 10°C milk is added to 200 g of 70°C coffee in an insulated cup. The coffee and milk reach a final temperature T_f.			
A 200-g aluminum block at +150°C is placed in a large insulated cup with 40 g of ice at −10°C. The aluminum and ice reach a final temperature T_f (the ice has completely melted).			

12.4.7 Equation Jeopardy Each of the equations below describes a thermodynamics process. Fill in the table that follows to describe the process in other ways.

Mathematical representation	Describe the process in words
$Q = (0.40\text{ kg})(4186\text{ J/kg}\cdot{}^\circ\text{C})(100^\circ\text{C} - 30^\circ\text{C}) + (0.10\text{ kg})(2.56 \times 10^6\text{ J/kg})$	
$(0.40\text{ kg})(2090\text{ J/kg}\cdot{}^\circ\text{C})[0^\circ\text{C} - (-8^\circ\text{C})] + (0.40\text{ kg})(33.3 \times 10^4\text{ J/kg}) + (0.40\text{ kg})(4186\text{ J/kg}\cdot{}^\circ\text{C})(T_f - 0^\circ\text{C}) + (2.0\text{ kg})(4186\text{ J/kg}\cdot{}^\circ\text{C})(T_f - 50^\circ\text{C}) = 0$	

12.4.8 Evaluate the solution A group in your physics lab has produced a graph that group members think describes the following process.

A: Heating causes a solid to warm to its melting temperature.
B: Continued heating causes the solid to melt, thus producing a liquid.
C: With more heating, the liquid warms slowly to its boiling temperature.
D: Continued heating causes the liquid to convert to a gas.
E: The heating then warms the gas.

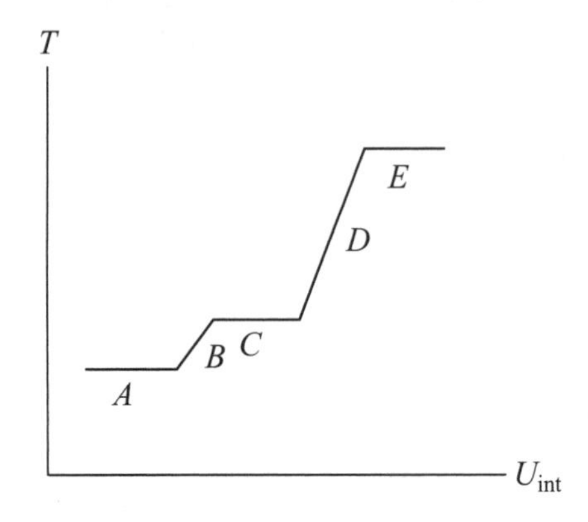

Indicate any flaws in the group's graphical description of this process. Use words and different graphical representations to explain and correct errors.

12.4.9 Design an experiment You just bought a new immersion heater to be used to heat water for your tea. The instruction booklet that came with the heater says that the power rating is 200 W. Design an experiment to test whether the manufacturer provided an accurate rating.

a. Fill in the table that follows to help you.

Describe in words an experiment to test the power rating.	Draw a picture of the experimental setup. Include quantities you will measure and calculations you will make.	Complete the calculations to predict the outcome of the experiment.	List assumptions and how they affect the result. List experimental uncertainties and how they will affect the result.
			Assumptions: How they affect the result: Uncertainties: How they affect the result:

b. Perform the experiment; record the outcome of the experiment, and decide whether the manufacturer's rating is reasonable.

12.4.10 Design an experiment You have an object made of an unknown material. Design two independent experiments to determine the specific heat of the material. Compare the outcomes of the experiments and account for the differences. In the space below, describe the experimental designs, your analysis procedure, the outcomes of the experiments, and your conclusions.

12.4.11 Represent and reason Fill in the table that follows, which shows the changing pressure and volume for two different processes involving an ideal gas.

***P*-versus-*V* graph for a process**	**Determine the work done on the system during the process.**	**Describe in words a possible process.**	**Explain the process using your knowledge of gas particles.**
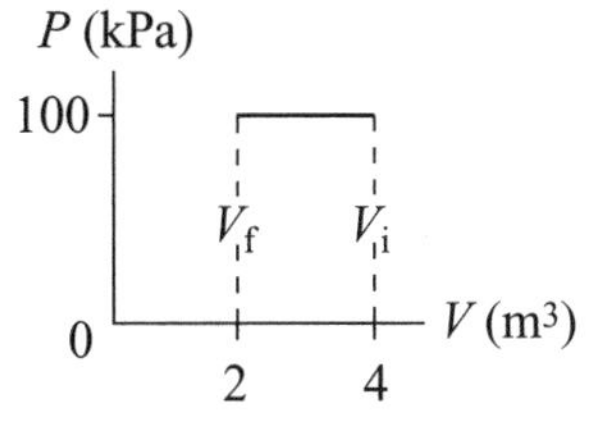 P (kPa); 100; 0; V_f 2; V_i 4; V (m³)			
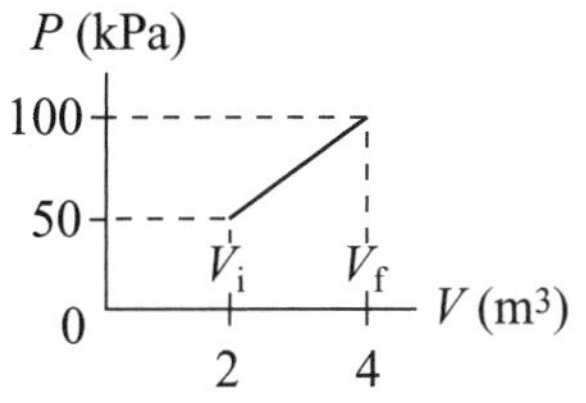 P (kPa); 100; 50; 0; V_i 2; V_f 4; V (m³)			

12.4.12 Represent and reason The first row of the table graphically represents an isochoric (constant volume) process, an isothermal (constant temperature) process, and an isobaric (constant pressure) process. Each process involves 1.0 mol of an ideal gas. Complete the table that follows for each process.

Isochoric process	**Isothermal process**	**Isobaric process**
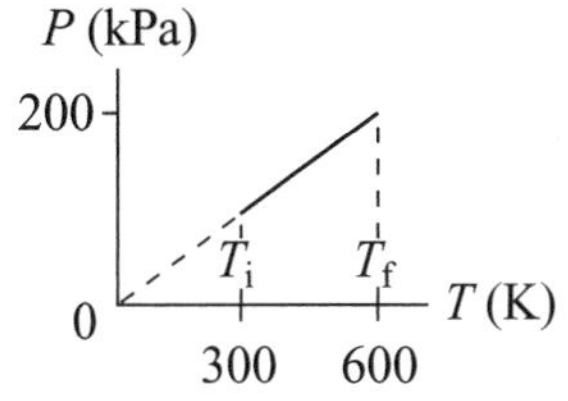 P (kPa); 200; 0; T_i 300; T_f 600; T (K)	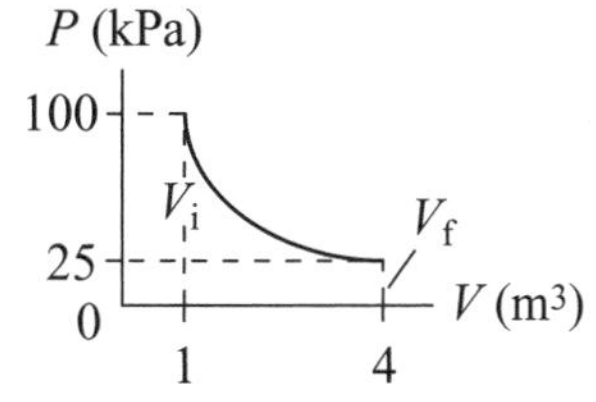 P (kPa); 100; 25; 0; V_i 1; V_f 4; V (m³)	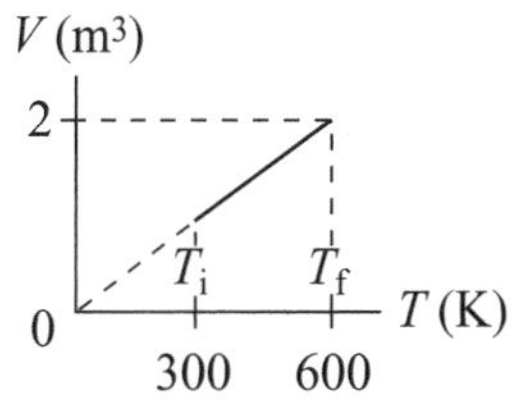 V (m³); 2; 0; T_i 300; T_f 600; T (K)

(continued)

Isochoric process	Isothermal process	Isobaric process
Use the ideal gas law to calculate the volume of the gas. Then determine the work done by the environment on the gas during the process.	Use the ideal gas law to calculate the temperature of the 1.0 mol of gas. Then use the temperature to determine the *change* of the internal energy of the gas.	Use the ideal gas law to determine the gas pressure. Then determine the work done by the environment on the gas during the process.
Use the initial and final temperatures to determine the change in internal energy of the 1.0 mol of gas during the process.	Use the ideal gas to write the pressure of the gas in terms of its volume (and other constant quantities).	Use the initial and final temperatures to determine the change in internal energy of the 1.0 mol of gas during the process.
Use these results to construct a qualitative work–heating–internal energy change bar chart for the process.	Use these results to construct a qualitative work–heating–internal energy change bar chart for the process.	Use these results to construct a qualitative work–heating–internal energy change bar chart for the process.
$W + Q = \Delta U_{int}$ + 0 –	$W + Q = \Delta U_{int}$ + 0 –	$W + Q = \Delta U_{int}$ + 0 –

12.4.13 Reason Answer the questions concerning the processes described in Activity 12.4.12.

Isochoric process	Isothermal process	Isobaric process
Explain why the line in the isochoric graph passes through the origin.	Explain why the graph line is not a straight line.	Explain why the line in the isobaric graph passes through the origin.
Explain the process using the knowledge of the molecules and their motion.	Explain the process using the knowledge of the molecules and their motion.	Explain the process using the knowledge of the molecules and their motion.

13 Second Law of Thermodynamics

13.1 Qualitative Concept Building and Testing

13.1.1 Reason For each process, describe the energy changes that occur and indicate what would have to happen for the process to reverse itself—that is, to go naturally from the final state back to the initial state.

a. Fill in the table that follows.

Process	**Describe the types of energy that change, the work done on the system, and the heating during the processes.**	**Describe the reverse process and decide if it ever occurs.**
Water at the top of Niagara Falls drops onto the rocks below. Slight warming occurs. The water, rocks, and Earth are the system.		
Water at the top of Niagara Falls drops onto the blades of an electric generator near the bottom of the falls, causing the blades to rotate and to generate an electric current that in turn causes a lightbulb to glow for a short time interval. The water, generator, lightbulb, and Earth are the system.		
Logs, paper, and kindling are stacked in your fireplace. You light them on fire and soon have a blaze. The next morning you have ashes in your fireplace. The paper, kindling, logs, and air are the system.		
A pendulum bob is held initially to the side and then released. It swings back and forth with slowly decreasing amplitude and eventually stops because of air resistance and friction in its bearing. The system is the pendulum, air, bearing, and Earth.		
Each second your body converts 100 J of metabolic energy to thermal energy that is transferred to the air surrounding your body. (The thermal energy is produced by converting the chemical energy stored in complex molecules from food to simpler molecules.) The system is your body and the surrounding air.		

(*continued*)

Hot gas in a cylinder pushes a piston, which causes the blades of an electric generator to turn, which in turn causes a lightbulb to glow for a short time interval. The system is the original hot gas (which cools while pushing the piston), the generator, and the lightbulb (which first glows and then stops glowing and cools down).		

b. Decide if any of the reverse processes described in the table are prohibited because energy is *not* conserved. Explain.

13.1.2 Reason and explain

a. Below is a list of different types of matter with the same amount of energy. Order the types with respect to their potential to do work on some system, with the most useful for doing work listed first.

- 1000 J of the energy stored in the molecules of gasoline
- 1000 J of thermal energy in the air of your bedroom
- 1000 J of gravitational potential energy of water at the top of a waterfall
- 1000 J of thermal energy in very hot gas in a cylinder closed by a movable piston
- 1000 J of thermal energy in cold gas in a cylinder closed by a movable piston
- 1000 J of thermal energy in water at 65°C in the rocks at the bottom of a waterfall
- 1000 J of thermal energy in the water in a boiler at 700°C
- 1000 J of chemical energy stored in the bonds of a piece of wood

b. Look at your list in part a and try to identify a type or types of energy that seem the most useful for doing work on some system and a type or types that seem the least useful. Explain.

13.1.3 Reason and explain Based on the analysis in Activities 13.1.1 and 13.1.2, devise in words (using everyday language) a qualitative explanation for what happens to the energy in a system during real-world processes that are not reversible. Consider in particular the nature of the initial type or types of energy (see examples in Activity 13.1.2), the nature of the final type or types of energy, and the ability of systems to do work on another system when they have different types of energy in the initial state.

13.1.4 Test your idea

a. Fill in the table to describe the processes given below.

Process	**Describe the reverse process.**	**Use the rule you formulated in Activity 13.1.3 to predict if the process is reversible. Explain.**	**Based on everyday experiences, is your prediction correct?**
A block skids to a stop while moving across a tabletop. The block and the table surface are the system.			
A golf ball is thrown vertically upward at speed v and reaches a maximum height $h = \sqrt{2gv}$ above its launch point. Ignore the influence of air. The ball and Earth are the system.			
A large foam ball is moving vertically up at speed v and reaches a maximum height h' somewhat less than $h = \sqrt{2gv}$. The ball, Earth, and air are the system.			
Two identical metal balls at the end of strings swing toward each other at equal speeds, collide elastically, and then bounce back at the same speed. The balls and Earth are the system.			
Two identical carts with chewing gum stuck to their fronts roll toward each other at equal speeds and have an inelastic collision. They stick together and stop. The carts are the system.			
Two cups of water, one cold and the other hot, are mixed in an insulated bowl. The mixture reaches an intermediate temperature. The water is the system.			

b. If any of your predictions are incorrect, how should your rule be revised so that its use leads to a correct prediction?

13.1.5 Reason Describe three real-world, nonreversible phenomena (without repeating those given in previous activities). Discuss the types of energy produced in the final state compared to the types of energy lost in the initial state.

13.2 Conceptual Reasoning

13.2.1 Observe and explain Imagine an hourglass with all of its sand in the top bulb. Over the course of an hour, the sand slowly spills into the bottom bulb.

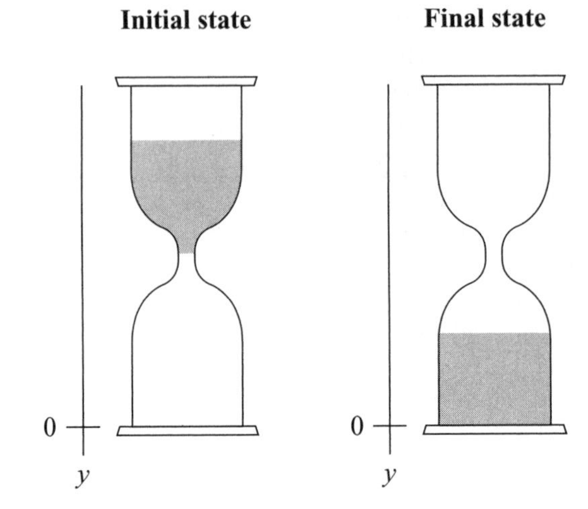

a. Consider the sand, hourglass, and Earth as the system. Describe the energy changes that occurred in this system during that hour.

b. Fill in the table that follows.

Process	Water at the top of Niagara Falls drops onto the rocks below. The system is the water, the rocks, and Earth.	A car has 1 gal of gasoline and travels 30 mi on a level road when it runs out of the gasoline. The system is the initial gasoline, the car, and the atmosphere surrounding the car.	A person uses 250,000 J of metabolic energy (converting the chemical energy in complex molecules to thermal energy) while running 1 mi. The system is the person and the surrounding air.
Describe the analogies between the process described and the hourglass process.			

13.2.2 Reason For each process described, indicate if work is done on the system, if energy is transferred through heating, and whether the system's energy changes. If the system's energy changes, specify what kind of energy decreases and what kind increases. Then indicate the signs of the work W done on the system by the environment $(+, -, \text{or } 0)$, the heating Q, and the energy changes $(U_f - U_i) = \Delta U$ for each type of energy that changes. Fill in the table that follows.

Process	**Indicate what object does the work on the system and whether heating occurs. Identify types of energy change.**	**Indicate the signs of *W, Q*, and ΔU (for each type of energy).**
Your car slides on ice and runs into a tree, causing the front of the car to become slightly hotter and crumpled. The car and road surface are the system. The tree is outside the system.	Work: Heating: Types of energy change:	*W*: *Q*: ΔU for each energy change:
You compress a gas by pushing down on the handle of a piston inside a cylinder that holds the gas. The cylinder is in a pool of water that keeps the cylinder and gas temperature constant. The gas is the system.	Work: Heating: Types of energy change:	*W*: *Q*: ΔU for each energy change:
While jogging on a hot summer day, moisture on your skin at temperature 35°C evaporates. Consider the system as a small amount of that moisture from just before it evaporates to just after it evaporates.	Work: Heating: Types of energy change:	*W*: *Q*: ΔU for each energy change:
You place a kettle of water on the burner of a stove. The water warms from room temperature to a somewhat higher temperature. The system is the water and kettle.	Work: Heating: Types of energy change:	*W*: *Q*: ΔU for each energy change:

13.3 Quantitative Concept Building and Testing

13.3.1 Observe and explain Place four nickels on a tray.

a. Give the tray a good vertical shake so that the coins turn over randomly. Note the distribution of coins in terms of the number of heads and the number of tails. Repeat the shake and note the new distribution. Think of how many different combinations or distributions of heads and tails you can get; we call these *macrostates*. Draw the possible combinations in the table, as shown in the first cell. Note that the table has more cells than you need.

3 heads and 1 tail (H) (T) (H) (H)			

b. Repeat the experiment, but this time mark all the coins themselves with different color dots or some other distinguishing marks. Record the distribution of the coins after the first shake. Think of how many different or unique ways you can get the marked coins to distribute (we call each state a *microstate*). For example, there are four different microstates to get three heads and one tail—each coin can be the tail and the other three coins the heads. Draw all possible combinations (microstates). Again, the table has more cells than you need.

(H) (H) (H) (T)	(H) (H) (T) (H)		

c. Add the total number of microstates in part b. Then divide the number of microstates for each macrostate (for example, the macrostate with one head and three tails has four possible microstates) by the total number of microstates and then multiply by 100. This is the percentage chance to get a particular macrostate after one shake. Record the macrostates and the percentage chance for each in the table that follows.

Macrostates (combinations)	**Record the number of microstates for the macrostate.**	**Record the percentage chance of getting the macrostate after one shake.**
Four heads, zero tails	One	
Three heads, one tail	Four	25%

13.3.2 Reason Suppose that the universe is made up of four coins and that its life starts with four heads. Each shake of the universe causes it to move to a new state with a different combination of coins—most likely a state with greater probability than the starting combination. However, there is a 1-in-16 shakes chance that the universe will return to its starting combination—four heads. What about the real world? In 1889, the French mathematician Henri Poincaré attacked this problem using the statistical approach to entropy. He concluded, "To see heat transfer from a cold body to a warm one … it will suffice to have a little patience." (The word *little* was a playful exaggeration.) Justify this comment, the so-called *recurrence paradox*, in everyday words so a nonscience student can understand.

13.3.3 Reason

a. Apply your knowledge of probability to explain why a drop of food coloring in a glass of clear water spreads out so that all of the water has an even color after some time.

b. Discuss whether the food coloring could condense back to an original droplet after it spreads evenly in a glass of clear water.

c. Explain using your knowledge of probability why a gas always occupies the whole volume of the container.

13.3.4 Reason Water at the top of Niagara Falls can do physics-type work on the blades of an electric generator near the bottom of the falls. The generator can light an electric lightbulb. The thermal energy of the water after hitting the rocks below has lost much of its ability to do work—to turn the generator blades. And the change in the *type* of energy has caused an irreversible change in the world's ability to do work. When this happens we say the entropy of the universe has increased. For each of the processes below, indicate which state has increased entropy, which we now consider as a decreased ability to do work. For the lower entropy state of the system, indicate how the system's energy in that state might be used to do work on some other system. In this case system 1 does work on some system 2, which is considered the environment.

Process	**Does the initial or final state of system 1 have lower entropy—that is, a greater ability to do work? Explain.**	**For the lower entropy state, indicate one way system 1 can do work.**
Coal is placed in the burner of an old steam engine freight train. After burning, ashes and combustion products remain (such as CO_2 in the atmosphere). The coal, train, and atmosphere are system 1.		
Each second your body converts 100 J of metabolic energy (converting complex molecules from food) to thermal energy transferred to the air surrounding your body. System 1 is your body and the surrounding air.		
Extremely hot gas in a cylinder pushes against a piston, which in turn can push against other things. System 1 is the original hot gas, the piston and cylinder, and other things against which the piston can push.		

13.3.5 Reason In Activity 13.3.1 you found that flipped coins had the greatest probability to land in disorganized states with equal numbers of heads and tails and the least probability to be in organized states with all heads or all tails. In Activity 13.3.3 you found that processes tended to move in a direction toward states with random kinetic energy (thermal energy) and away from states with more organized forms of energy that can do work. Are these two trends consistent? Explain.

13.3.6 Reason Imagine that you have two containers: one filled with 100 g of water at 80°C and the other filled with 200 g of water at 20°C.

a. What will the final temperature be of the mixture in a calorimeter?

b. How can you explain this process using the second law of thermodynamics?

13.3.7 Represent and reason A thermodynamic engine uses the thermal energy in the gas or liquid in a hot reservoir to warm the engine fluid. The engine then does work pushing a turbine or a generator. The engine gets rid of some of its thermal energy by heating the gas or liquid in a cool reservoir. Finally, the environment does work on the cooled engine fluid so that it returns to its starting state. This process is represented in several different ways in the table that follows.

Sketch of the process	Flowchart	P-versus-V diagram	Mathematical representation
1 → 2 Heating engine 2 → 3 Engine does work 3 → 4 Cool engine (Cool) 4 → 1 Work on engine	T_{hot}, Q_{hot}, W, Q_{cool}, T_{cool}	P (N/m²); V (m³); 1, 2, 3, 4; Q_{hot}, $W_{by\ engine}$, Q_{cool}, $W_{on\ engine}$	$Q_{hot} = W_{by\ engine} + Q_{cool} - W_{on\ engine}$ or $Q_{hot} = W_{net} + Q_{cool}$ where W_{net} = area inside the graph.

Are these representations consistent? Explain.

The temperatures must be in kelvins (K). With this in mind, determine the maximum efficiencies of the thermodynamic engines described in the table that follows.

Burning coal heats the gas in the turbine of an electric power plant to 700 K. After turning the blades of the generator, the gas is cooled in cooling towers to 350 K.	Near Bermuda, ocean water at the surface is about 24°C, and at a depth of 800 m under the water, it is about 10°C. Consider the efficiency of a thermodynamic engine working between these two depths.	An inventor claims to have a thermodynamic engine that attaches to a car's exhaust system and thereby "recycles" the thermal energy in the hot exhaust system. The temperature of the exhaust gas is 160°C and the temperature of the output of this proposed heat engine is 20°C.

13.4 Quantitative Reasoning

13.4.1 Represent and reason Below we show a cyclic process involving 1 mol of an ideal gas. Fill in the table that follows for this process.

P-versus-V graph for a cyclic process P (kPa) 120, 60, 0 1, A, 2, D, B, 4, C, 3 1, 4 V (m³)	Determine the work done on the gas by the environment during each leg (A, B, and C) of the cycle. Determine the net work done on the gas $W_{\text{environment on gas}} =$ Find the work that the gas does on the environment; it is equal to $W_{\text{gas on environment}} =$ $-W_{\text{environment on gas}} =$	Use the ideal gas law to determine the temperature of the gas at each corner of the process (1, 2, and 3).
Use the temperatures listed in the previous cell of the table to determine the thermal energy of the gas at each corner of the process and the change in internal thermal energy during each leg of the process.	Use the results recorded in the previous cells of the table and the first law of thermodynamics to determine the heating during each leg of the process.	Determine the efficiency of the process: $e = \dfrac{\text{net work done by the gas on the environment}}{\text{sum of all the positive heating legs of process}}$

13.4.2 Represent and reason Below we show a cyclic process involving an ideal gas. Complete the table for this process.

P-versus-*V* graph for a cyclic process *P* (kPa) 120, 60, 0; 1 *A* 2, *C*, 3 *B*; *V* (m^3) 1, 4	Determine the work done on the gas by the environment during each leg (*A*, *B*, and *C*,) of the cycle. Determine the net work done on the gas $W_{\text{environment on gas}} =$ Find the work that the gas does on the environment; it is equal to $W_{\text{gas on environment}} =$ $-W_{\text{environment on gas}} =$	Use the ideal gas law to determine the temperature of the gas at each corner of the process (1, 2, and 3).
Use the temperatures listed in the previous cell of the table to determine the thermal energy of the gas at each corner of the process and the change in internal thermal energy during each leg of the process.	Use the results recorded in the previous cells of the table and the first law of thermodynamics to determine the heating during each leg of the process.	Determine the efficiency of the process: $e = \dfrac{\text{net work done by the gas on the environment}}{\text{sum of all the positive heating legs of process}}$

13.4.3 Equation Jeopardy The following equations represent the four parts (*A*, *B*, *C*, and *D*) of a cyclic process with an ideal gas. In this case we consider the work done by the system on the environment, and consequently $Q + (-W_1) = \Delta U_{\text{int}}$.

$$W_1 = (3.0 \times 10^5\,\text{N/m}^2)(0.020\,\text{m}^3 - 0.010\,\text{m}^3) + 0 + (1.0 \times 10^5\,\text{N/m}^2)(0.010\,\text{m}^3 - 0.020\,\text{m}^3) + 0$$

$$\Delta U_{\text{int}} = (3/2)(1.0\,\text{mol})(8.3\,\text{J/mol}\cdot\text{K})[(700\,\text{K} - 360\,\text{K}) + (480\,\text{K} - 700\,\text{K}) + (240\,\text{K} - 480\,\text{K}) + (360\,\text{K} - 240\,\text{K})]$$

$$Q = Q_A + Q_B + Q_C + Q_D$$

a. Draw a P-versus-V graph for the process with labeled axes (including a scale).

b. Determine the net change in the internal energy during the entire cycle.

c. Determine the energy transferred to the system through heating for each of the four parts of the process.

d. Indicate the temperature of the gas at each corner of the cycle and decide if the temperatures are consistent with the ideal gas law.

e. Determine the efficiency of the cycle in doing work. In this case, define *efficiency* as the net work done by the system on the environment divided by the heating of the system during any parts of the cycle where the heating is positive (thermal energy is added to the system).

13.4.4 Evaluate the solution

The problem: A heat engine operates between temperature reservoirs at 250°C and 50°C. The hot reservoir through heating adds 5000 J of thermal energy to the engine during each cycle. If it operates at maximum efficiency, determine the maximum net work the engine can do during each cycle.

Proposed solution: The maximum efficiency is:

$$e_{\text{max}} = 1 - \frac{50°\text{C}}{250°\text{C}} = 0.80$$

Thus, the maximum work that the engine can do is:

$$W_{\text{net max}} = e_{\text{max}} Q_{\text{hot}} = 0.80(5000\ \text{J}) = 4000\ \text{J}$$

a. Identify any missing elements or errors in the solution.

b. Provide a corrected solution if there are errors.

13.4.5 Regular problem Automobiles convert the useful energy stored in the chemical bonds of gasoline to less useful thermal energy. This conversion causes an irreversible degradation of useful energy (an increase in its entropy). Consider the following example.

a. The effective friction force caused by rolling resistance for car tires is approximately 50 N. To counter this rolling resistance, the car tires have to push back against the road, exerting a 50-N force made possible by the engine turning the driveshaft. The road in turn pushes the car forward, exerting a 50-N force. This forward push of the road balances the resistive rolling friction force—hence, the car moves at constant velocity. How much work is done during a 100-mi trip by this 50-N force of the road on the car tires in the forward direction?

b. Noting that burning 1.0 gal of gasoline converts about 1.2×10^8 J of chemical energy to other forms, how much gasoline is needed to overcome rolling resistance during the 100-mi trip if the gasoline's energy is 100% efficient in overcoming rolling resistance?

c. Because of inefficient transmissions, friction in engines, and inefficient combustion processes, the car is only about 30% efficient. The energy needed to overcome rolling resistance divided by the total energy used by the car for this purpose is about 0.3. With this in mind, how much gasoline is used to overcome rolling resistance during the 100-mi trip?

d. Was energy conserved during this 100-mi trip? Explain. Why should we worry less about energy conservation and more about energy "degradation"?

13.4.6 Regular problem A 60-kg woman walking on level ground at 1 m/s metabolizes energy at a rate of 230 W. When she walks up a 5° incline at the same speed, her metabolic rate increases to 370 W. Determine her efficiency at converting chemical energy into gravitational potential energy.

13.4.7 Reason While applying the second law of thermodynamics, keep in mind that natural processes lead to an increase in entropy in closed systems. In an open system, the net entropy of the universe increases, but some parts may have entropy decreases at the expense of other parts that have larger entropy increases. Consider the following example. Use the data to help your reasoning. A person consumes and converts complicated molecules with low entropy, such as carbohydrates, fats, and proteins, into much smaller molecules (CO_2, H_2O) with high entropy. A healthy adult consumes about 500 kg, or a half ton, of these foods each year. The degradation of this food results in the transfer to the environment of 2 billion joules of thermal energy that have little use. Consider this: Human beings gradually grow from a relatively helpless state to independent participants in a complex world. Is the development of humans an exception to the second law of thermodynamics with large low-entropy organized molecules? Explain.

13.4.8 Design an experiment Describe an experiment that can be performed to estimate the average efficiency of a human body. Fill in the table that follows to help you.

Describe how you can estimate the energy that the body gets from food.	Describe how you can estimate the "useful" energy that the body produces.	Describe the mathematical procedure to estimate the efficiency of the human body.

14 Electric Charge: Force and Energy

14.1 Qualitative Concept Building and Testing

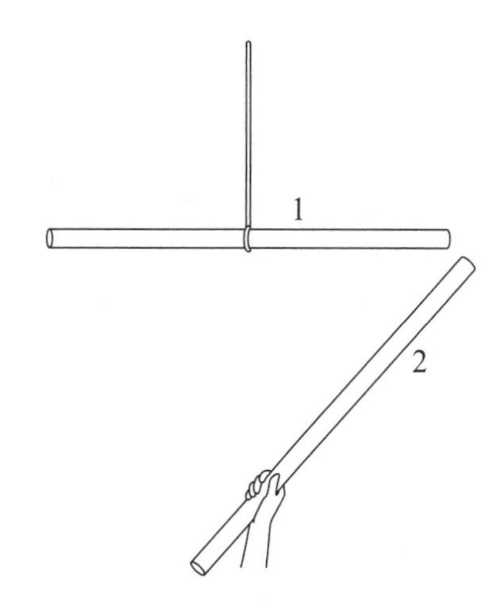

14.1.1 Observe and find a pattern For the experiments that follow, you need two foam insulation tubes, a small piece of felt or wool, string, and plastic wrap. Suspend one tube from a string, as shown in the illustration. Before starting the experiments that follow, bring one end of tube 2 near one end of hanging tube 1. Is there any interaction? Now rub one end of each tube vigorously with different materials, as described below.

a. Bring the rubbed end of tube 2 or the material doing the rubbing near the rubbed end of the suspended tube 1 and record the behavior of tube 1.

Object 1	Object 2	Record your observation
Tube 1 rubbed with felt	Tube 2 rubbed with felt	
Tube 1 rubbed with felt	The felt that was used to rub tube 2	
Tube 1 rubbed with plastic wrap	Tube 2 rubbed with plastic wrap	
Tube 1 rubbed with plastic wrap	The wrap that was used to rub tube 2	
Tube 1 rubbed with plastic wrap	Tube 2 rubbed with felt	

b. Identify patterns in these observations. Devise an explanation.

14.1.2 Test your explanation Assemble two long pieces of nylon (stockings work fine) and a plastic grocery bag. Design an experiment to test how two pieces of nylon rubbed with the plastic bag interact with each other. Fill in the table that follows to make predictions about the outcomes of your experiments based on the patterns that you found in Activity 14.1.1.

Design an experiment to test how nylon pieces rubbed with plastic interact.	Predict the outcome of your experiment.	Perform the experiment and then describe the outcome.	Was the reasoning leading to your prediction correct? Explain.

14.1.3 Explain In Activities 14.1.1 and 14.1.2 you found a consistent pattern: Identical objects rubbed with a second material repel each other. The second material in turn attracts the objects it rubbed. Think of a mechanism that might explain why rubbing objects makes them attract or repel each other.

14.1.4 Test the explanation Your friend Gaurang says that electric interactions are the same as magnetic interactions because magnets also attract and repel each other. Consequently, he believes that when you rub objects, they become magnetized. Assemble a magnet on a swivel, two foam tubes, felt, and plastic wrap. Design an experiment whose outcome will allow you to decide whether Gaurang is correct or whether rubbing objects makes them participate in a different type of interaction.

Describe an experimental setup to test the idea that rubbing causes materials to become magnetic (Gaurang's idea).	Predict the outcome of your experiment based on Gaurang's idea.	Perform the experiment and then describe the outcome.	Make a judgment about Gaurang's idea based on the outcome.

14.1.5 Observe and explain You have two foam tubes; one tube is suspended at the center from a string, and the other is free. Vigorously rub one end of each tube with felt. Slowly bring the rubbed end of the free tube closer and closer to the rubbed end of the hanging tube. Describe your observations. What can you infer about how the electric force depends on the separation of the objects?

14.1.6 Observe and explain

a. Perform the experiments described and fill in the table that follows.

Experiment	**Rub one end of foam tube 2 with felt and bring it close to a hanging foam tube 1 that has not been rubbed. Repeat, but this time rub the other end of tube 2 with plastic wrap and bring it near the end of unrubbed tube 1.**	**Rub one end of foam tube 2 with felt and bring it close to the end of a hanging metal rod 1 that has not been rubbed. Repeat, but this time rub the other end of foam tube 2 with plastic wrap and bring it near the end of the hanging metal rod 1.**
Record what you observed.	Tube 2 rubbed with felt: Tube 2 rubbed with plastic wrap:	Tube 2 rubbed with felt: Tube 2 rubbed with plastic wrap:

b. Devise an explanation involving a possible internal structure of the foam material that might explain why the rubbed tubes attract the unrubbed tubes. Draw a charge diagram for the inside of the unrubbed tube.

c. Devise an explanation involving a possible internal structure of the metal rod that might explain why the rubbed tubes attract it. Draw a charge diagram for the inside of the metal rod.

d. Devise an explanation for why the metal rod responds differently than the foam tube.

e. List everyday experiences that are consistent with these observations.

14.1.7 Test your ideas Hang a small piece of aluminum foil from a 30-cm-long piece of thread, which is tied at the top to a plastic or wooden rod (for example, a ruler). Use the ideas that you devised in Activity 14.1.6 to predict what happens when you bring the end of a foam tube rubbed with fur near the piece of foil. Then repeat the procedure using a foam "peanut" (standard packing material) hanging from the thread. Fill in the table that follows.

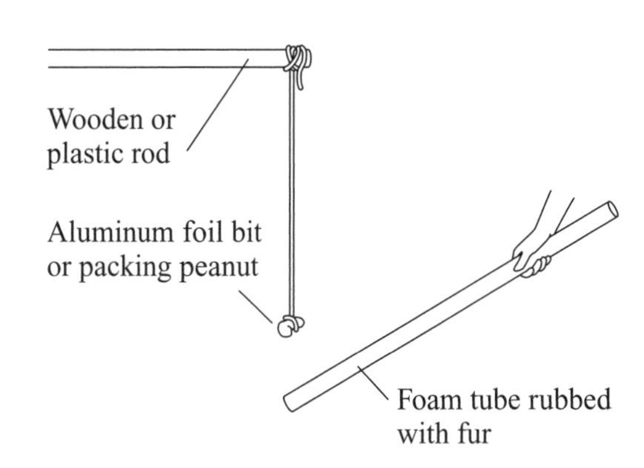

Experiment: Aluminum foil suspended on a string

Predict in words what you will observe.	Explain your prediction using a charge diagram.	Perform the experiment and record your observations.	Revise the explanation if necessary.

Experiment: Packing peanut suspended on a string

Predict in words what you will observe.	Explain your prediction using a charge diagram.	Perform the experiment and record your observations.	Revise the explanation if necessary.

14.1.8 Observe and explain An electroscope is made up of a vertical metal rod with a metal sphere top that sticks out of a glass enclosure; a cork-shaped piece of plastic prevents electric charge from going from the metal rod and metal leaves at the bottom onto the cylindrical metal enclosure. A thin metal leaf hangs down against the bottom of the rod inside the enclosure.

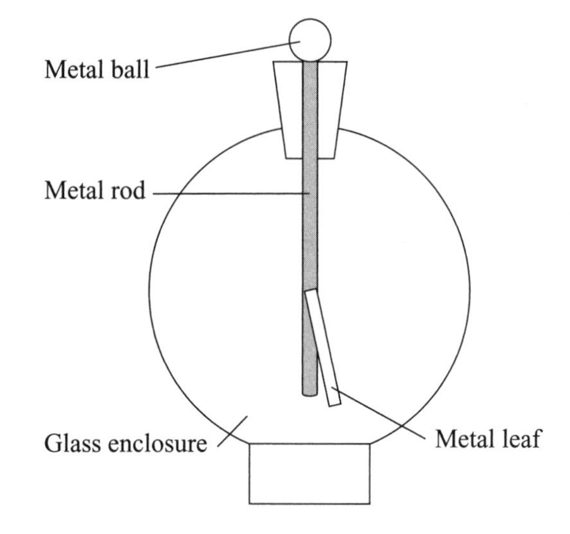

a. Perform the experiments described and fill in the table that follows.

Experiment	Record your observation.	Explain your observation.
Rub a foam tube with any material and then rub the top of the electroscope with the foam tube.		
Touch the top of the electroscope with your hand.		
Rub the top of the electroscope with a rubbed foam tube, remove the tube, and then touch the top of the electroscope with your hand while wearing a plastic glove.		

b. Modify the explanation you devised in Activity 14.1.7 to account for the outcomes of the third experiment.

14.1.9 Test your ideas Charge one electroscope by rubbing its top with a foam tube that has been rubbed with wool (the foam tube rubbed with wool is negatively charged). Place a second uncharged electroscope near the first charged electroscope.

a. Predict what happens if you touch a metal rod from the top of one electroscope to the top of the other. Wear a plastic glove to make sure you do not touch the rod with your bare skin. Explain your prediction, then perform the experiment and reconcile predictions with the outcome.

b. Discharge the electroscopes (touch them with your fingers) and predict what happens if you charge one electroscope and repeat the experiment in part a, only this time you touch a wooden rod from the top of the charged electroscope to the top of the uncharged electroscope. Fill in the table that follows to explain your predictions and then reconcile them with the observations.

Experiment: Touch the metal rod to the tops of the charged and uncharged electroscopes.

Predict in words what you will observe.	Explain your prediction.	Perform the experiment; record your observations.	Revise the explanation if necessary.

Experiment: Touch the wooden rod to the tops of the charged and uncharged electroscopes.

Predict in words what you will observe.	Explain your prediction.	Perform the experiment; record your observations.	Revise the explanation if necessary.

14.2 Conceptual Reasoning

14.2.1 Represent and reason Several experiments are described below. Complete the table that follows.

Sketch the situation.	Describe the situation in words.	Draw a microscopic representation (charge diagram) of the charges inside the specified object.	Perform the experiment and describe the outcome.
	Rub a foam tube with fur and bring it near one end of an empty plastic bottle placed on a swivel.	Plastic bottle:	
	Rub a foam tube with fur and bring it *near* an electroscope without touching the electroscope.		

Sketch the situation.	Describe the situation in words.	Draw a microscopic representation (charge diagram) of the charges inside the specified object.	Perform the experiment and describe the outcome.
	Move the tube in the last experiment away from the electroscope.		
	A piece of aluminum foil rolled in a ball hangs vertically from a string near a charged foam tube.		The ball swings and touches the rubbed foam tube and then immediately swings back away from the tube.

14.2.2 Represent and reason Two positively charged objects are held near each other in the muzzle of a cannon (see part a of the illustration). When the "trigger" holding the cannonball is released, the positively charged cannonball flies out the end of the muzzle in part b. Certain types of energy have increased. Describe some type of energy decrease that you think might compensate for the increase in these other energies. *Note:* The situation shown in part a of the illustration is similar to that of a compressed spring; instead of the coils of the spring being squeezed together, two like charges are squeezed or pushed together. In part b of the illustration, this compressed electric "spring" is more relaxed. Fill in the table that follows.

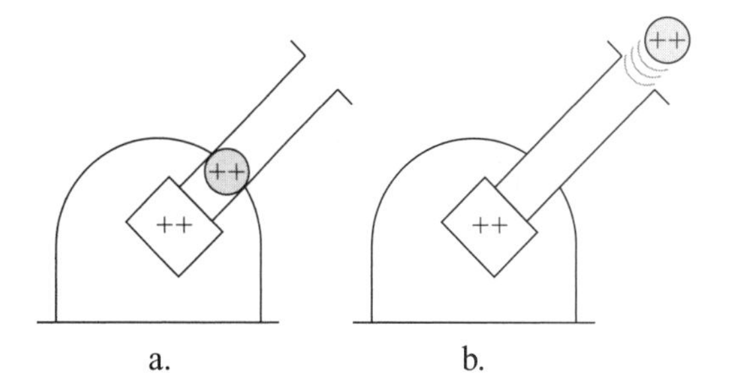

a. b.

Using the language of energy, explain in words the described process, going from the initial state to the final state. Indicate your system choice.	Draw an energy bar chart representing the initial and final states.
	K_i U_{gi} U_{si} U_{qi} $+W=$ K_f U_{gf} U_{sf} U_{qf} $\Delta U_{\delta U(\text{int})}$ + 0 −

14.2.3 Represent and reason Imagine the energy changes of two opposite-sign charged objects used as a nutcracker, as illustrated in the figure to the right. What happens when the negatively charged block shown in a is released and moves near the nut, as shown in b? What type of energy decreases to make up for the increase in kinetic energy? Fill in the table that follows to answer these questions. *Note:* The situation shown in part a is similar to that of a stretched spring. Instead of the coils of a spring being stretched, the two opposite charges are pulled apart—like stretching a spring. In part b, this electric stretched "spring" is in the process of relaxing.

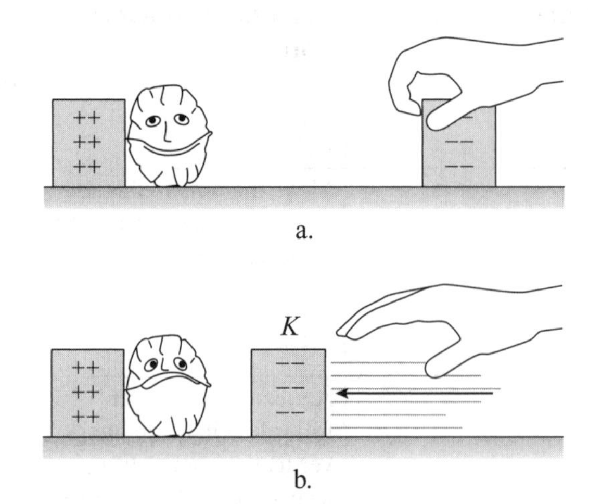

Using the language of energy, explain in words the process described above going from the initial state to the final state. Indicate your system choice.	Draw an energy bar chart representing the initial and final states.
	K_i U_{gi} U_{si} U_{qi} $+W=$ K_f U_{gf} U_{sf} U_{qf} $\Delta U_{\delta U(\text{int})}$ + 0 −

14.2.4 Reason

a. Does the analogy of a compressed spring for a system consisting of two similarly charged objects pushed close to each other make sense? Explain.

b. Does the analogy of a stretched spring for a system consisting of two oppositely charged objects pulled far apart make sense? Explain.

c. Discuss the limitations of both analogies.

14.3 Quantitative Concept Building and Testing

14.3.1 Find a pattern Charles Coulomb used a torsion balance to measure the force that one charged ball exerts on another charged ball to find out how the force between two electrically charged objects depends on the magnitudes of the charges and on their separation. Coulomb could not measure the absolute magnitude of the electric charge on the metal balls. However, he could divide charges in half by touching a charged metal ball with an identical uncharged ball. The table that follows provides data that resemble what Coulomb might have collected. Find patterns in the data and devise a mathematical relationship based on these observations. Use graph paper to help. Remember to decide which are the independent variables and which is the dependent variable. Then analyze the changes in the dependent variable as you change only *one* independent variable at a time.

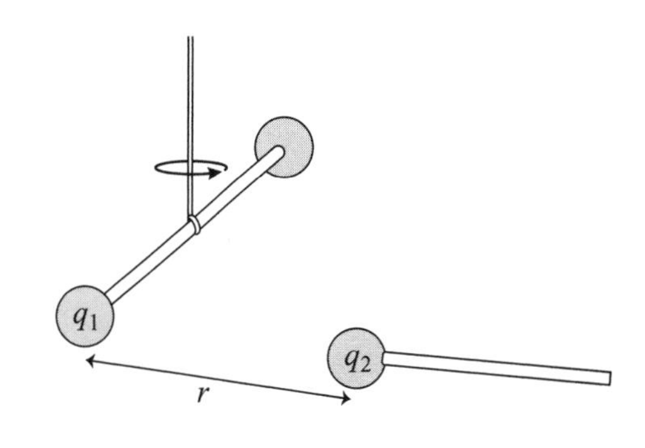

Charges (q_1, q_2)	Distance	Force
1, 1 (unit)	1 (unit)	1 (unit)
1/2, 1	1	1/2
1/4, 1	1	1/4
1, 1/2	1	1/2
1, 1/4	1	1/4
1/2, 1/2	1	1/4
1/4, 1/4	1	1/16
1, 1	2	1/4
1, 1	3	1/9
1, 1	4	1/16

14.4 Quantitative Reasoning

14.4.1 Represent and reason The metal balls on the cart in the illustration have equal magnitude charges and are very light. The rods supporting and connecting them are made of an insulating material and are also light. The cart rests on a smooth table.

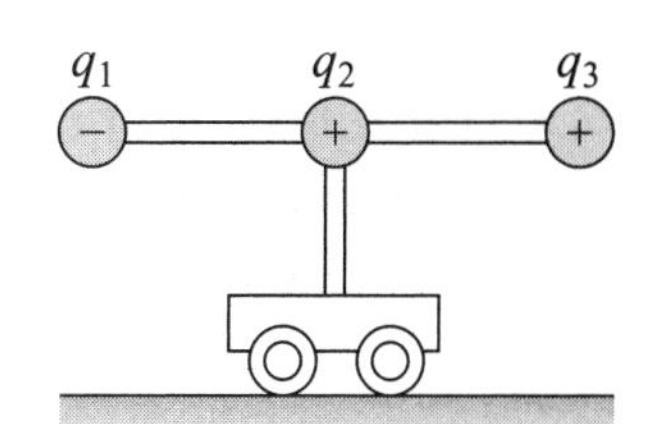

a. Fill in the table that follows. In this instance we consider only electric forces—not other types of force.

Draw labeled arrows representing electric forces exerted on the left metal ball. Represent the ball with a dot.	**Draw labeled arrows representing electric forces exerted on the center metal ball.**	**Draw labeled arrows representing electric forces exerted on the right metal ball.**	**Draw labeled arrows representing electric forces exerted on the whole cart (a system with three charged balls).**

b. Will the cart tend to accelerate either to the left or to the right? Explain your answer.

14.4.2 Represent and reason A positively charged ball of mass m hangs at the end of a string. Another positively charged ball is secured at the top end of the string to a wooden support. Fill in the table that follows.

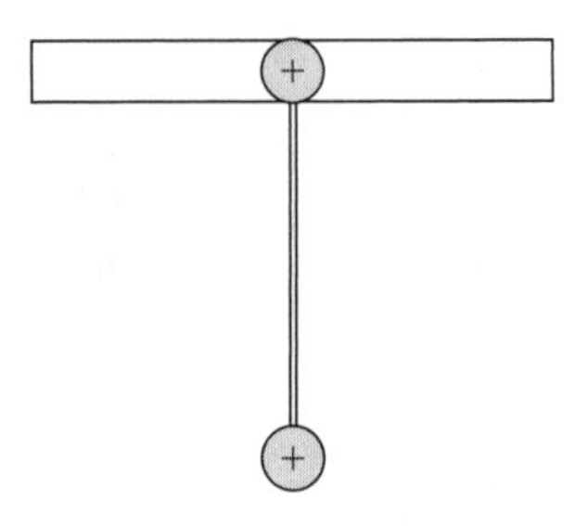

Draw a force diagram for the hanging ball if both balls are positively charged.	**Represent the diagram mathematically using Newton's second law.**	**Draw a force diagram for the hanging ball if the top ball is negatively charged.**	**Represent the diagram mathematically using Newton's second law.**

14.4.3 Represent and reason Two equal-mass stationary balls hang at the end of strings, as shown at the right. The ball on the left has electric charge $+5Q$, and the ball on the right has electric charge $+Q$. The strings make angles less than 45° with respect to the vertical plane. Fill in the table that follows.

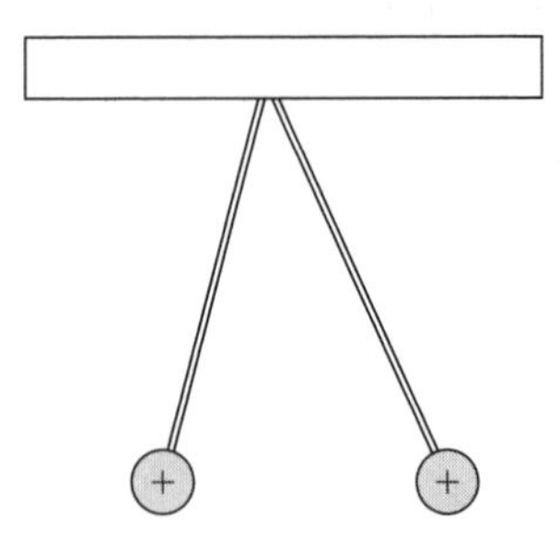

Draw a force diagram for the left ball.	Draw a force diagram for the right ball.	Decide which string makes a bigger angle with the vertical plane or if they make the same angle.
Apply Newton's second law in component form for the right ball. Horizontal x axis: Vertical y axis:	Based on your analysis, rank the forces $T_{\text{S on }Q}$, $F_{5Q\text{ on }Q}$, and $F_{\text{E on }Q}$, listing the largest force first. Explain the ranking:	

14.4.4 Equation Jeopardy The application of Newton's second law for a positively charged object at one instant of time is shown in the equation that follows. Other charged objects are along a horizontal line. Complete the table.

$$(9.0 \times 10^9\ \text{N} \cdot \text{m}^2/\text{C}^2)\left[-\frac{(2.0 \times 10^{-4}\text{C})(3.0 \times 10^{-5}\text{C})}{(2.0\ \text{m})^2} - \frac{(9.0 \times 10^{-4}\text{C})(3.0 \times 10^{-5}\text{C})}{(3.0\ \text{m})^2}\right]$$

$$= (4.0\ \text{kg})a_x$$

Draw a force diagram for the object at the instant the equation applies.	**Sketch a situation the equation might describe at that particular instant.**	**Write in words a problem for which the equation is a solution (it applies at only one instant in time).**	**Determine one change that could be made in the situation so that the net force exerted on the object of interest is zero.**

14.4.5 Evaluate the solution

The problem: A 2.0-kg cart with a $+2.0 \times 10^{-5}$ C charge on it sits at rest 1.0 m to the right of a fixed dome with charge $+1.0 \times 10^{-4}$ C. The cart is released. Determine how fast it is moving when it is 3.0 m from the fixed-charged dome.

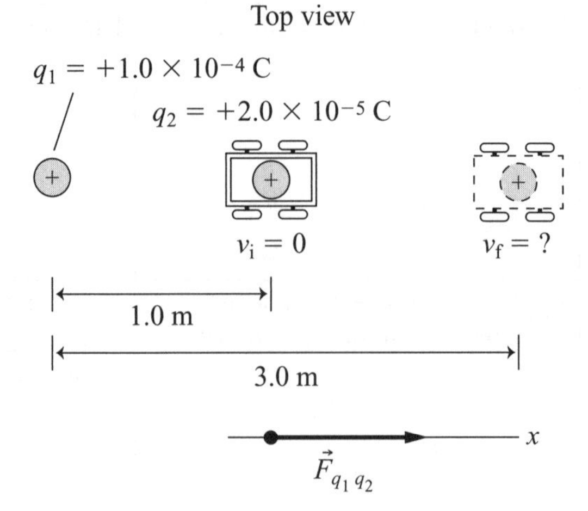

Proposed solution: The situation is shown at the right.

Simplify and diagram

We assume that the dome and cart are point particles.

See the force diagram to the right.

Represent mathematically and solve

$$\Sigma F_x = kq_1q_2/r = ma_x$$

$$a_x = kq_1q_2/rm$$

$$= (9 \times 10^9 \text{ N}\cdot\text{m}^2/\text{C}^2)(1.0 \times 10^{-4}\text{ C})(2.0 \times 10^{-5}\text{ C})/(1.0\text{ m})(2.0\text{ kg}) = 9.0\text{ m/s}^2$$

$$v^2 = 0^2 + 2(9.0\text{ m/s}^2)[(3.0\text{ m}) - (1.0\text{ m})] \quad \text{or} \quad v = 18\text{ m/s}$$

a. Identify any missing elements or errors in the solution.

b. If there are errors, provide a corrected solution or the missing elements.

14.4.6 Represent and reason A negatively charged ball, initially at rest, falls until it hits a massless spring, which it compresses while stopping. The bottom of the spring rests on a positively charged block.

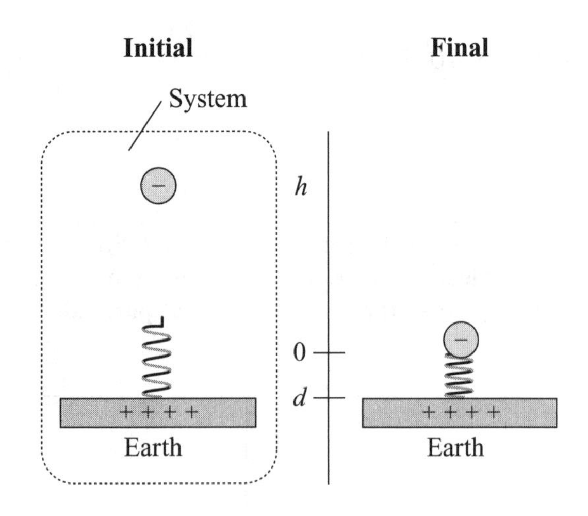

Draw a bar chart consistent with the process.	Apply the generalized work–energy equation to the process.
K_i U_{gi} U_{si} U_{qi} $+W=$ K_f U_{gf} U_{sf} U_{qf} $\Delta U_{\delta U(int)}$ + 0 –	

14.4.7 Represent and reason

a. Chris releases the trigger on an electric cannon. The cannonball with charge $+q$ and mass m fires vertically upward due to its repulsion from the stationary ball with a charge $+Q$. The cannonball reaches the apex of its flight at distance h above its starting position. Represent the process physically with a bar chart and mathematically in part a of the table that follows.

b. Now, suppose that the charge $+Q$ is reduced to $+Q/2$. Represent this process with a bar chart and mathematically. Describe in words how reducing the charge affects the process in part b of the table.

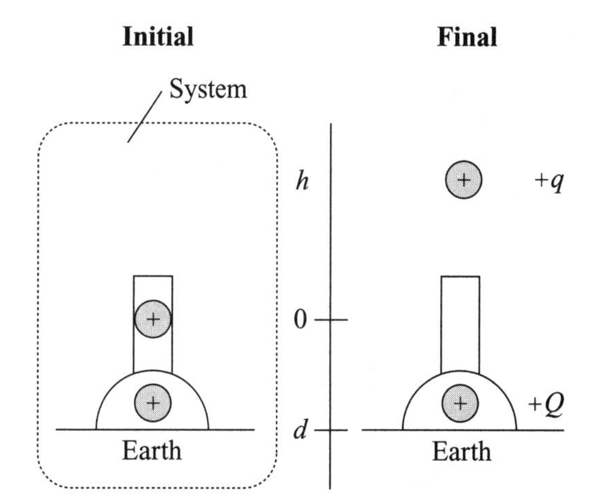

Draw a bar chart consistent with the process.	Apply the generalized work–energy equation.
a. K_i U_{gi} U_{si} U_{qi} $+W=$ K_f U_{gf} U_{sf} U_{qf} $\Delta U_{\delta U(int)}$ + 0 –	
b. K_i U_{gi} U_{si} U_{qi} $+W=$ K_f U_{gf} U_{sf} U_{qf} $\Delta U_{\delta U(int)}$ + 0 –	

14.4.8 Bar–chart Jeopardy The bar charts below could represent many processes. Complete the table for each bar chart.

Bar charts for unknown processes	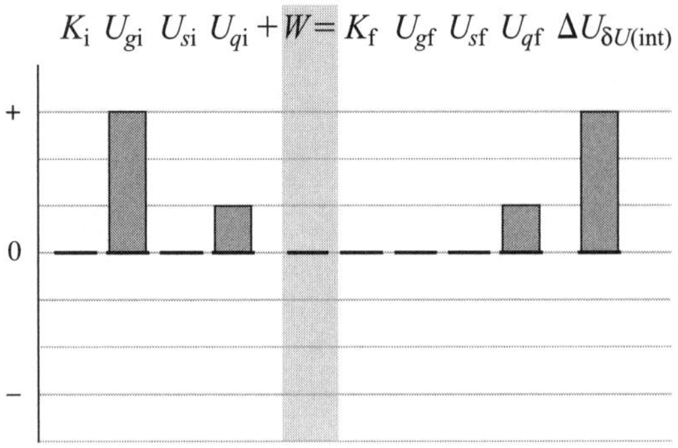 K_i U_{gi} U_{si} U_{qi} $+W=$ K_f U_{gf} U_{sf} U_{qf} $\Delta U_{\delta U(int)}$	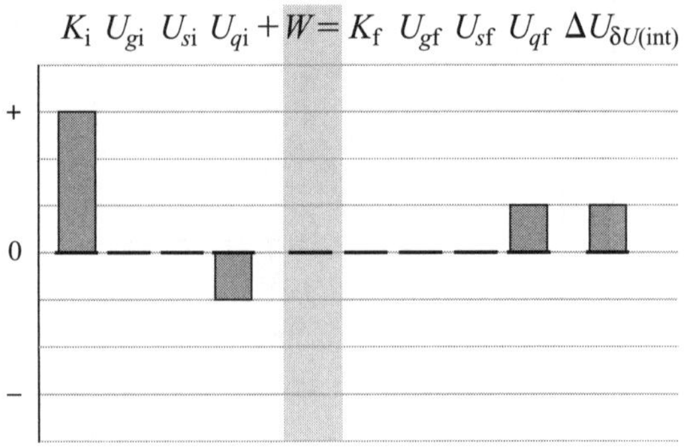 K_i U_{gi} U_{si} U_{qi} $+W=$ K_f U_{gf} U_{sf} U_{qf} $\Delta U_{\delta U(int)}$
Draw an initial–final sketch of one possible process described by the bar chart.		
Describe the process in words.		
Convert the bar chart into the work–energy relationship as applied to this process.		

14.4.9 Regular problem Consider a simplified version of a real situation that occurs in the center of our sun. A proton of charge $+e$ moves directly toward a stationary deuterium nucleus, also of charge $+e$. (We assume that the deuterium does not move.)

a. Determine the speed that the proton must move toward the deuterium when far from it so that it is able to get within 1.0×10^{-15} m before stopping. (At this distance, there is a good chance that the proton and deuterium will fuse to form a helium nucleus. The fusion releases considerable energy.)

Sketch and translate	Simplify and diagram (construct a bar chart)
Sketch the initial and final problem states. Choose a system.	
Represent mathematically	**Solve and evaluate**
Use the bar chart to apply a generalized work–energy equation.	

b. Estimate the temperature of hydrogen when this fusion can occur with good probability. The proton mass is 1.67×10^{-27} kg. What assumptions did you make?

14.4.10 Equation Jeopardy

The equation below describes one or more physical processes.

$$\frac{1}{2}(1.67 \times 10^{-27}\,\text{kg})v_i^2 + \frac{1}{2}(1.67 \times 10^{-27}\,\text{kg})v_i^2 = \frac{(9.0 \times 10^9\,\text{N}\cdot\text{m}^2/\text{C}^2)(1.6 \times 10^{-19}\,\text{C})^2}{1.0 \times 10^{-15}\,\text{m}}$$

Draw a bar chart that is consistent with the equation.	**Sketch the initial–final states that the equation might describe.**	**Write in words a problem for which the equation could be a solution.**

14.4.11 Evaluate the solution

The problem: A 2.0-kg cart with a $+2.0 \times 10^{-5}$ C charge on it starts at rest 1.0 m from a fixed dome with charge $+1.0 \times 10^{-4}$ C. The cart is released. Determine how fast it is moving when it is 3.0 m from the fixed-charged dome.

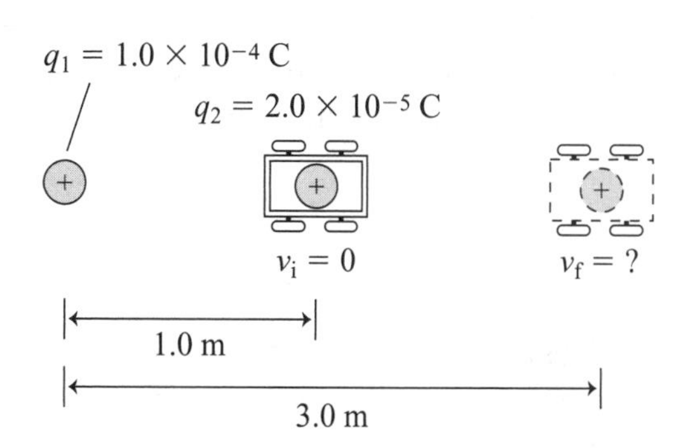

Proposed solution:

Sketch and translate

The situation is depicted in the illustration to the right.

Simplify and diagram

We assume that the dome and cart are point particles.

See the bar chart to the right.

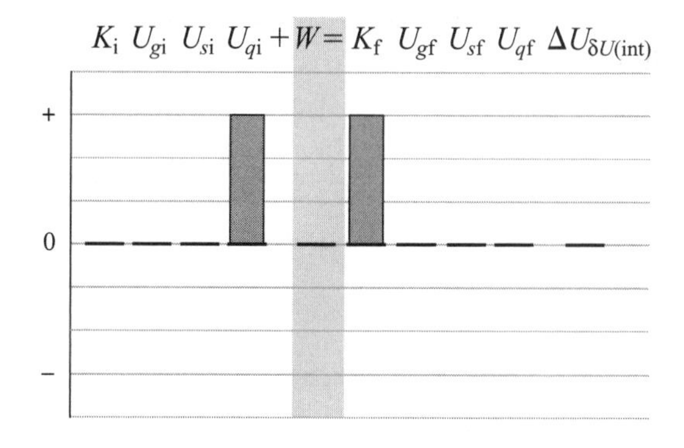

Represent mathematically and solve

$$
\begin{aligned}
U_{qi} &= K_f \\
kq_1q_2/r &= (1/2)mv^2 \\
v &= [2kq_1q_2/rm]^{1/2} \\
&= [2(9 \times 10^9)(1 \times 10^{-4})(2 \times 10^{-5})/(1.0)(2.0)]^{1/2} \\
&= 18 \text{ m/s}
\end{aligned}
$$

a. Identify any missing elements or errors in the solution to the problem.

b. If there are errors, provide a corrected solution or missing elements.

14.4.12 Regular problem One simple and productive model of a hydrogen atom (although rarely used in modern physics) is a positive nucleus (a proton) and a negatively charged electron moving around it in a circular orbit. Estimate the electron's speed in this model. The radius of the atom is 0.51×10^{-10} m.

14.4.13 Regular problem Use the result of Activity 14.4.12 to determine the minimum energy that the proton nucleus–electron system needs to gain for the electron to become free (to remove the electron far from its proton nucleus).

14.4.14 Design an experiment Your group is working on a static electricity project. You need to use a nonconducting string from which you will hang pieces of aluminum foil. Your friend brings two kinds of dental floss. Design an experiment to find out which floss is conducting and which one is not. Describe the experiment and explain how you will make a decision based on its outcome.

15 Electric Fields

15.1 Qualitative Concept Building and Testing

15.1.1 Represent and reason Imagine two pointlike charged objects of mass m_1 and m_2 that have electric charges q_1 and q_2, respectively. Complete the table that follows and analyze the objects' gravitational and electrostatic interactions using the filled cells as hints.

Problem	Gravitational interaction	Electrostatic interaction
What property of objects determines whether they participate in the interaction?		Electric charge
What is the direction of the force between the interacting objects?	It is an attractive force.	
Write an expression for the magnitude of the force between interacting objects.		
How does the magnitude of the force depend on properties of the objects?	It is directly proportional to m_1 and directly proportional to m_2.	
How does the magnitude of the force depend on the distance between the objects?		
Write an expression for the potential energy of the interacting objects.	$U_g = -G\frac{m_1 m_2}{r}$	$U_e = k\frac{q_1 q_2}{r}$

15.1.2 Reason Discuss similarities and differences between gravitational and electrostatic interactions. Suggest possible mechanisms for how these interactions can occur at a distance—without direct contact between objects.

15.1.3 Observe and explain Assemble an uncharged electroscope; a foam tube; fur; a small, clear, glass beaker (or plastic cup); and a small metal cup. Perform the experiments described in the table below and explain the outcomes using the idea of an electric field.

Experiment	Result	Sketch the situation.	Explain the experiment using the idea of an electric field.
Bring the charged end of a foam tube down toward the top of an uncharged electroscope without touching the electroscope.	The needle of the electroscope deflects.		
Take the charged end of the foam rod away from the top of the uncharged electroscope.	The needle of the electroscope goes back to its original position.		
Repeat the first two experiments, this time using a small glass beaker to cover the metal ball on the electroscope.	The needle behaves the same way in both cases, but the deflection is a little smaller.		
Repeat the first two experiments, only this time using a small metal cup to cover the metal ball on the electroscope.	There is no deflection of the needle, as if the charged tube was not brought close.		

15.1.4 Design an experiment Use any equipment you have to design an experiment that allows you to observe that the electric interaction between two objects can be blocked or shielded. Perform the experiments and record the outcomes.

15.2 Conceptual Reasoning

15.2.1 Represent and reason

a. For each situation pictured in the table that follows, represent at the dots the gravitational force or the electric force that the source mass or source charge exerts on a test mass or test charge. Fill in the table that follows.

Word description	Represent with arrows the gravitational force that the Earth (the source mass) exerts on small objects (called *test masses*) at the points shown. Draw the arrows with the correct relative lengths.	Represent with arrows the electric force that the object with a large negative charge (the source charge) exerts on small positively charged objects (called *test charges*) at the points shown.	Represent with arrows the electric force that the object with a large positive charge (the source charge) exerts on small positively charged objects (called *test charges*) at the points shown.
Picture description	• (Earth) •	• (−) •	• (+) •

b. Use a field approach to explain in words how the source object can exert a force on test objects without directly touching them at each of these points. Discuss how the magnitude of this force might depend on the magnitude of the gravitational field, on the magnitude of the mass of the test objects in the gravitational field, on the magnitude of the electric field, and on the test charges in the electric field.

c. Discuss how the presence of a source mass or a source charge alters the space. How far do you think this alteration extends?

d. Discuss whether a system that includes Earth and a test object possesses gravitational potential energy. Discuss whether a system that includes the source-charged object and the test-charged object possesses electric potential energy. Does each energy depend on the magnitude of the source mass or the test mass? On the source charge or the test charge?

15.2.2 Represent and reason Estimate the direction and the magnitude of the $\vec{E}$ field at points A, B, and C in the figure that follows.

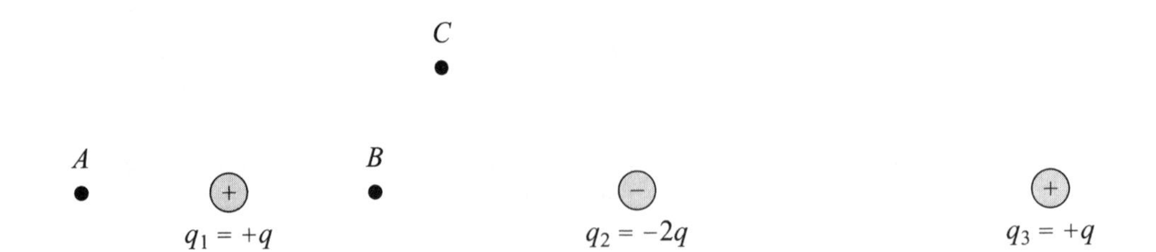

15.2.3 Represent and reason Estimate, by drawing on the figures that follow, the direction and relative magnitude of the $\vec{E}$ field at points A, B, and C due to the electric dipole on the heart (shown at one instant during a heartbeat cycle).

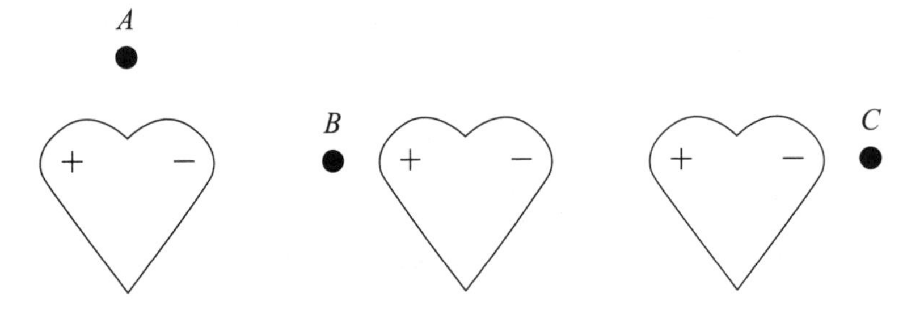

15.2.4 Reason Use the analogy between the electric field and the gravitational field to estimate the Earth's gravitational field $\vec{g}$ at the points shown in the figure.

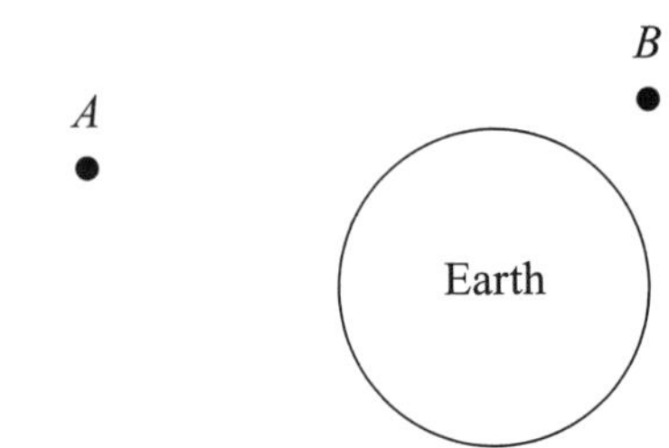

a. What would you choose as the source-mass object? What would you use as the test-mass object?

b. Draw $\vec{g}$ field vectors at points A, B, and C.

c. Discuss how the magnitude and direction of the $\vec{g}$ field are related to the acceleration of free-falling objects placed at these points.

15.2.5 Diagram Jeopardy $\vec{E}$ field vectors due to one or more electrically charged objects are shown below. Indicate with circles, including + or – signs, the locations of the charged objects causing the fields.

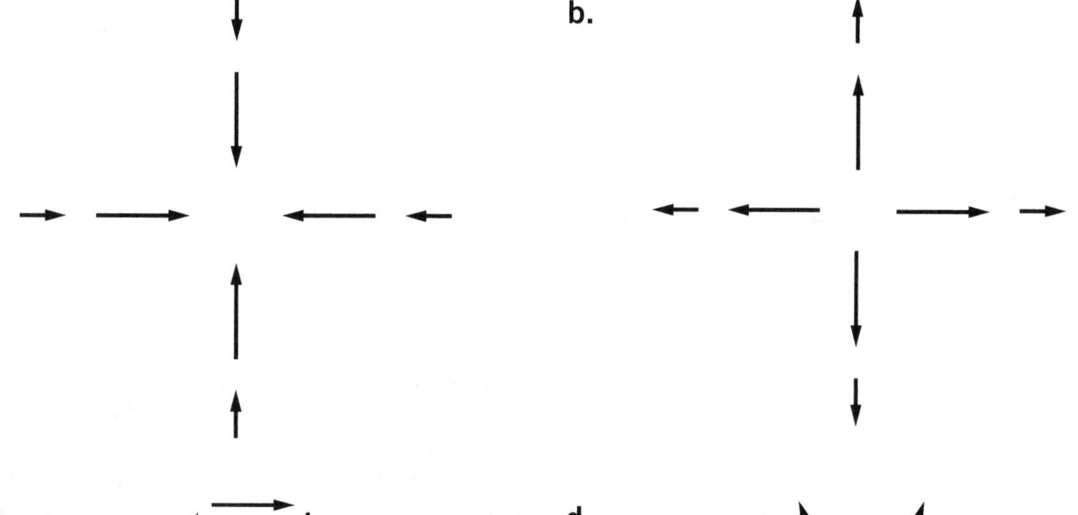

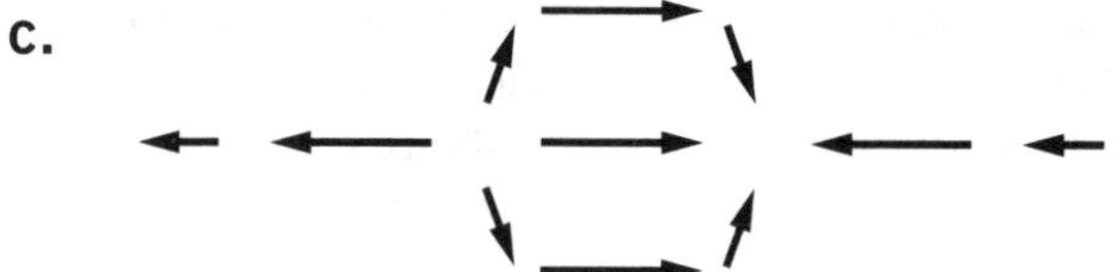

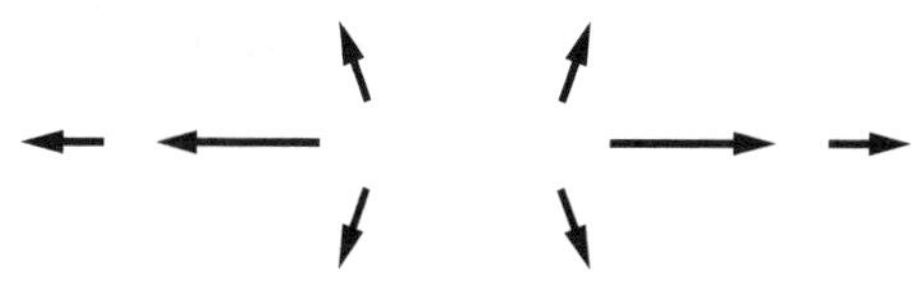

15.2.6 Reason Answer the following questions:

a. Can electric field lines cross? Explain.

b. What is the direction of the $\vec{E}$ field line at a point midway between two equal-magnitude, positively charged objects? How do you know?

c. What is the direction of the $\vec{E}$ field line at a point midway between two equal-magnitude, oppositely charged objects? How do you know?

15.2.7 Represent and reason Draw $\vec{E}$ field lines for the electric field created by the source-charged objects described in the table that follows. Show the vectors that are tangent to the lines.

a. A pointlike positively charged object	**b.** A pointlike positively charged object with twice the magnitude of charge as in part a	**c.** A pointlike negatively charged object	**d.** A pointlike negatively charged object with twice the magnitude of charge as in part c
e. Two positively charged pointlike objects of equal-magnitude charge, separated by a distance *s*	**f.** Two negatively charged pointlike objects of equal-magnitude charge, separated by a distance *s*	**g.** A small positively charged object and a small negatively charged object of equal-magnitude charge, separated by a distance *s*	**h.** A small positively changed object and a small negatively charged object with twice the magnitude of electric charge, separated by a distance *s*

15.2.8 Represent and reason Imagine that a small, positively charged object moving toward the top of the page enters an electric field with the lines shown below.

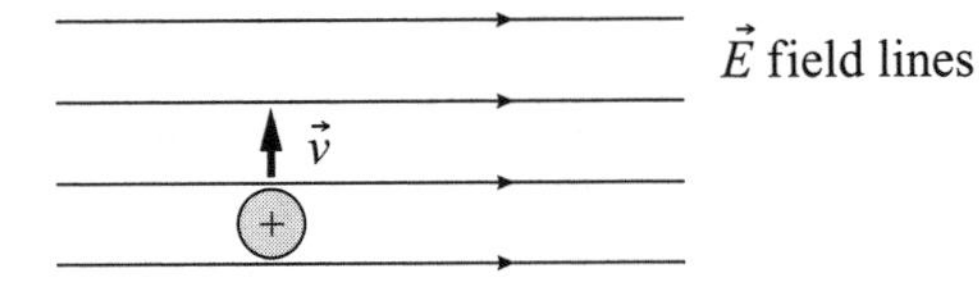

a. Sketch on the illustration an approximate path of the object as it moves through the field. The direction of the initial velocity of the object as it enters the field is shown in the figure.

b. Discuss whether the lines represent the paths that a charged object follows after it enters the field.

15.2.9 Represent and reason The figure on the right shows the instant when a hollow metal box is placed in a uniform electric field (its effect on the external field is not shown).

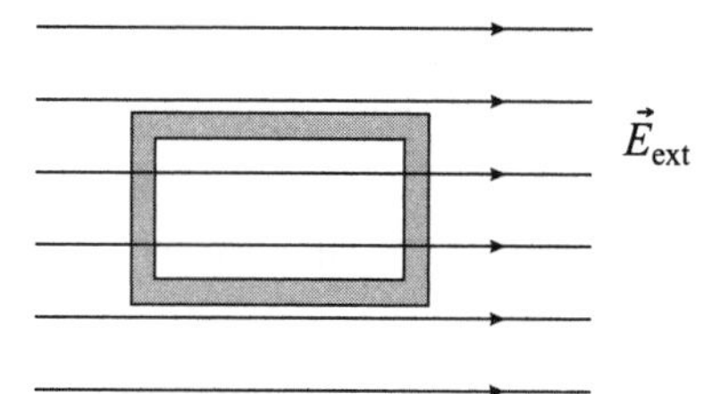

a. Indicate the electric charge distribution in the metal due to the external electric field. *Note:* If electrons move from one part of the box to another, the part with a deficiency of negatively charged electrons is now positively charged, and the part with an excess of electrons is negatively charged.

b. Draw electric field lines caused by this induced-charge distribution on the surface of the box and discuss the magnitude of the total $\vec{E}$ field inside the box.

c. Discuss how your reasoning in parts a and b helps explain why it is safe to sit in a car during a lightning storm.

d. Draw the new shape of the field lines outside the box. How does the redistribution of electrons inside the box affect the electric field outside?

15.2.10 Reason Complete the table that follows.

Compare the electric potentials at points *A, B,* and *C* due to the positive source charge, the largest potential listed first. Show at least one other point at which electric potential is the same as at point *A*.	Compare the electric potentials at points *D, E,* and *F*, the largest potential listed first. (*Note:* A large negative number is less than a small negative number.) Show at least one other point at which electric potential is the same as at point *E*.
• *C* • *B* • *A* (+)	• *F* • *E* • *D* (−)

15.2.11 Represent and reason In the table that follows, draw lines of equal gravitational potential caused by Earth, a mass source for the gravitational potential. Think of how you can write an expression for the gravitational potential that is analogous to the expression for the electric potential.

Far away from the surface of the Earth, when the Earth is modeled as a sphere	**Close to the surface of the Earth, when the surface of the Earth can be modeled as a plane**
Earth	Earth

15.2.12 Represent and reason Using the same source charges as in Activity 15.2.7 a–d and the electric field lines that you drew there, draw the surfaces of equal potential. Find a pattern between the direction of the lines and the change of electric potential (whether the electric field lines point in the direction of increasing or decreasing potential).

15.2.13 Explain Sometimes physicists use the analogy between lines of equal electric potential and lines of equal altitude on topographical maps.

a. Explain how this analogy works and why it is useful.

b. Describe how the closeness of the equal-altitude lines relates to the steepness of a mountain.

c. Use this analogy to draw electric field lines for the electric field whose equal potential surfaces look as follows. Explain.

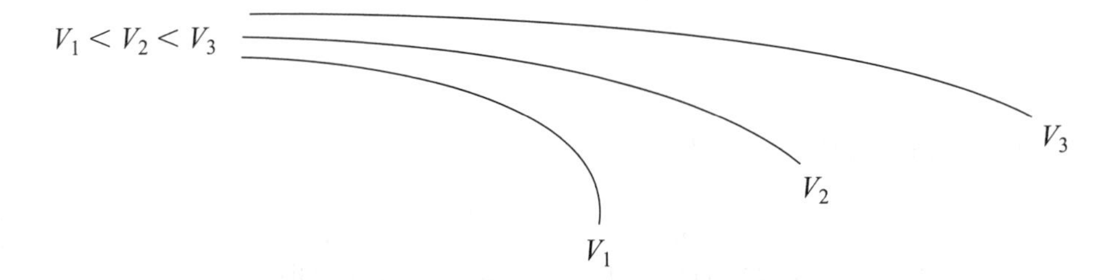

15.2.14 Reason

a. Draw equal potential surfaces and $\vec{E}$ field lines for a negatively charged, infinitely large metal plate. For help, complete the activities in the table that follows.

Sketch the plate with electric charges (shown below).	Draw lines of equal electric potential. Indicate where the potential is higher.	Draw $\vec{E}$ field lines.
− − − − − − − − − − − −	− − − − − − − − − − − −	− − − − − − − − − − − −

b. Why are the $\vec{E}$ field lines perpendicular to the surface of the plate and to the equal potential surfaces? Explain.

c. Are both the $\vec{E}$ field lines and equal potential surfaces equally spaced? Explain.

d. Examine the locations of the $\vec{E}$ field lines and the equal potential surfaces with respect to each other. Discuss any patterns that you find. Explain.

15.3 Quantitative Concept Building and Testing

15.3.1 Derive A pointlike object of charge q (a source charge) is located at the origin of a coordinate system.

a. Use the definition of $\vec{E}$ field to derive a relationship for the magnitude of the $\vec{E}$ field caused by this charged object at some point A a distance r from the charged object. Represent the relationship graphically (the magnitude of $\vec{E}$ -versus-r).

b. Use the definition of electric potential to write an expression for the electric potential at point A due to the pointlike charge q at the origin. Point A is a distance r from the charge q. Represent the relationship graphically (V-versus-r).

c. Discuss whether there is any relationship between the $\vec{E}$ field and electric potential.

15.3.2 Reason

a. Imagine that you have a positively charged, solid metal ball. Complete the table that follows.

Indicate in the drawing how the charge is distributed inside the ball and on the ball's surface and explain your drawing.	R	Draw electric field lines outside the ball and compare their distribution with the distribution of electric field lines of a pointlike charge.	R
Draw electric field lines inside the ball (if any). Explain.	R	Draw a graph of the magnitude of the $\vec{E}$ field versus the distance r from the center of the ball.	E, R, r
		Draw a graph of the V field versus the distance r from the center of the ball.	V, R, r

b. Discuss how you can have a situation in which the electric field at some location is zero but the electric potential is not. Does this seem reasonable? Explain.

15.3.3 Reason You have two metal spheres of radii R_1 and R_2. The sphere on the left has a charge $+q_1$, and the sphere on the right is not charged.

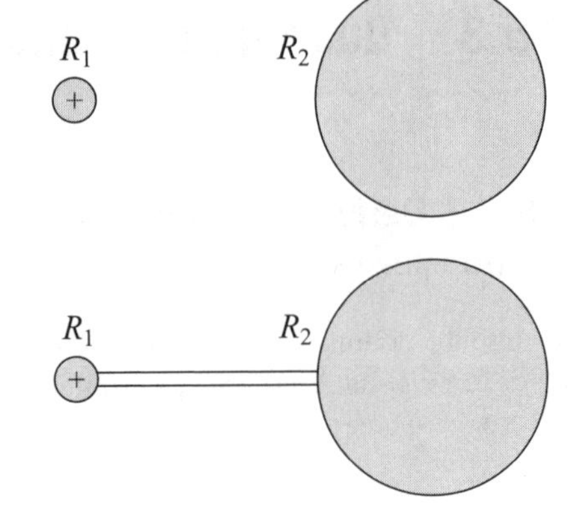

a. Explain whether the electric potential on the surfaces of the spheres will be the same or different before they are connected with a metal rod.

b. Will the electric potential of the surfaces be the same or different after they are connected?

c. Which sphere will have a larger charge after the connection? How does the total charge of the two spheres after the connection compare to the initial charge of the left sphere?

d. Discuss how the situation in parts a–c is related to the method of discharging objects by connecting them to Earth—so-called *grounding*. Earth is a huge conductor, like a huge metal sphere.

15.4 Quantitative Reasoning

15.4.1 Represent and reason A small aluminum ball is charged with $+1.0 \times 10^{-9}$ C. Determine the magnitude of the $\vec{E}$ field due to this source charge at the four points shown below. Represent the $\vec{E}$ field at each point using an arrow.

5 cm to the right of the ball	7 cm to the left of the ball	1 cm above the ball	10 cm below the ball
⊕ •	• ⊕	• ⊕	⊕ •

15.4.2 Represent and reason Charged objects $q_1 = +q$ and $q_2 = -q$ are separated by a distance r, as shown in the art below. Determine an expression for the magnitude of the $\vec{E}$ field at points $(r, 0)$ and $(0, r)$ due to these two charges. Fill in the table that follows.

Sketch	Draw an electric field diagram.	Determine the component fields E_x and E_y.	Determine the magnitude of the net electric field.
$-q$ $+q$ ← r → ← r →			
r $-q$ $+q$ ← r →			

15.4.3 Represent and reason A 0.060-kg electrically charged ball hangs at the end of a string oriented 53° outward from a charged vertical plate. The $\vec{E}$ field produced by the plate at the position of the ball is 1.0×10^5 N/C and points away from the plate. Complete the table that follows to determine the charge on the ball (sign and magnitude). Assume that the gravitational constant is 10 N/kg.

Sketch the situation.	Draw a force diagram for the hanging ball.	Represent the diagram mathematically using Newton's second law.	Determine the charge on the ball.
$E = 1.0 \times 10^5$ N/C 53° q? $m = 0.060$ kg		x: y:	

15.4.4 Equation Jeopardy An electrically charged block moves on a horizontal surface. The application of Newton's second law and kinematics to this situation is shown in the equations that follow. Complete the table. Assume that the gravitational constant is 10 N/kg. *Note: n* is the magnitude of the normal force, and N is the unit of force. The zeros are for forces that have zero components along that axis.

$$\Sigma F_x = (+1.0 \times 10^{-5}\,\text{C})(+2.0 \times 10^{6}\,\text{N/C}) + 0 - 0.40\,n + 0 = (4.0\,\text{kg})\,a_x$$

$$\Sigma F_y = 0 + n + 0 - (4.0\,\text{kg})(10\,\text{N/kg}) = 0$$

$$2a_x(24\,\text{m} - 8.0\,\text{m}) = v^2 - 0$$

Draw a force diagram for the block.	Sketch a situation the equation might describe.	Write in words a problem for which the equation is a solution.	Determine one change that could be made in the situation so that the net force exerted on the object of interest is zero.

15.4.5 Evaluate the solution

The problem: A 0.040-kg cart is moving at a speed of 6.0 m/s when it enters a 1.8×10^4 N/C electric field that stops the cart in 0.40 m. Determine the electric charge on the cart.

Proposed solution:

Sketch and translate

The situation is shown in the illustration.

$\vec{E}$

q_i q_f

0.40 m

Simplify and diagram

We assume that the cart and its charge are a point-like object and that there is no friction.

A force diagram for the cart is shown at the right.

x

Represent mathematically, solve and evaluate

$$a = (v_f^2 - v_i^2)/2(x_f - x_i) = (6.0\,\text{m/s})^2/2(0.40\,\text{m}) = 90\,\text{m/s}^2$$

$$qE = ma$$

$$q = ma/E = (0.040\,\text{kg})(90\,\text{m/s}^2)/(1.8 \times 10^4\,\text{N/C}) = 2.0 \times 10^{-4}\,\text{C}$$

a. Identify any errors or missing elements in the solution to this problem.

b. Provide a corrected solution if there are errors.

15.4.6 Regular problem A 0.50-kg cart with charge $+3.0 \times 10^{-6}$ C slides on a horizontal surface that exerts a 2.0-N friction force on the cart. The cart moves in a 1.0×10^{-6} N/C horizontal constant $\vec{E}$ field that points toward the right. If the cart starts at rest, determine its speed after moving 2.0 m.

Sketch the situation.	Draw a force diagram for the cart.	Represent the situation mathematically.	Determine the cart's speed.

15.4.7 Represent and reason The same situation is represented in different ways in the table.

Words
Determine the electric potential V at a distance r above a point directly between an electric dipole, which consists of an object of charge $+Q$ on the left and a second object of charge $-Q$ that is a distance r to the right.
Sketch 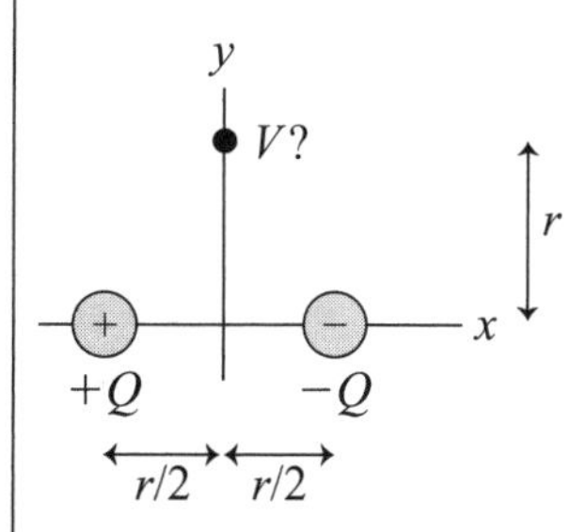
Mathematical representation and solution $V = \frac{k(+Q)}{\left[\left(\frac{r}{2}\right)^2 + r^2\right]^{\frac{1}{2}}} + \frac{k(-Q)^2}{\left[\left(\frac{r}{2}\right)^2 + r^2\right]^{\frac{1}{2}}} = 0$

Are the representations consistent with each other? Explain.

15.4.8 Represent and reason Determine the speed of the dust particle illustrated in the table that follows.

Words	Sketch
A 0.0010-g dust particle charged to $+1.0 \times 10^{-7}$ C encounters a uniform 1.0×10^{4} N/C electric field, as shown. Determine the change in its speed after it moves a vertical distance of 1.0 m.	
Represent mathematically and solve.	
Indicate any assumptions you made.	

15.4.9 Represent and reason Determine the electric potential at point A. Specify your choice of the reference point where the potential is zero.

Words
Three charged objects are shown. Determine the electric potential V at point A—that is, at position $x = 0.9$ m.
Sketch $q_1 = +2 \times 10^{-6}$ C $q_2 = -4 \times 10^{-6}$ C $q_3 = +2 \times 10^{-6}$ C $x_1 = 0$ $x_2 = 0.3$ m $x_3 = 0.6$ m $x_A = 0.9$ m
Represent mathematically and solve.

15.4.10 Evaluate the solution

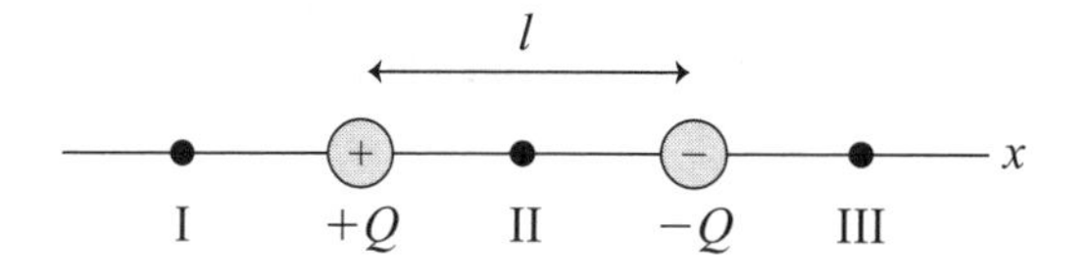

The problem: Imagine two electrically charged objects $+Q$ and $-Q$ connected by a plastic rod (not shown in illustration) of length l. Write an expression for the electric potential at position I (distance r to the left of $+Q$), position II (halfway between the charged objects), and position III (distance r to the right of $-Q$). Identify and correct any errors in the solutions in the table that follows.

Proposed solution	Identify errors in the solution (if there are any).	Provide a corrected solution (if there are errors).
$V_I = kQ/r^2 + k(-Q)/(2r)^2$ $= 3kQ/4r^2$		
$V_{II} = kQ/(r/2) + k(-Q)/(r/2)$ $= 0$		
$V_{III} = kQ/2r + kQ/r$ $= 3kQ/2r$		

15.4.11 Represent and reason Complete the table that follows. Do not solve for anything.

Description in words of the process	Rocket launch: A 0.010-kg rocket with a $+1.0 \times 10^{-3}$ C charge rests on a very large horizontal plate. Another very large horizontal plate above the first has a hole in it directly above the rocket. When a switch is closed, the lower plate and upper plates become charged so that there is a +10,000-V electric potential on the lower plate relative to zero potential on the upper plate. The rocket shoots up through the hole. Determine the maximum height that the rocket reaches above its starting position. (*Note:* The electric potential does not change after the rocket passes through the hole in the top plate.)
Sketch the process.	

(*continued*)

Construct a qualitative bar chart.	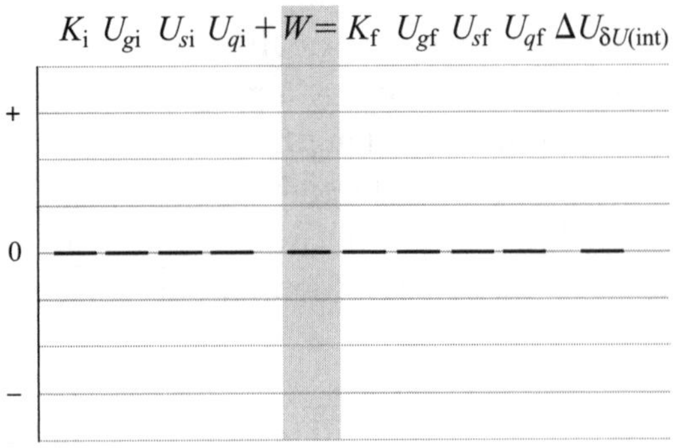 K_i U_{gi} U_{si} U_{qi} $+W=$ K_f U_{gf} U_{sf} U_{qf} $\Delta U_{\delta U(int)}$ + 0 −
Write a mathematical description of the process.	

15.4.12 Equation Jeopardy

The generalized work–energy equation is applied to a physical process in the last row of the table that follows. Complete the table for this process.

Write a word description of the process.	
Sketch the process.	
Construct a qualitative bar chart.	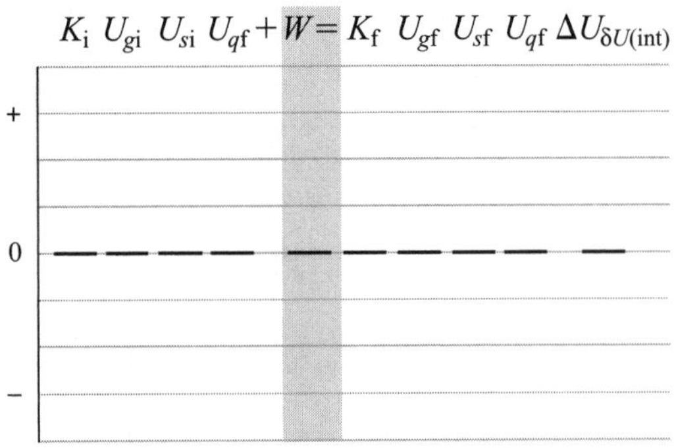 K_i U_{gi} U_{si} U_{qf} $+W=$ K_f U_{gf} U_{sf} U_{qf} $\Delta U_{\delta U(int)}$ + 0 −
Mathematical description of the process	$(2.0 \times 10^{-5}\text{ C})(40{,}000\text{ V})$ $= (1/2)(1.0 \times 10^{-3}\text{ kg})v^2$ $+(1.0 \times 10^{-3}\text{ kg})(10\text{ N/kg})(40\text{ m})$

15.4.13 Regular problem To cleanse the air of dust and pollen, some homes have electrostatic precipitators in their heating and air-conditioning systems. These units work by moving particle-laden air through an ionizing area, in which the particles acquire a charge, and then into an area in which an electric field is present. Because the particles are charged, they experience a force and are attracted to a collector that home owners need to clean from time to time. Such units are also used on a larger scale as industrial scrubbers in smokestacks, where particle-laden air rises through the stack, acquires a charge by an industrial ionizing source, and is filtered via electrostatic attraction. Imagine that you are a member of a team from the Environmental Protection Agency, which is trying to determine whether the school's heating smokestack is effective in removing most particulate matter. You find that the smoke particles move up the 20-m-high stack at a constant 5-m/s speed. By checking the ionizing equipment, you deduce that the specific charge (charge per unit mass) imparted to each particle is 1×10^{-5} C/kg. Most particles have a mass of 1 μg (10^{-6} g), but some are as large as 100 μg. You see that two plates separated by 0.30 m are on opposite sides of the chimney, with a 3000-V/m $\vec{E}$ field between them. Will you recommend that the operating license for the smokestack be renewed? Support your ruling with a careful analysis.

16 DC Circuits

16.1 Qualitative Concept Building and Testing

16.1.1 Observe and explain You have two electroscopes, initially uncharged, situated near each other. Explain the outcomes of each experiment described in the table that follows.

Experiment	Rub a foam tube with wool and then rub the top of one electroscope with the rubbed part of the tube. Observe that the needle of the electroscope deflects.	Then while wearing latex gloves connect a metal wire between the top of the first electroscope and the top of the second uncharged electroscope. The leaves of the second electroscope instantly separate, and the leaves of the first electroscope come closer to each other but do not go down completely.	Now rub the first electroscope with the charged tube again. The leaves of both electroscopes deflect more.
Explain using the language of *V* field (electric potential).			

16.1.2 Observe and explain You again have two electroscopes situated near each other; one is charged, and the other is not. This time the leads of a neon lightbulb connect the electroscopes (you may need an extra wire connected to one of the bulb's leads). Fill in the table that follows.

Experiment	Use the bulb to connect a charged electroscope and an uncharged electroscope. You observe a short flash of light when you first connect them.	Then charge both electroscopes and touch the neon bulb leads across the tops of the electroscopes. You observe no light.
Explain using the language of *V* field (electric potential).	Why does the flash last just a short time interval?	

16.1.3 Observe and explain You have a Wimshurst generator and a neon lightbulb. Complete the table that follows.

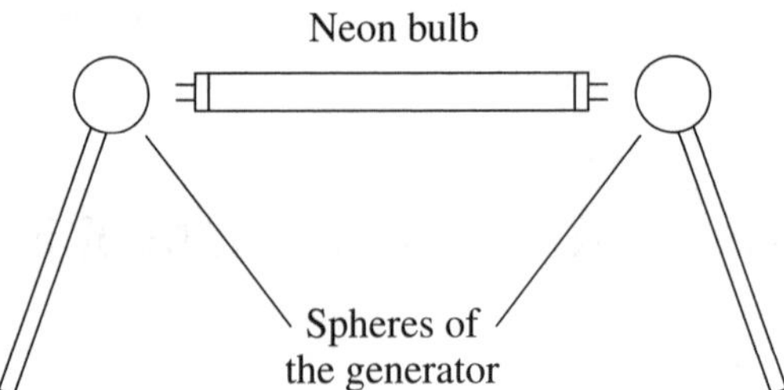

Experiment	Crank the handle of the Wimshurst generator. When you bring the metal spheres of the generator near each other, there is a strong spark.	Crank the handle of the generator with the metal spheres about 5-cm apart. Touch the lead wires of a neon lightbulb to the metal spheres of the generator. You see a bright flash of light from the neon bulb.
Explain		

16.1.4 Observe and explain You charge a Wimshurst generator by cranking the handle and then hang a Ping-Pong® ball with a metal coating from a thin thread between the charged spheres of the generator. Fill in the table that follows.

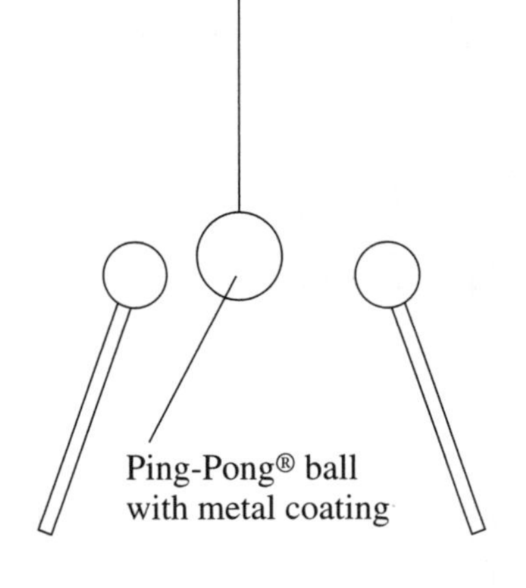

Ping-Pong® is a registered trademark of Parker Brothers, Inc.

Experiment	The metal-coated ball swings rapidly back and forth between the oppositely charged spheres of the generator for about 30 s but eventually slows down and stops.
Explain using the language of *V* field and energy	

16.1.5 Observe and explain You have a battery, a wire, and a lightbulb. Try different arrangements of these three elements to make the lightbulb glow.

Draw pictures of the arrangements where the bulb lights and several where it does not. Explain how this experiment is similar to 16.1.3 and 16.1.4 and how it is different.

16.1.6 Explain Use your observations and explanations for the previous activities to answer the following questions.

a. Summarize the conditions that are necessary for the continuous flow of electric charge through a metal wire.

b. What properties of the device connected to the wire are necessary to maintain a continuous electric charge flow through the wire? Think of such a device in your everyday experience.

16.1.7 Test your ideas One of the devices that maintains a continuous potential difference across different points of a circuit is a battery. Use a 45-V battery for the following testing experiment.

a. Fill in the table that follows.

Experiment	**Predict the outcome.**	**Perform the experiment and record the outcome. Explain.**
What happens if you connect the poles of the battery to parallel metal plates placed near each other with a metal-coated Ping-Pong® ball hanging from a thin thread between the plates?		

b. Explain ways in which this experiment is analogous to a battery connected to a lightbulb (see Activity 16.1.5). Indicate the corresponding parts of the two different systems described in the two different activities. How are the systems different?

16.1.8 Observe and design Draw circuit diagrams according to the word descriptions below. Build the circuits, and observe the relative brightness of the lightbulbs.

Circuit description	**Draw a circuit diagram.**	**Discuss the brightness of the lightbulb (is it more bright or less bright than in the first experiment).**
One 1.5-V battery, one lighted light-bulb, and wires		
Two 1.5-V batteries arranged so that the positive side of one touches the negative side of the other, forming a chain (in physics they are said to be *in series*); one lighted lightbulb; and wires		
Two 1.5-V batteries arranged so that their positive sides are together and negative sides are together, forming a ladder (in parallel); one lighted light-bulb; and wires		

16.2 Conceptual Reasoning

16.2.1 Reason You learned in Section 16.1 that for a lightbulb to glow, the two poles of a battery must be connected to the lightbulb with conducting wires. You also observed experiments in which several batteries were connected to lightbulbs in series or in parallel. Use your knowledge of the internal structure of conductors and the understanding of the role of a battery to explain these observations using two analogies: one involving flowing water and the other involving a group of people running on a track. Remember that an analogy does not need to account for all aspects of a phenomenon. However, if you find similar aspects, make a note of them.

Parts of the electric circuit	Parts of the water system	Parts of the running people system
Moving electrons		
Battery		
Connecting wires	Pipes with water in them	
Lightbulb		Muddy patch on the track

Observed properties of the electric current	Observed properties of the water system	Observed properties of the running people system
When batteries are in series, the lightbulb is brighter.		
When identical batteries are in parallel, the lightbulb is the same brightness.		

16.2.2 Reason In Chapter 15 you learned the concept of *V* field or potential difference. Use this concept to explain the role of a battery in a circuit.

a. Describe an analogy between some part of a system with water flow in a pipe caused by a pump and the potential difference provided by a battery in a circuit.

b. Describe an analogy between some part of a system with water flow in a pipe caused by a pump and the physical quantity *electric current* in a circuit.

16.2.3 Reason Use the flowing-water system and running-people system to find analogies for the quantities *potential difference* and *electric current*. Fill in the table that follows.

Electric circuit	**Water system**	**Running people**
Potential difference between two points ΔV		
Current I through a wire		

16.2.4 Observe and explain

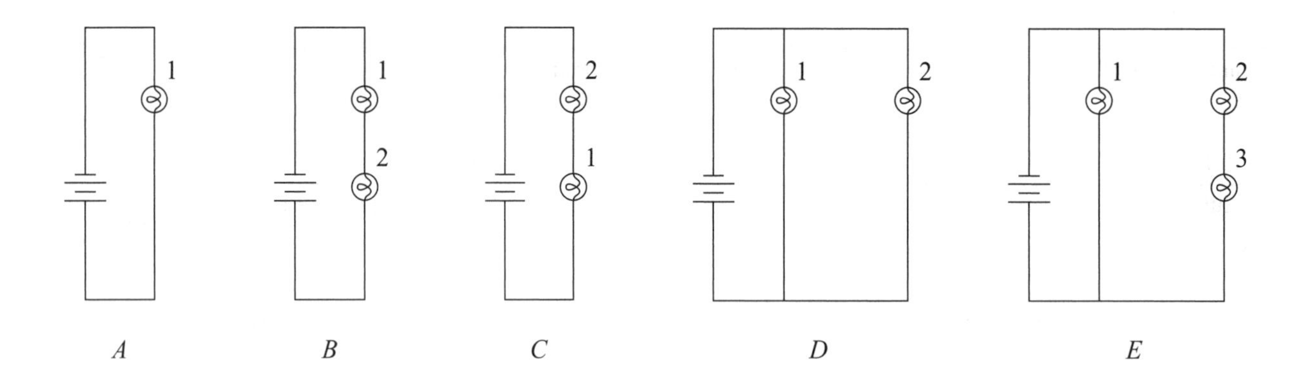

a. Build circuit *A* using a battery, wires, and identical bulbs (see the figures above) and then notice the brightness of the bulb. Then build circuit *B* and notice the brightness of the bulbs. Explain the differences in your observations using the concept of *V* field (potential difference) or any of the analogies.

b. Build circuit *C* and notice the brightness of the bulbs. Explain your observations using the concept of current and any analogies.

c. Build circuit *D* and notice the brightness of the bulbs. Now build circuit *E* and notice the brightness of the bulbs. Explain the differences in your observations using the concepts of potential difference and current.

d. Can you say that a battery is a source of constant current? Explain your answer.

16.2.5 Reason Use the analogies you discussed in Activity 16.2.2 and the ideas of potential difference and current to rate the bulbs in the circuit shown at the right according to their brightness, listing the brightest bulb first. Indicate whether any bulbs are equally bright. Explain your ratings.

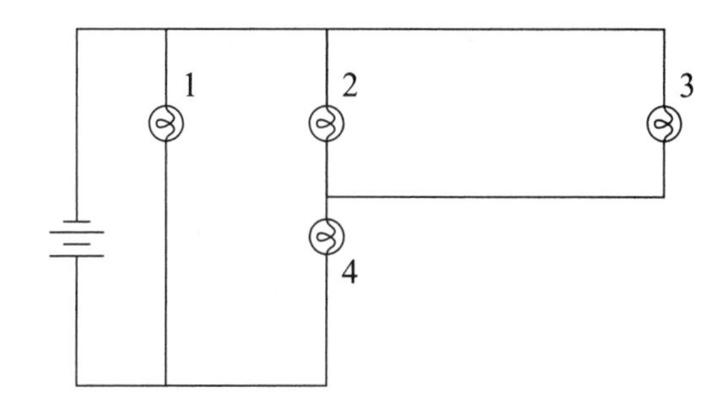

16.2.6 Reason

a. Rate the bulbs in the circuit shown to the right according to their brightness when the switch is open.

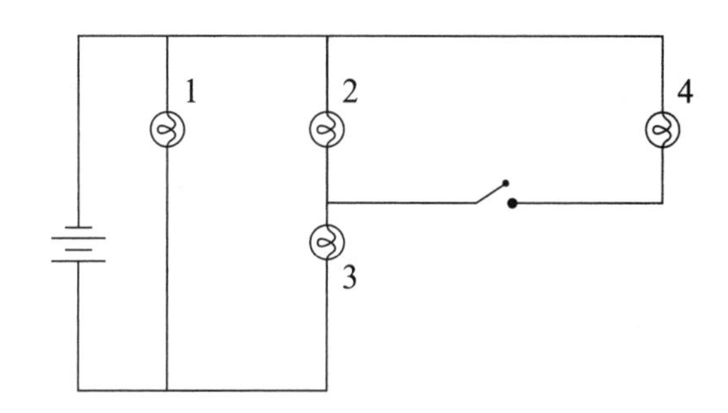

b. Now rate the bulbs in the circuit when the switch is closed.

c. Indicate how the brightness of the first three lightbulbs changes after the switch is closed.

16.2.7 Represent and reason

a. Predict how the brightness of the top bulb shown in the illustration to the right changes when you close switch 1.

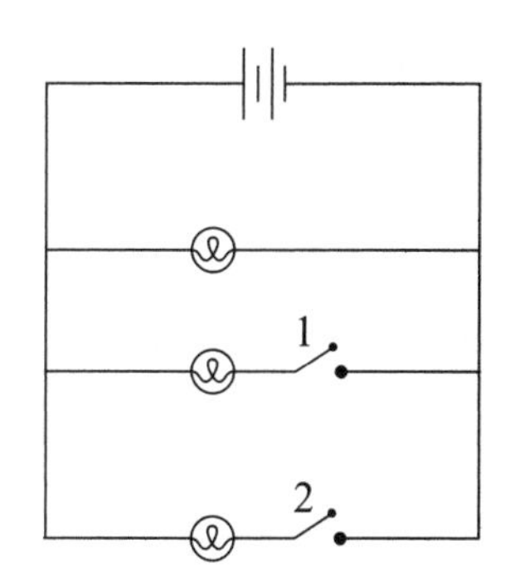

b. Predict how the brightness of the top bulb changes when you close switch 2 (switch 1 is open).

c. Rate the brightness of the three bulbs when both of the switches are closed. Explain your ratings.

d. Perform the experiment, record your observations, and compare them to your predictions. Explain the discrepancies

16.3 Quantitative Concept Building and Testing

16.3.1 Observe, find a pattern, and explain You connect a resistor in series with an ammeter and a variable potential difference source. In the table, the electric current I through the resistor is shown as you vary the potential difference ΔV across the resistor.

Potential difference ΔV (V)	0.0	2.0	4.0	6.0	8.0	10.0
Current I (A)	0.000	0.020	0.038	0.061	0.079	0.105

Complete the table that follows.

Construct a circuit diagram.	Represent data graphically.	I (A) / ΔV (V)
Describe the pattern mathematically.	Explain the relationship between the current through the resistor and the potential difference across it.	

16.3.2 Test your idea Design an experiment to test whether the linear relationship for current and potential difference holds for different resistors. Use commercial resistors and lightbulbs.

Describe experiments in words and specify what type of resistive device you will study.	**Draw an electric circuit and measure the value of the resistance of the resistor using an ohmmeter (an electric device that measures the electric resistance of an object).**	**Write your prediction using the relationship $I = (1/R)\Delta V$.**	**Perform the experiment, record the outcome, and decide whether the relationship holds for this particular type of resistor.**

16.3.3 Observe and explain The readings of three ammeters are shown at the right. What can you say about the magnitude of the current through each of the lightbulbs? Explain the pattern.

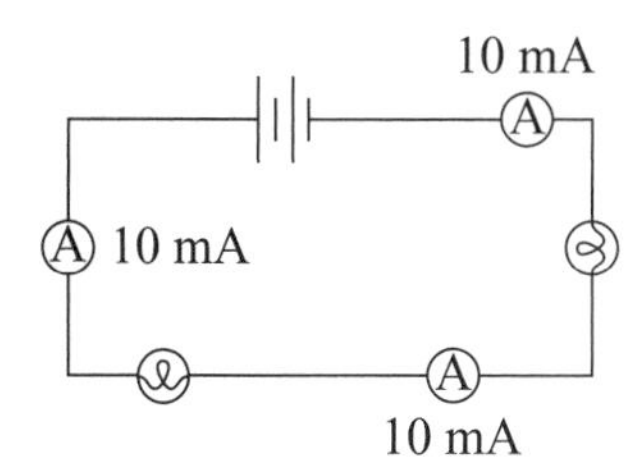

16.3.4 Evaluate the reasoning Your friend says that when two identical lightbulbs are connected in series to each other and then to a battery, the lightbulb connected closest to the negative pole of the battery will be brighter. He explains this by claiming that the second bulb will get fewer electrons because the first bulb will use up some of the electrons. Do you agree or disagree? How can you convince your friend of your opinion? You can use theoretical arguments or perform an experiment to test his suggestion.

16.3.5 Evaluate the reasoning Your friend says that when two identical lightbulbs are connected in series to each other and to the terminals of a battery, the lightbulb closest to the negative pole of the battery will have a greater potential difference across it. She explains it by saying that it will be harder for the electric field to push through to the second bulb after it has already pushed through the first. Do you agree or disagree? How can you convince your friend of your opinion? You can use theoretical arguments or perform an experiment to test her suggestion.

16.3.6 Design an experiment

a. Design an experiment to investigate the relationship between currents through resistors 1, 2, 3, and 4, as shown in the illustration to the right. Describe the experiment and record the results in any format you find appropriate.

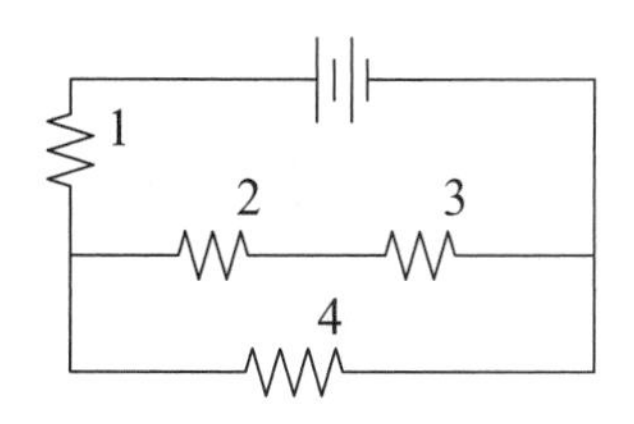

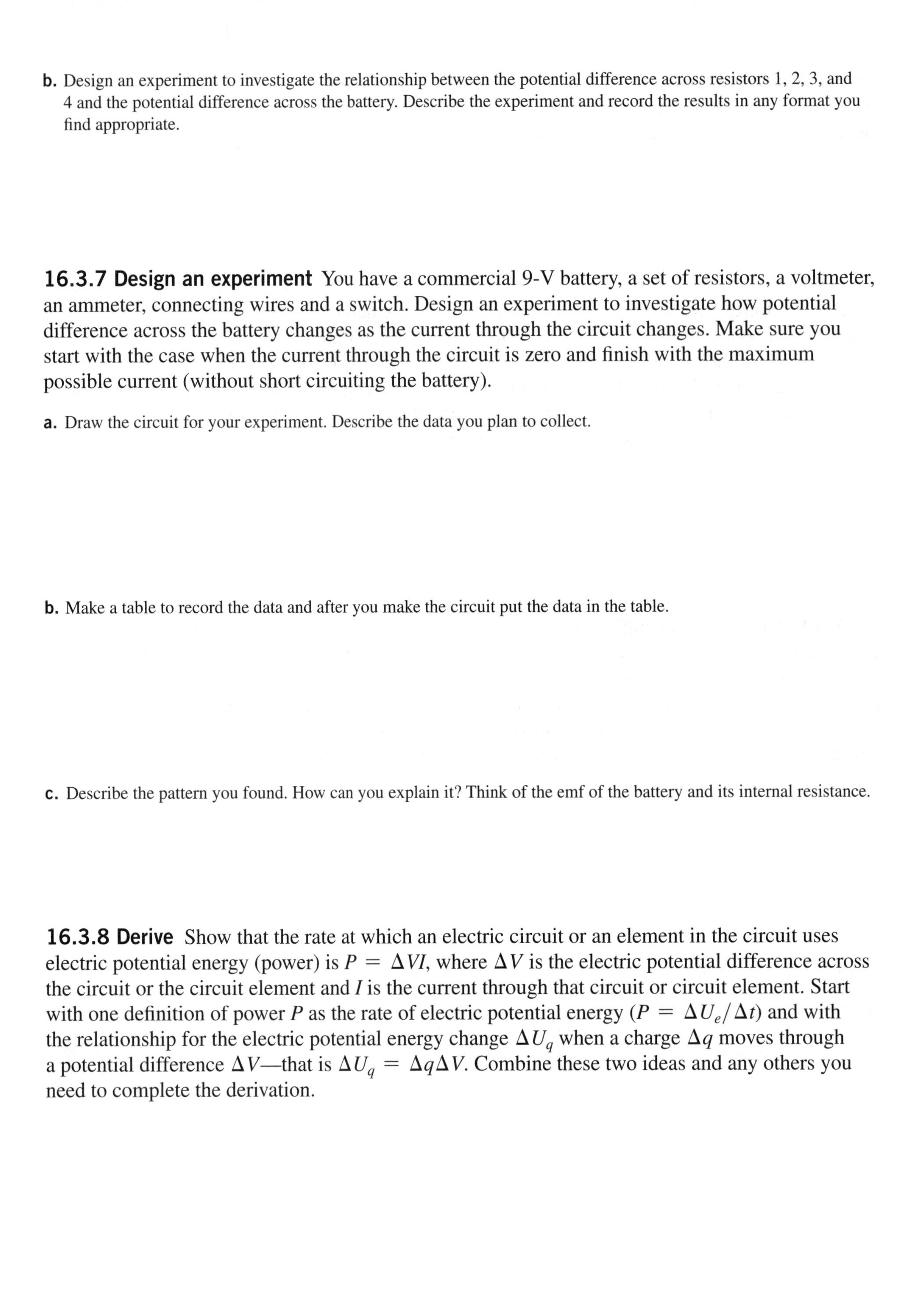

b. Design an experiment to investigate the relationship between the potential difference across resistors 1, 2, 3, and 4 and the potential difference across the battery. Describe the experiment and record the results in any format you find appropriate.

16.3.7 Design an experiment You have a commercial 9-V battery, a set of resistors, a voltmeter, an ammeter, connecting wires and a switch. Design an experiment to investigate how potential difference across the battery changes as the current through the circuit changes. Make sure you start with the case when the current through the circuit is zero and finish with the maximum possible current (without short circuiting the battery).

a. Draw the circuit for your experiment. Describe the data you plan to collect.

b. Make a table to record the data and after you make the circuit put the data in the table.

c. Describe the pattern you found. How can you explain it? Think of the emf of the battery and its internal resistance.

16.3.8 Derive Show that the rate at which an electric circuit or an element in the circuit uses electric potential energy (power) is $P = \Delta VI$, where ΔV is the electric potential difference across the circuit or the circuit element and I is the current through that circuit or circuit element. Start with one definition of power P as the rate of electric potential energy ($P = \Delta U_e/\Delta t$) and with the relationship for the electric potential energy change ΔU_q when a charge Δq moves through a potential difference ΔV—that is $\Delta U_q = \Delta q \Delta V$. Combine these two ideas and any others you need to complete the derivation.

16.3.9 Test your idea You have a 9-V battery and two different lightbulbs labeled bulb A and bulb B. You connect them in parallel and see that bulb A is much brighter than bulb B.

a. Use the ideas that you developed in Activity 16.3.8 to explain this observation and then predict what you will observe if you connect the bulbs in series to the same battery.

b. Perform the experiment and record the outcome. Did it match your prediction? If not, revise your explanation to account for the outcome.

16.3.10 Test your ideas Build a circuit consisting of a battery (rated 9 V), a lightbulb, and a switch connected in series. Keep the switch open.

a. Draw the circuit diagram of your circuit below:

b. Predict the potential difference across the battery, across the lightbulb, across a connecting wire and across the switch. Now use a voltmeter to check your predictions. Write down the readings. Discuss any surprising results you found and reconcile them with your prediction.

c. Now close the switch and repeat the experiment. Write down the readings. Do they make sense?

d. Discuss whether Ohm's law in the form of $I = \frac{\Delta V}{R}$ applies to a battery and to a switch in an open circuit. Discuss whether Ohm's law applies to a battery, a switch and a connecting wire in a closed circuit.

16.4 Quantitative Reasoning

16.4.1 Represent and reason Imagine that you have a 9-V battery connected by wires to a lightbulb. Fill in the table that follows and list the assumptions you make in the space below the table. Consider that the negative terminal of the battery is at zero potential.

Draw the circuit.	Draw a qualitative electric potential-versus-position graph.
	V (V) Negative battery terminal · Positive battery terminal · One side of the bulb · Other side of the bulb · Negative battery terminal

16.4.2 Represent and reason Complete the table that follows for the circuit in the illustration. List the assumptions you make in the space below the table.

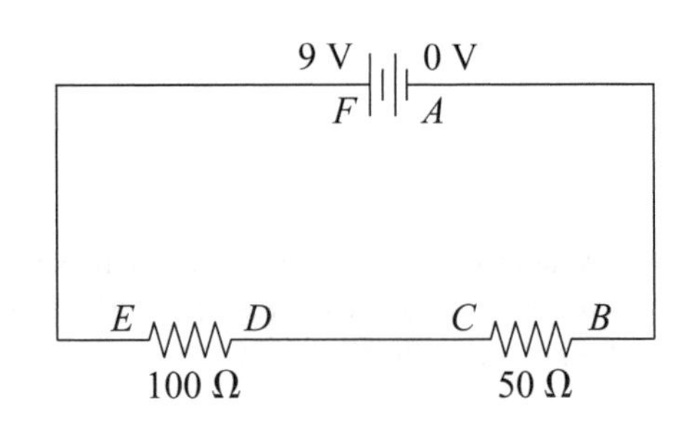

Find the current in the circuit and then calculate the electric potential at each lettered position in the circuit.	Plot the electric potential-versus-position for the circuit.
	V (V) 9 5 0 A B C D E F A

16.4.3 Represent and reason The application of Kirchhoff's loop rule for a circuit is shown in the equation that follows.

$$-1.0\text{ V} - I(2\,\Omega) + 4.0\text{ V} - I(6\,\Omega) - I(4\,\Omega) = 0$$

Draw a circuit that is consistent with the equation and label the resistors and batteries in the circuit. Draw an arrow and label the electric current.

16.4.4 Represent and reason Complete the last three cells of the table that follows for the circuit shown in the first cell. Do not solve for the currents.

The circuit	Identify currents with arrows and symbols.	Apply the junction rule for junctions *A* and *B*.	Apply the loop rule for three different loops.
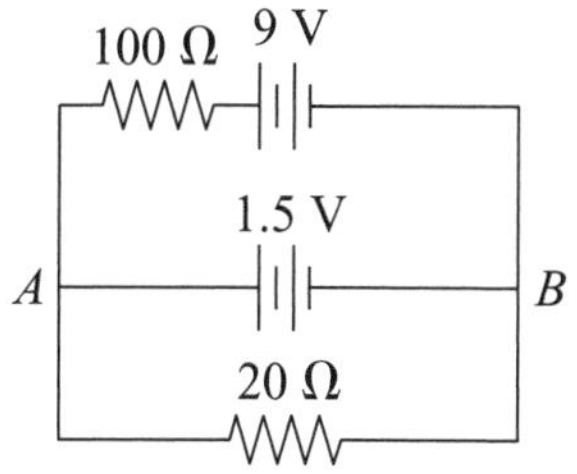 100 Ω, 9 V, 1.5 V, 20 Ω, *A*, *B*	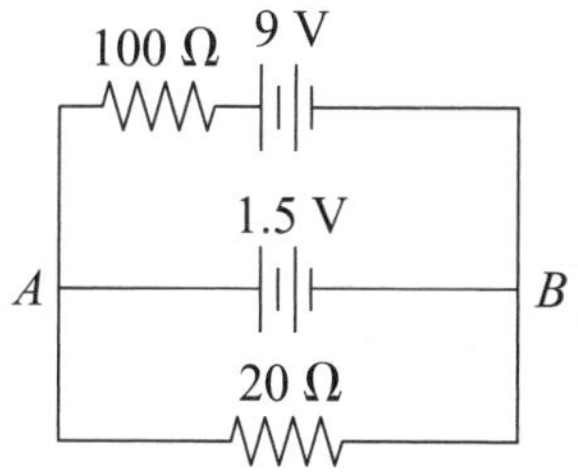 100 Ω, 9 V, 1.5 V, 20 Ω, *A*, *B*	Junction *A*: Junction *B*:	

16.4.5 Represent and reason For the circuit in Activity 16.4.4, determine the current through each branch. *Note:* A branch of a circuit is one part of the circuit where the same electric current flows through each element in that branch.

16.4.6 Represent and reason Complete the last three cells of the table that follows for the circuit shown in the first cell. Do not solve for the currents.

The circuit	Identify currents with arrows and symbols.	Apply the junction rule for junctions *A* and *B*.	Apply the loop rule for three different loops.
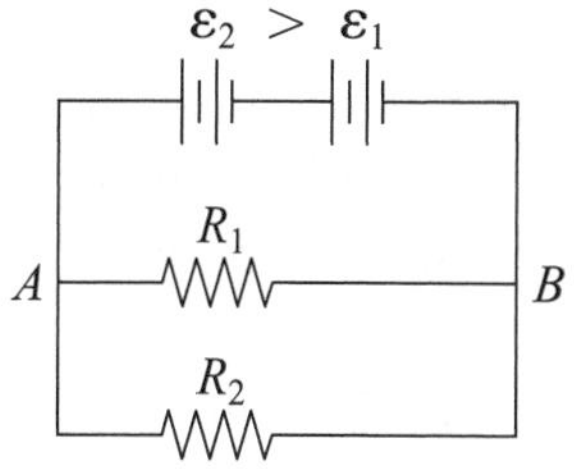 $\varepsilon_2 > \varepsilon_1$, R_1, R_2, *A*, *B*	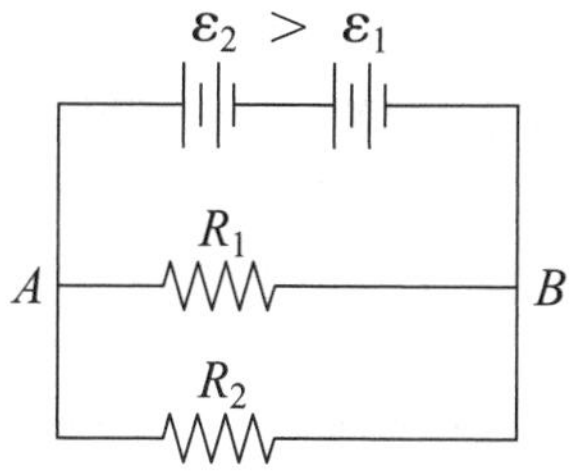 $\varepsilon_2 > \varepsilon_1$, R_1, R_2, *A*, *B*	Junction *A*: Junction *B*:	

16.4.7 Represent and reason Complete the last three cells of the table that follows for the circuit shown in the first cell. Do not solve for the currents.

The circuit	Identify currents with arrows and symbols.	Apply the junction rule for junctions *A* and *B*.	Apply the loop rule for three different loops.
9 V 1 Ω 4.5 V 0.5 Ω *A* *B* 7 Ω	9 V 1 Ω 4.5 V 0.5 Ω *A* *B* 7 Ω	Junction *A*: Junction *B*:	

16.4.8 Reason For the circuit in Activity 16.4.7, determine the current through and the potential difference across each resistor.

16.4.9 Evaluate the solution

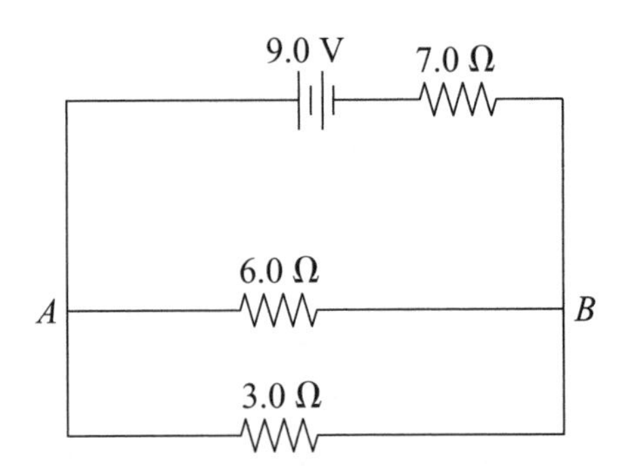

The problem: Apply Kirchhoff's loop rule for three loops and the junction rule for one junction for the circuit shown at the right. Do not solve for the electric currents.

Proposed solution:

$$+9.0\text{ V} - (7.0\,\Omega)I - (6.0\,\Omega)I = 0$$
$$+9.0\text{ V} - (7.0\,\Omega)I - (3.0\,\Omega)I = 0$$
$$-(6.0\,\Omega)I + (3.0\,\Omega)I = 0$$

a. Identify any errors in the proposed solution.	b. Provide a corrected solution if there are errors.

16.4.10 Electric circuit Jeopardy You have a circuit consisting of a variety of elements including a 9-V battery (measured as 9 V when you put a voltmeter across it without an external circuit), a switch, and several resistors. You measure current through different circuit elements and the potential difference across them (the same element has the matching voltmeter and ammeter numbers). The results are in the table below. Draw a picture of the circuit where these measurements could have been taken, determine the values of resistances if possible, and show where the voltmeters and ammeters could be located.

Element	Ammeter reading	Voltmeter reading
1	0.071 A	8.86 V
2	0.071 A	7.10 V
3	0.071A	0 V
4	0.035 A	1.76 V
5	0.035 A	1.76 V

16.4.11 Electric circuit Jeopardy You have a circuit with the same 9-V battery as in the previous activity, several resistors, and a switch. You measure current through different circuit elements and the potential difference across them (the same element has the matching voltmeter and ammeter numbers). The results are in the table below. Draw a picture of the circuit where these measurements could have been taken, determine the values of resistances if possible, and show where the voltmeters and ammeters could be located.

Element	Ammeter reading	Voltmeter reading
1	0	9.0 V
2	0	9.0 V
3	0	0

16.4.12 Regular problem What is the potential difference between points A and B if the emfs of the batteries are $\varepsilon_1 = 4.0$ V and $\varepsilon_2 = 1.0$ V and the resistances of the resistors are $R_1 = 10.0\ \Omega$ and $R_2 = 5.0\ \Omega$?

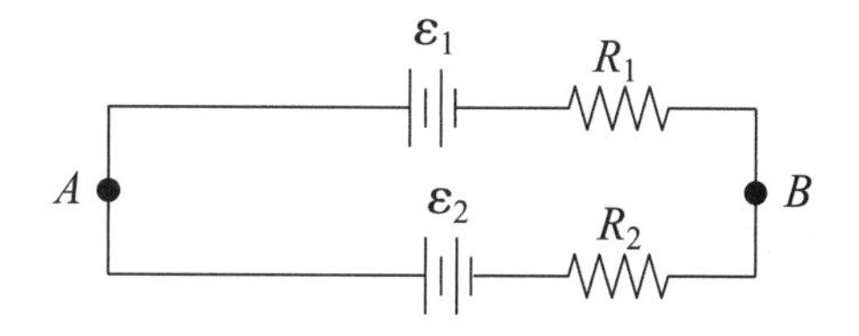

16.4.13 Design an experiment You have a spiral-shaped immersion water heater. Design two experiments to determine the power of the device. Describe the data that you will collect and the mathematical procedure you will use. Examine assumptions in your mathematical procedure. Will they lead to the power estimate higher or lower than actual? If you have the device, conduct the experiments and reconcile the difference between the ratings you obtained.

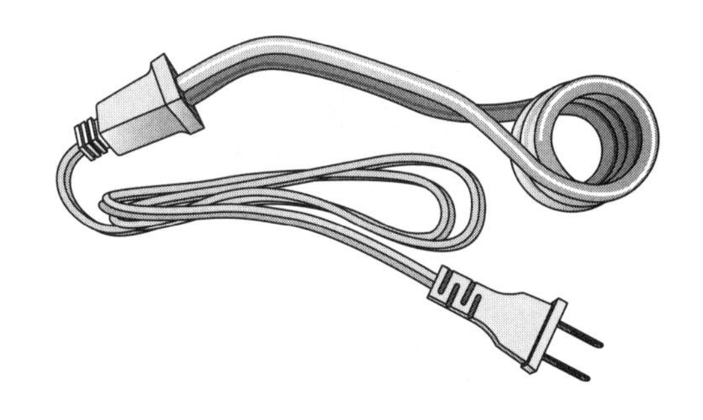

17 Magnetic Forces and Magnetic Fields

17.1 Qualitative Concept Building and Testing

17.1.1 Observe and find a pattern You have a bar magnet, a horseshoe magnet, and a compass.

a. Perform the experiments and complete the table that follows.

Experiment	Draw the relative orientations of the magnet with marked poles and the compass with marked poles.
Hold a bar magnet horizontally and slowly bring it closer to the compass.	
Repeat the first experiment but reverse the direction of the bar magnet.	
Hold the horseshoe magnet vertically and slowly bring the compass between the poles of the magnet.	

b. Describe in words a pattern between the orientation of the magnet's poles and the orientation of the compass.

c. Describe the same pattern representing a compass with an arrow S $\longrightarrow$ N, as illustrated.

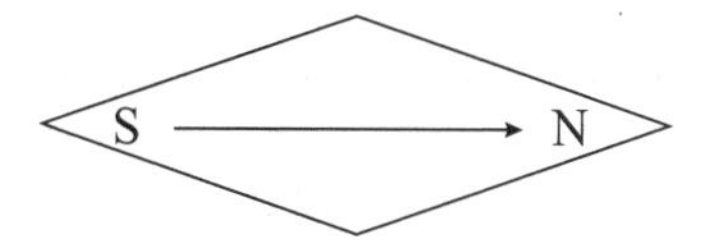

17.1.2 Observe and find a pattern Several experiments are described below.

a. Fill in the table that follows to indicate how magnetic poles are different from or the same as a positively or negatively charged object.

Observations with a magnet	Compare and contrast magnet poles with positively or negatively electrically charged objects.
The north pole of a bar magnet always attracts the south pole of another bar magnet and repels the other bar's north pole.	
Neither the north pole nor the south pole of a magnet exerts a force on a small aluminum ball hanging at the end of a thin, nonconducting thread.	
A foam tube is charged by rubbing so that one end is positive and the other is negative. You find that *both* ends attract the north pole of a bar magnet (and the south pole).	

b. Explain the results of the last experiment in the previous table. Note that a magnet is made of iron—an electric conductor. Start by drawing a picture of the charge distribution on the metal magnet when the positive end of the tube is near the magnet's N pole and again when the tube's negative end is near the magnet's N pole.

17.1.3 Observe and explain Connect a battery, a switch, some wires, and a lightbulb, as shown at the right. The bulb is an indicator of electric current. Make sure that one of the wires in the circuit is aligned along the geographical north–south direction. Place a compass under a north–south-oriented wire with the switch open; there is no current in the circuit. The needle points north.

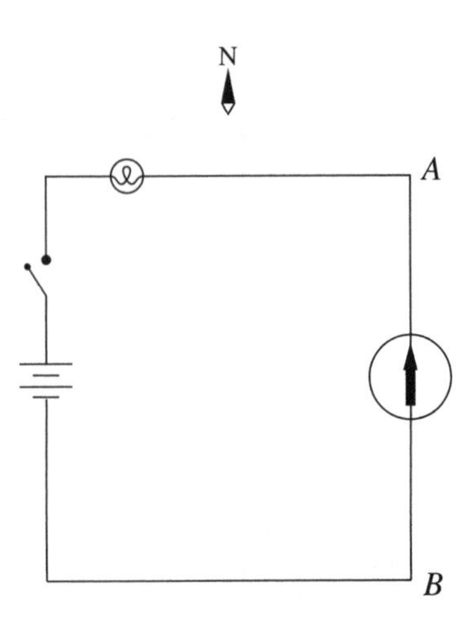

a. Perform the experiments described in the table that follows and record your observations.

Direction of the current	**Draw an arrow representing the orientation of the compass when placed below the wire *AB*.**	**Draw an arrow representing the orientation of the compass when placed above the wire.**
Current flows in the north–south ↓ direction in the wire, as shown in the figure.		
If you reverse the battery poles, the current now flows in a south–north ↑ direction in the wire above or below the compass.		

b. Use the thumb of your right hand to represent the direction of the current and your four fingers of the same hand to represent the direction of the compass. Does the orientation of your thumb and fingers describe a pattern between the direction of the current and the orientation of the compass for all of the previous experiments?

c. Come up with a reason why electric current (moving electrically charged particles) might affect the behavior of magnets differently than stationary charged objects would.

17.1.4 Design an experiment Your friend claims that a magnet simply consists of a positive electric charge and a negative electric charge locked at the ends of a metal bar. Design an experiment to test this claim. Predict the outcome.

17.1.5 Observe and explain Imagine that a wire passes up through the page you are reading. Iron filings are sprinkled on the page. We can think of the iron filings as small compasses. The top picture shows the filings when there is no current in the wire. The bottom picture is the arrangement of the filings when there is a significant current in the wire.

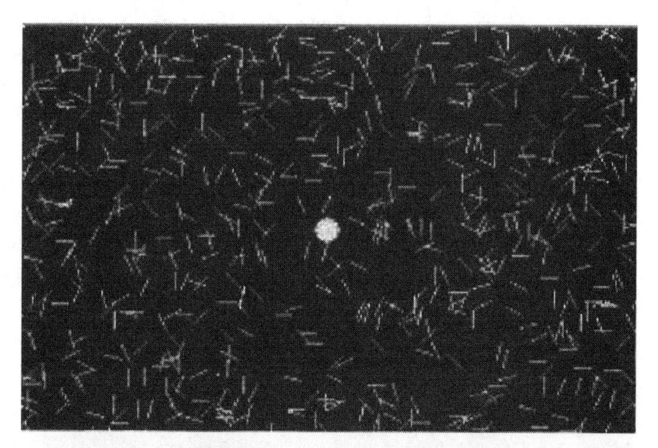

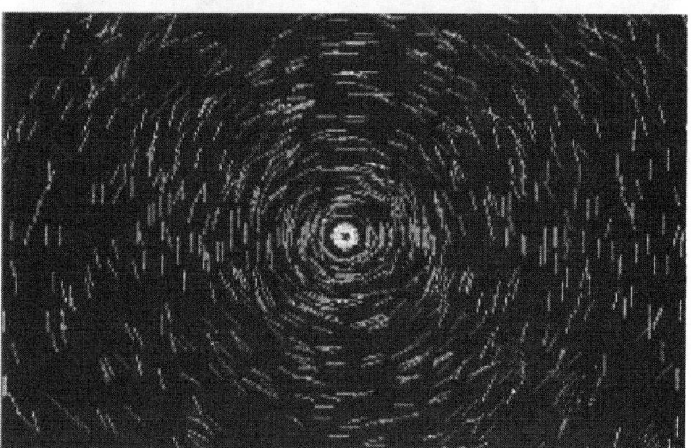

a. Is the picture at the bottom consistent with the results of the experiment in Activity 17.1.3? Explain your answer.

b. Draw a sketch that you think represents the orientation of magnetic field vectors produced by the electric current in the wire at five different points. (*Hint:* Choose the direction of the current as coming out of the page.)

17.1.6 Observe and find a pattern A cathode-ray tube (CRT) is part of a traditional television set or of an oscilloscope. Electrons "evaporate" from a hot filament called the cathode. They accelerate across a potential difference and then move at high speed toward a scintillating screen. The electrons form a bright spot on the screen at the point at which they hit it. A magnet held near the CRT sometimes causes the electron beam to deflect.

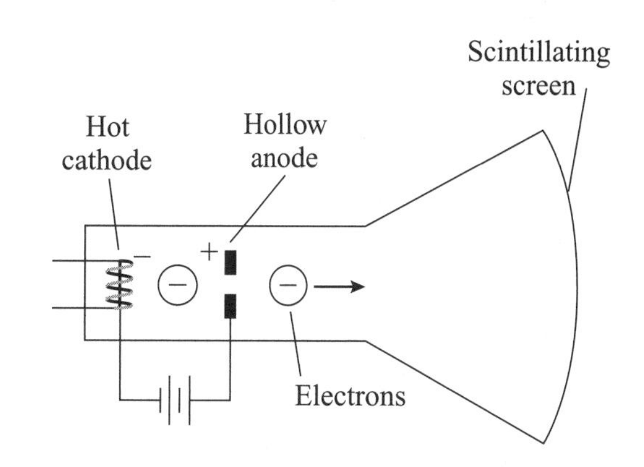

a. Devise a rule for the direction of the force $\vec{F}_{\text{B on E}}$ that the magnet exerts on the moving electrons relative to the direction of their velocity $\vec{v}$ and the direction of the magnetic field $\vec{B}$ produced by the magnet. Use the information provided in the table.

Experiment	Observation
Point the north pole of a magnet at the front of the scintillating screen—opposite the direction the electrons are moving.	Nothing happens to the beam.
Point the north pole of the magnet from the right side (as you face the coming beam) perpendicular to the direction the electrons are moving.	The beam deflects up.
Point the south pole of the magnet from the right side perpendicular to the direction the electrons are moving.	The beam deflects down.
Point the north pole of the magnet down from the top of the CRT, perpendicular to the direction the electrons are moving.	The beam deflects left.

b. Your friend says that the beam of electrons is deflected by the magnet because the electrons are charged particles and the magnet is made of iron. Because all conductors attract electrically charged particles, the experiment above is not related to magnetism. How can you convince your friend that she is mistaken?

17.1.7 Find a pattern A current-carrying wire is placed between the poles of an electromagnet. The direction of the B field lines produced by the magnet ($\vec{B}_{\text{external}}$) is shown in the figure. Invent a rule that relates the directions of the magnetic force $\vec{F}_{B\text{ on wire}}$, the directions of the $\vec{B}_{\text{field}}$, and the directions of the current I in the wire.

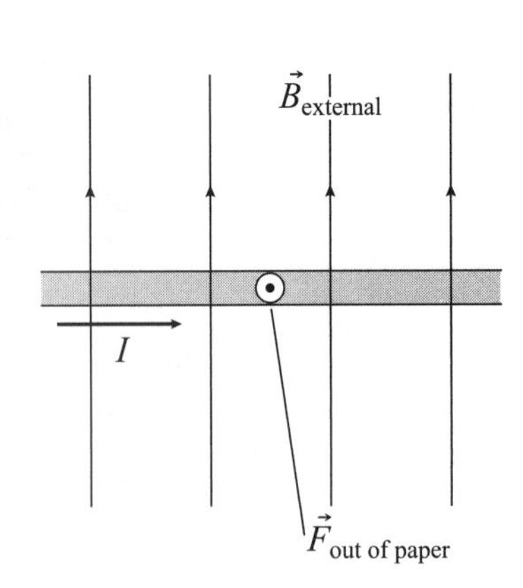

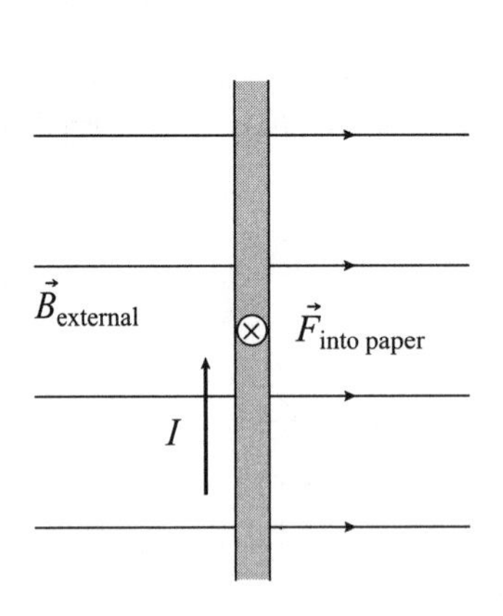

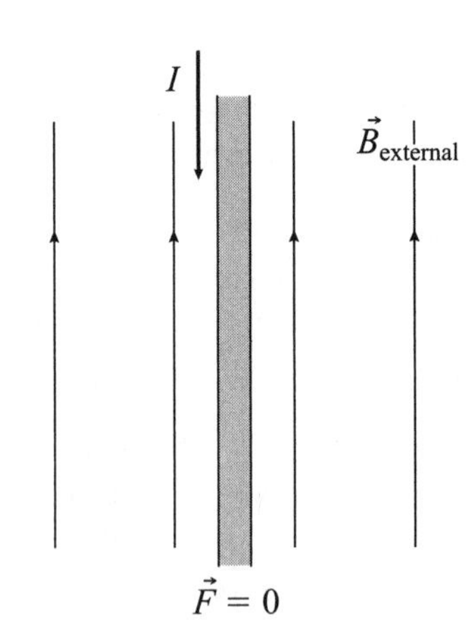

Come up with a rule that relates the directions of $\vec{F}_{B \text{ on wire}}$, $\vec{B}_{\text{external}}$, and I.

17.1.8 Test your idea You have a long wire, a power supply, and a horseshoe magnet. Design an experiment whose outcome you can predict using the rules you came up with in Activities 17.1.6 and 17.1.7. Draw a picture of the apparatus, write your prediction, perform the experiment, and record the outcome.

17.1.9 Represent and reason A rigid wire in the shape of a rectangular loop is shown in the illustration to the right. When the switch in the circuit is closed, there is current around the loop in a clockwise direction. The loop resides in a uniform external magnetic field. Decide the direction of the force exerted on the wires of the loop shown in the figures below and draw it in the figures.

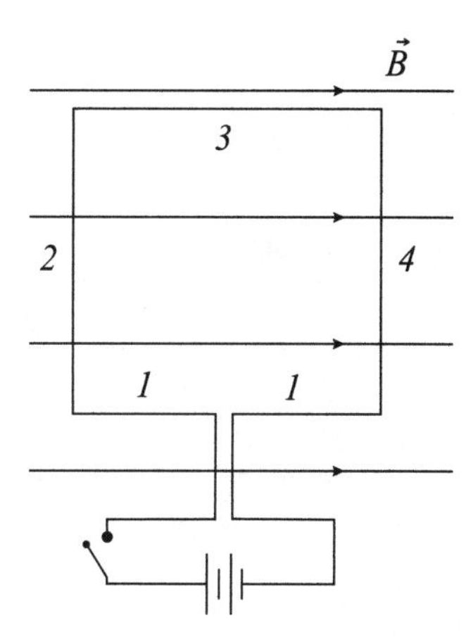

a. $\vec{F}_B$ on side *1*

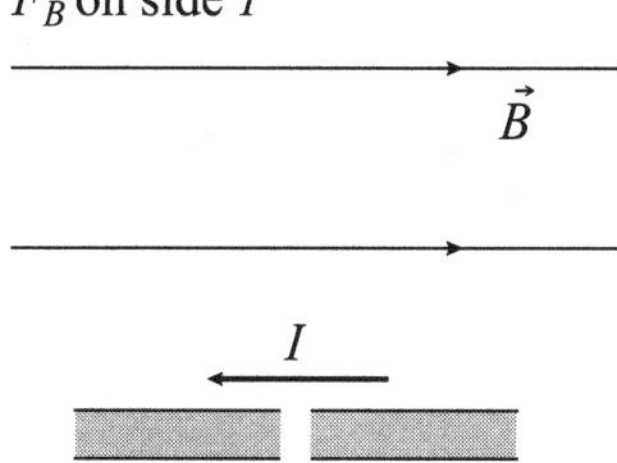

b. $\vec{F}_B$ on side *2*

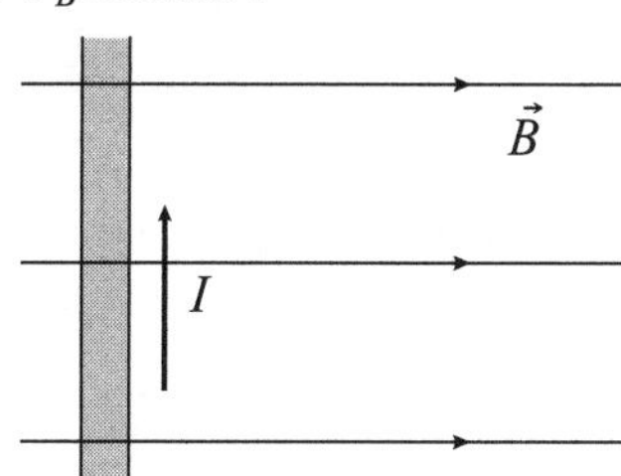

c. $\vec{F}_B$ on side *3*

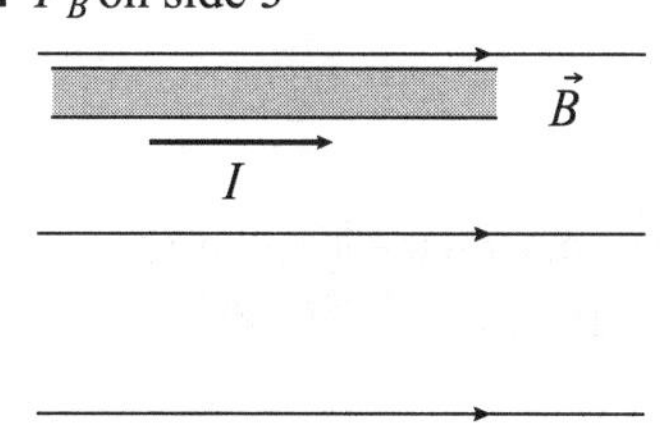

d. $\vec{F}_B$ on side *4*

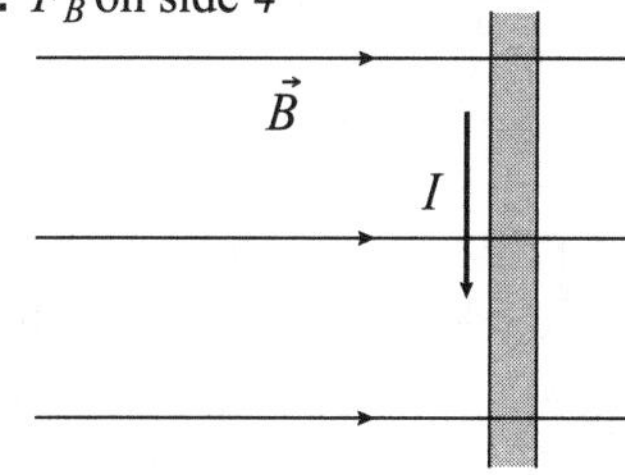

e. Does the magnetic field exert a net nonzero magnetic force on the loop? If so, what is the direction of the net force?

f. Does the field exert a net nonzero magnetic torque on the loop? If so, how does this torque tend to cause the orientation of the loop to change (assuming it is initially at rest)?

17.1.10 Represent and reason A rigid wire in the shape of a rectangular loop is shown at the right. When the switch in the circuit is closed, current flows up side *2*, across side *3*, and down side *4*. The loop's surface is perpendicular to the page and resides in an external magnetic field. The field's direction is parallel to the page and perpendicular to the surface of the loop. Decide the direction of the force exerted on the wires of the loop shown in the figures below and draw it in the figures.

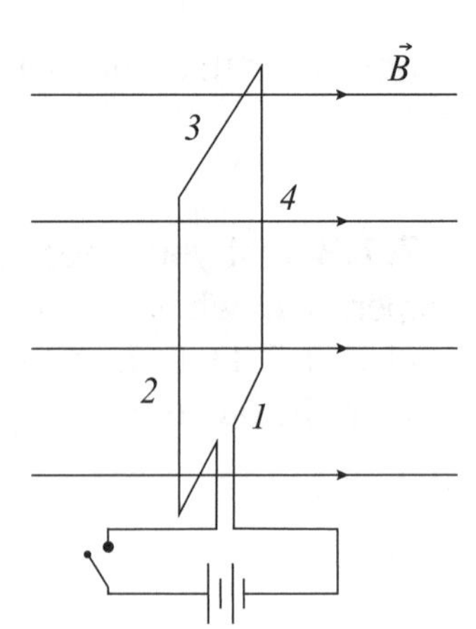

a. $\vec{F}_B$ on side *1*

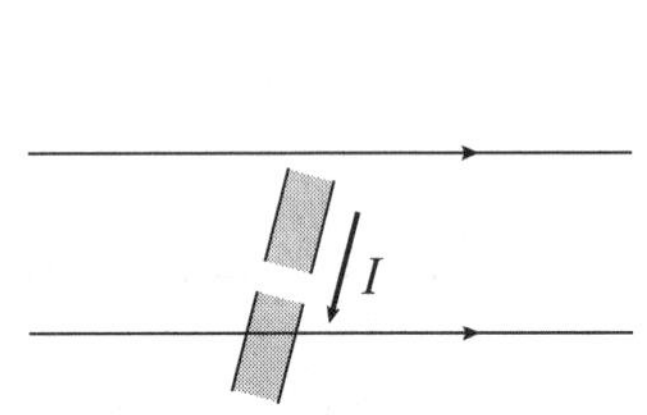

b. $\vec{F}_B$ on side *2*

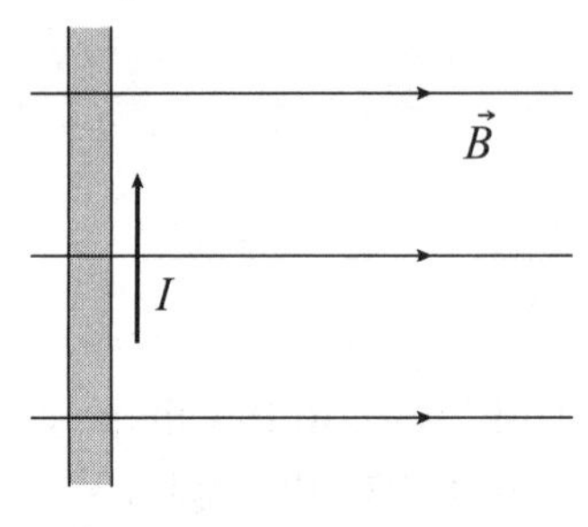

c. $\vec{F}_B$ on side *3*

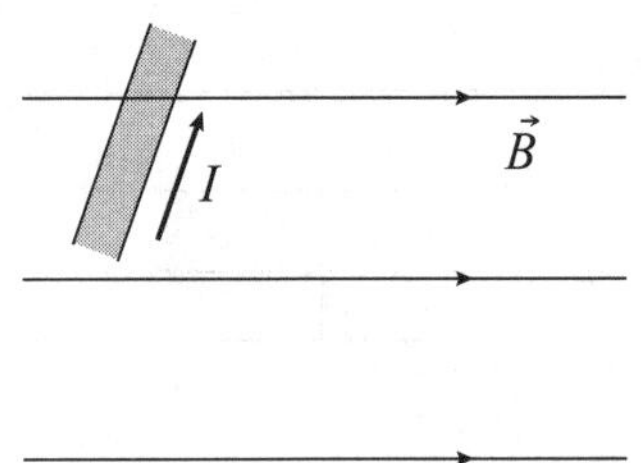

d. $\vec{F}_B$ on side *4*

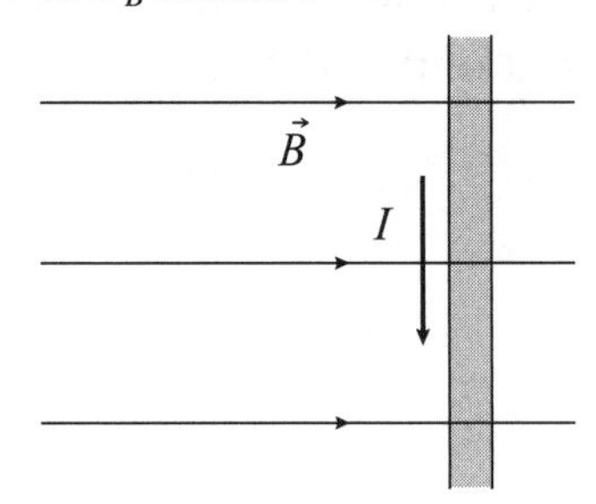

e. Does the magnetic field exert a net nonzero magnetic force on the loop? If so, what is the direction of the net force?

f. Does the field exert a net nonzero magnetic torque on the loop? If so, how does this torque tend to cause the orientation of the loop to change (assuming the loop is initially at rest)?

17.2 Conceptual Reasoning

17.2.1 Represent and reason For each situation represented in the illustration, decide if the external magnetic field (source) exerts a nonzero magnetic force on the short section of the current-carrying wire (test object). If the magnetic force is not zero, indicate the direction of the magnetic force by drawing it in the figures that follow. (*Note:* There must be other sections of wire, not shown, that are connected to these wires.)

a.

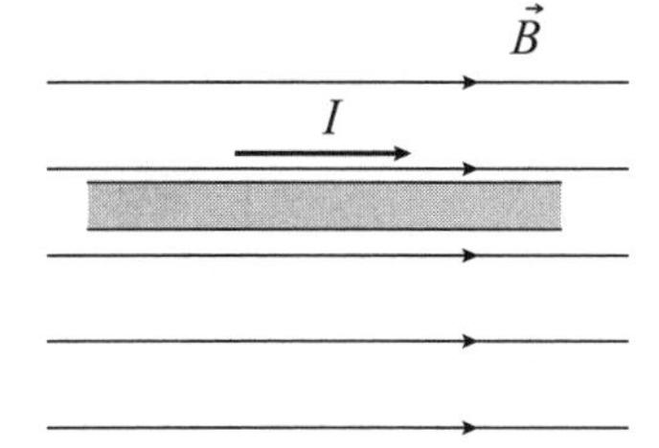

b.

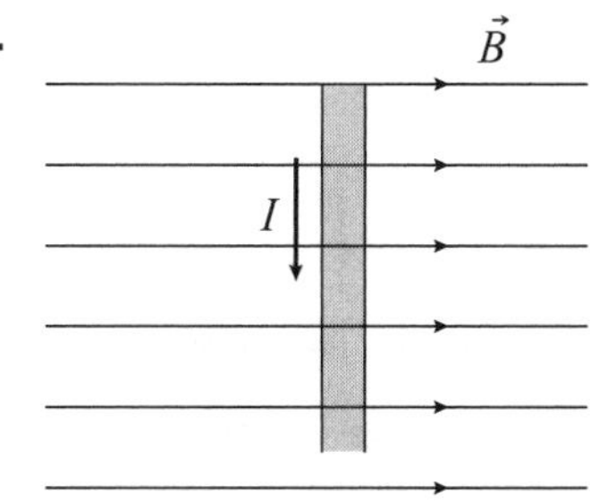

c.

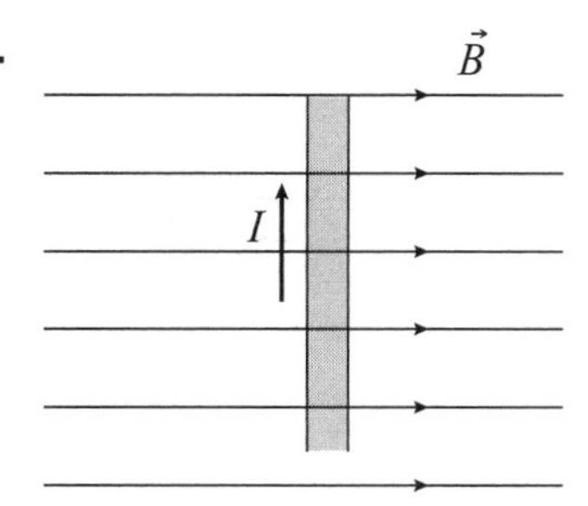

d.

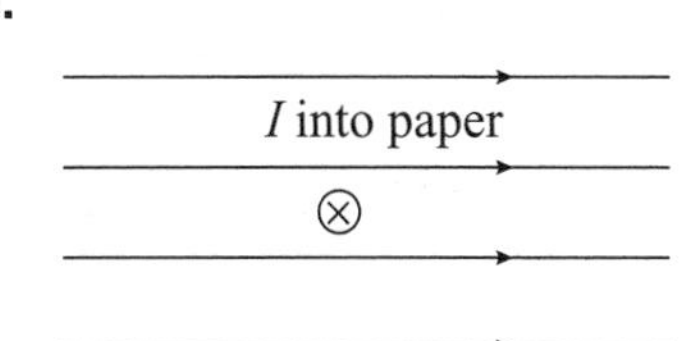

e.

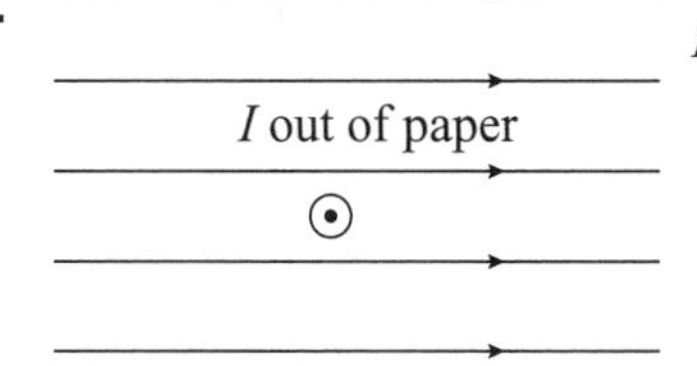

f.

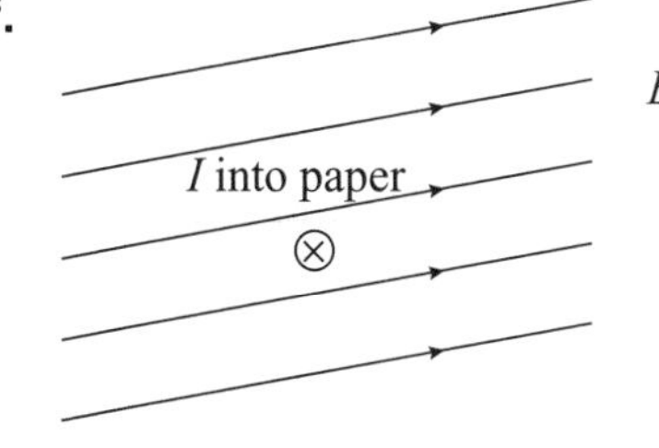

g. $\vec{B}$ into paper

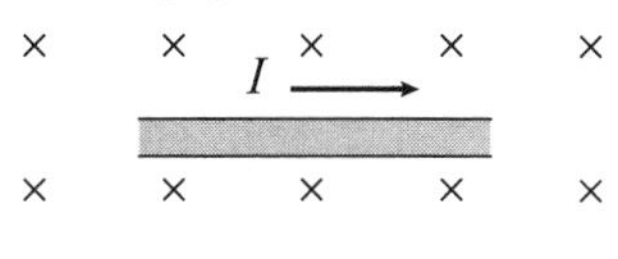

h. $\vec{B}$ out of paper

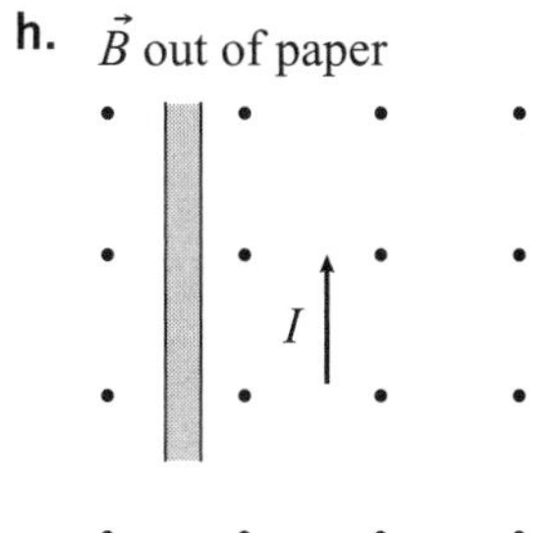

17.2.2 Represent and reason For each situation depicted in the table that follows, find the direction of the unknown physical quantity. Draw in the directions in the figures.

Situation 1	Situation 2	Situation 3
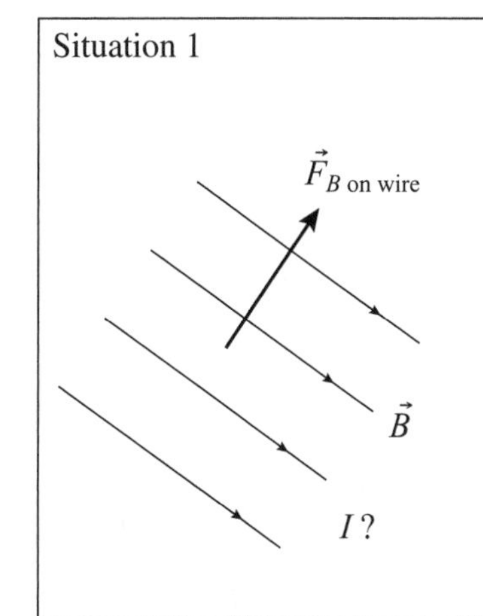	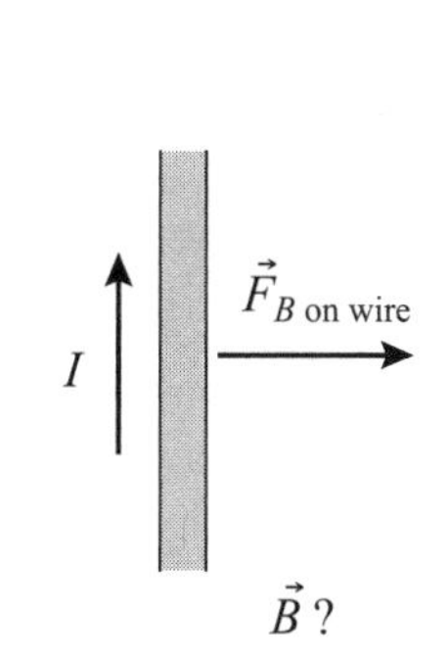	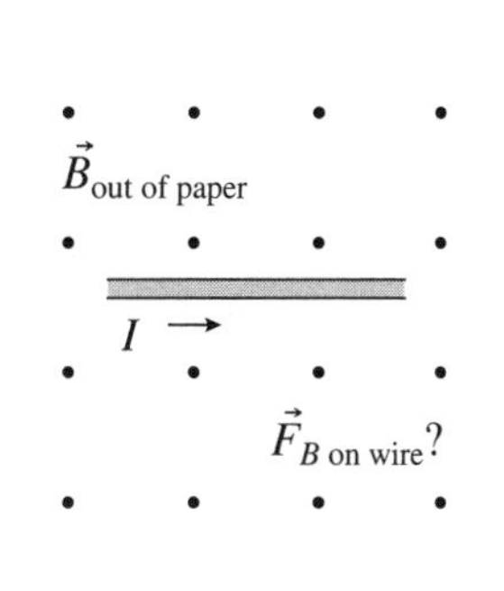

17.2.3 Represent and reason For each situation below, decide if a nonzero magnetic force is exerted on the moving electric charge (test object). If the force is not zero, draw in the direction of the magnetic force on the figures that follow.

a.

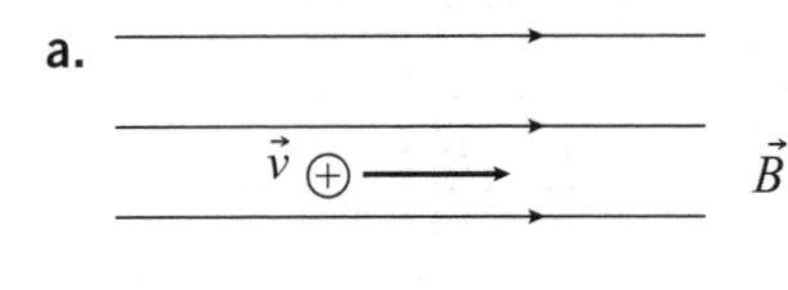

b.

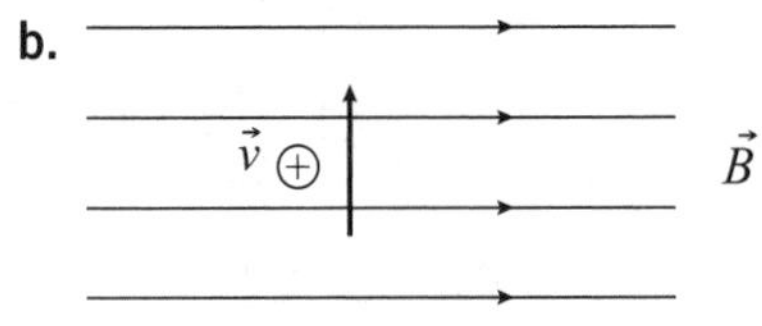

c.

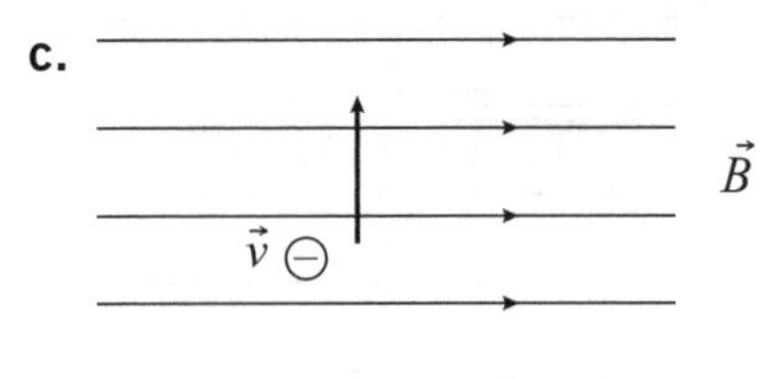

d.

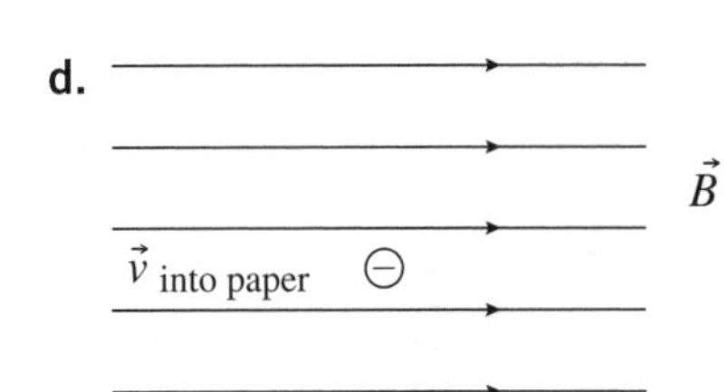

e.

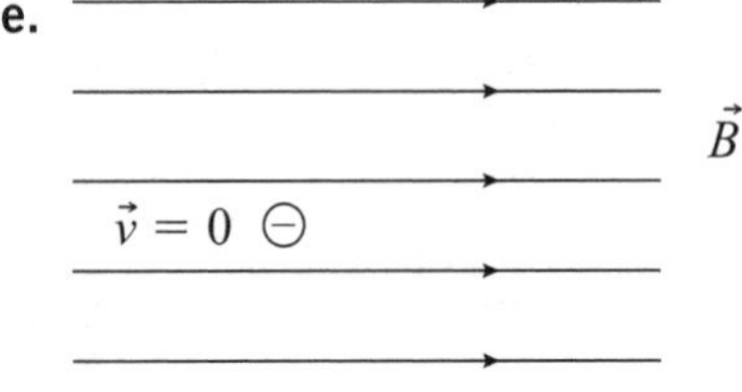

f.

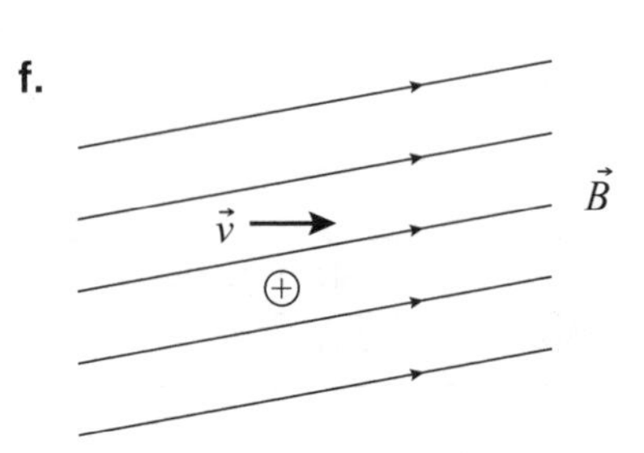

g.

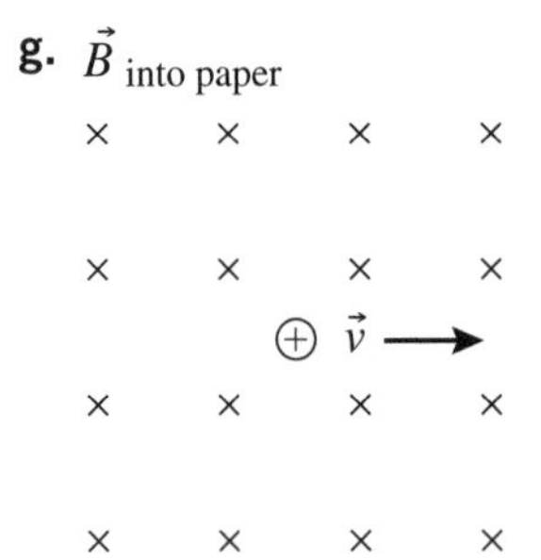

h.

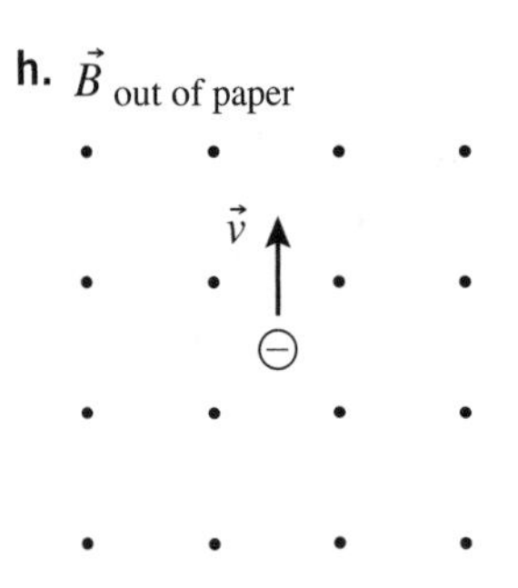

17.2.4 Represent and reason There is current in each of the wires shown in the illustration. Determine the direction of the magnetic field created by the current at the points indicated and draw it with an arrow, a dot (out of the page), or a cross (into the page). The objects indicated in the illustration are sources of a magnetic field.

a.

A B C

I

b.

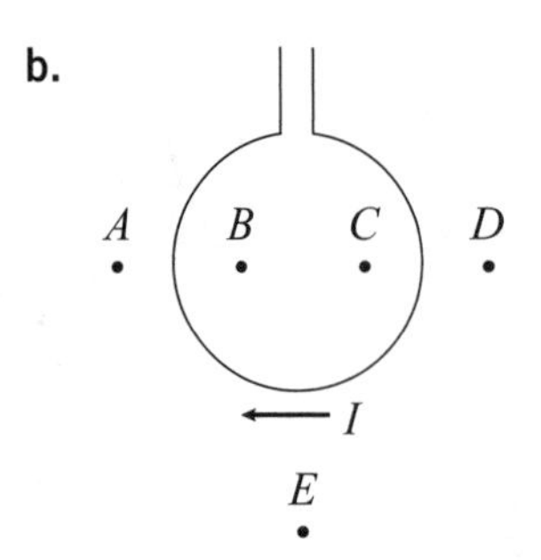

c.

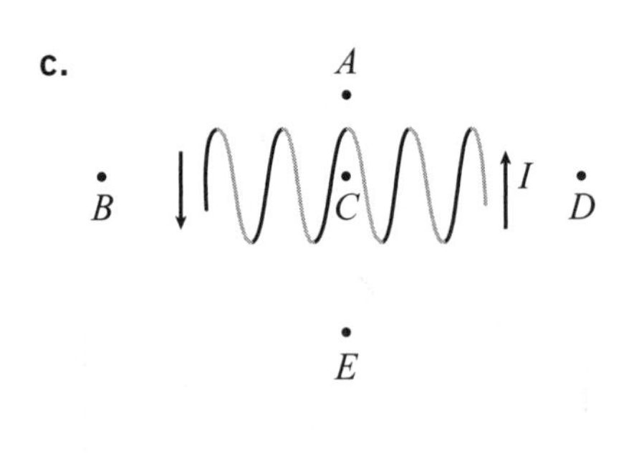

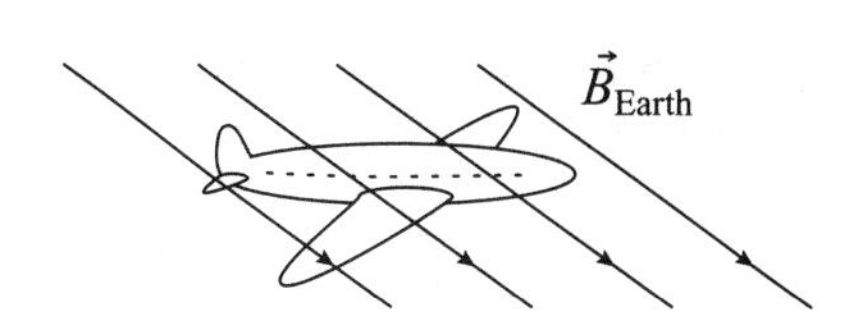

17.2.5 Pose a problem Pose a qualitative problem based on the situation shown in the illustration. The airplane is flying through Earth's magnetic field.

17.2.6 Represent and reason Two wires are parallel to each other. Wire 1 has electric current going into the page, and wire 2 has electric current coming out of the page. Fill in the table that follows to emphasize the importance of choosing a source of a field and a system on which the field exerts a force.

1 ⊗ 2 ⊙

Draw in B field lines $\vec{B}_1$ produced by the current I_1 in wire 1 (the source of the field). Be sure to include a line passing through wire 2.	Noting the field $\vec{B}_1$ passing through wire 2, draw the direction of the magnetic force $\vec{F}_{1\text{ on }2}$ that wire 1's magnetic field $\vec{B}_1$ exerts on wire 2 (the system).	Draw in B field lines $\vec{B}_2$ produced by the current I_2 in wire 2 (the source of the field). Be sure to include a line passing through wire 1.	Noting the field $\vec{B}_2$ passing through wire 1, draw the direction of the magnetic force $\vec{F}_{2\text{ on }1}$ that wire 2's magnetic field $\vec{B}_2$ exerts on wire 1 (the system).
1 ⊗ I_1 in 2 ⊙	1 ⊗ 2 ⊙ I_2 out	1 ⊗ 2 ⊙ I_2 out	1 ⊗ I_1 in 2 ⊙

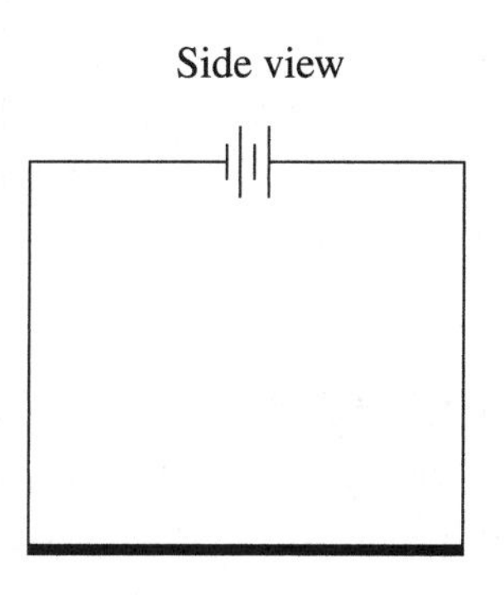

17.2.7 Represent and reason There is an electric current through a horizontal bar that hangs from two thin side wires (see the side view at the right). In what direction should an external magnetic field point so that the magnetic force that the magnetic field exerts on the bar helps support the bar? Explain. Draw a force diagram.

17.2.8 Represent and reason A word description and sketch of a hypothetical model of an atom is shown below.

a. Fill in the table that follows. Construct a physical representation that treats the atom like a circular loop of wire with an electric current in a magnetic field. *Hint:* The electron moves about the nucleus very fast, in effect producing a circular electric current about the nucleus.

Word description	Sketch	Construct a physical representation.
In the Bohr model of the hydrogen atom, a negatively charged electron moves in a circular orbit around a positively charged proton nucleus. Suppose one of these hydrogen atoms is in a magnetic field.	$\vec{B}$ Electron ⊖ ⊕	

b. Does the field produce a torque on the atom? If so, indicate the axis of rotation and the direction the atom would tend to turn about that axis.

17.2.9 Represent and reason Does the magnetic field exert a nonzero torque on the current loop in each case pictured below? If so, and if the loop is initially at rest, which way would the magnetic torque cause the loop to start turning? For each case draw in the forces on two opposite sides of the loop and show the direction of the net torque. (Current loops in a., b., and d. are perpendicular to the page.)

a.

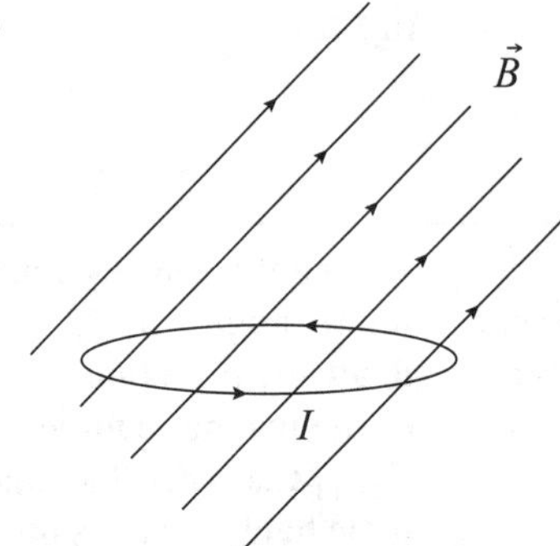

b. $\vec{B}$ into page

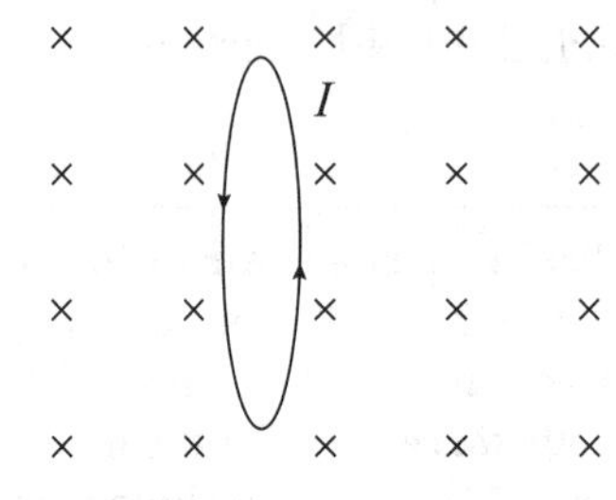

c. $\vec{B}$ into page

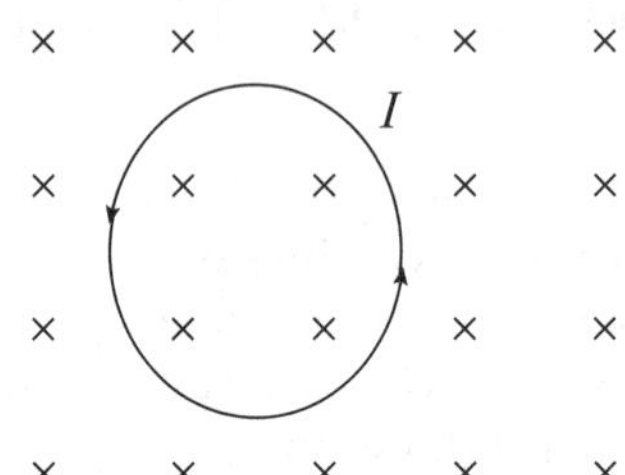

d.

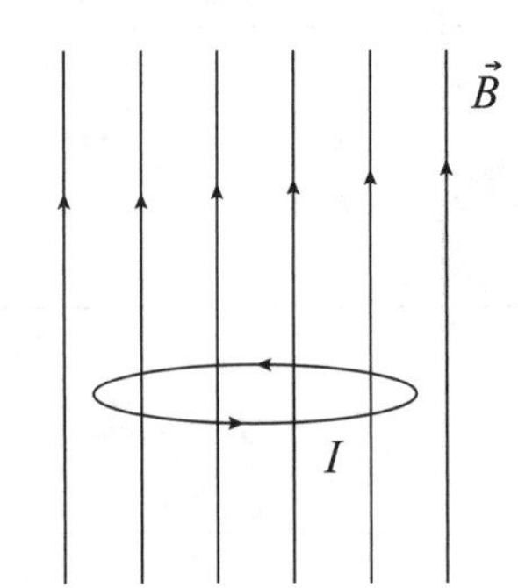

e.

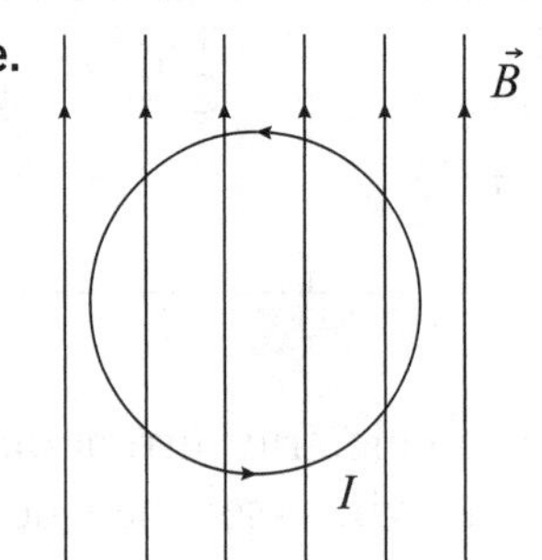

17.2.10 Reason A positively charged proton is ejected into space during a supernova, an explosion that occurs during the gravitational collapse of a large star near the end of its life. The proton travels through space for millions of years and finally reaches the magnetic field that surrounds Earth; we call the proton a cosmic ray. Sketch the path of the proton as it travels in Earth's magnetic field. Show its path as it might be seen from below Earth (see the person in the drawing).

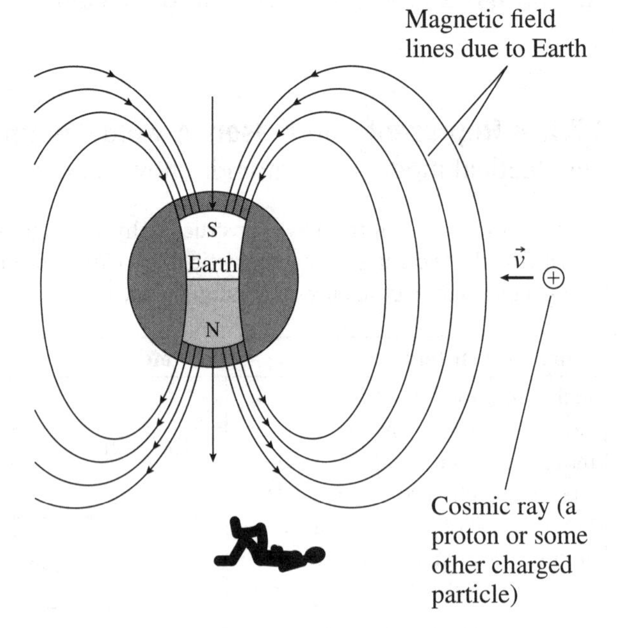

17.3 Quantitative Concept Building and Testing

17.3.1 Find a pattern The table below provides data concerning the magnitude of the magnetic force $\vec{F}_{\text{B on W}}$ exerted on a segment of a current-carrying wire by an external magnetic field as the following quantities are changed: (1) the magnitude of the external magnetic field $\vec{B}$, (2) the magnitude of electric current I, (3) the length of the segment of the current-carrying wire L, and (4) the direction of the electric current relative to the direction of the magnetic field.

Magnitude of the magnetic field $\vec{B}$ (T)	Current I in the wire (A)	Length L of the wire (m)	Angle θ between current direction and $\vec{B}$ field	Magnitude of the magnetic force F_B exerted on the wire (N)
$1B$	I	L	90°	F
$2B$	I	L	90°	$2F$
$3B$	I	L	90°	$3F$
B	$1I$	L	90°	F
B	$2I$	L	90°	$2F$
B	$3I$	L	90°	$3F$
B	I	$1L$	90°	F
B	I	$2L$	90°	$2F$
B	I	$3L$	90°	$3F$
B	I	L	0°	0
B	I	L	30°	$0.5F$
B	I	L	90°	F

Devise a rule relating the magnitude of the magnetic force F_B to these quantities.

17.3.2 Find a pattern The table below provides data concerning the magnitude of the magnetic force exerted on a moving charged particle by a magnetic field as the following quantities are changed: (1) the particle's speed, (2) the magnitude of the magnetic field, and (3) the direction of the particle velocity relative to the magnetic field.

Magnitude of the magnetic field $\vec{B}$ (T)	Charge of the moving particle	Speed v of the moving particle (m/s)	Angle θ between the velocity $\vec{v}$ and the $\vec{B}$ field	Magnitude of the magnetic force F_B exerted on the particle (N)
$1B$	q	v	90°	F
$2B$	q	v	90°	$2F$
$3B$	q	v	90°	$3F$
B	q	v	90°	F
B	$2q$	v	90°	$2F$

(continued)

Magnitude of the magnetic field $\vec{B}$ (T)	Charge of the moving particle	Speed v of the moving particle (m/s)	Angle θ between the velocity $\vec{v}$ and the $\vec{B}$ field	Magnitude of the magnetic force F_B exerted on the particle (N)
B	$3q$	v	90°	$3F$
B	q	v	90°	F
B	q	$2v$	90°	$2F$
B	q	$3v$	90°	$3F$
B	q	v	0°	0
B	q	v	30°	$0.5F$
B	q	v	90°	F

Devise a rule relating the magnitude of the force to these quantities.

17.3.3 Predict and test Assemble the apparatus shown in the illustration (a thick horizontal wire hanging from support wires on each side) and place a horseshoe magnet on a scale. Hang the apparatus between the poles of the magnet. Observe what happens to the reading of the scale when you turn on the current. Use the rule that you devised in Activity 17.3.2 to predict what will happen to the reading of the scale when you double the magnitude of the current. Fill in the table that follows.

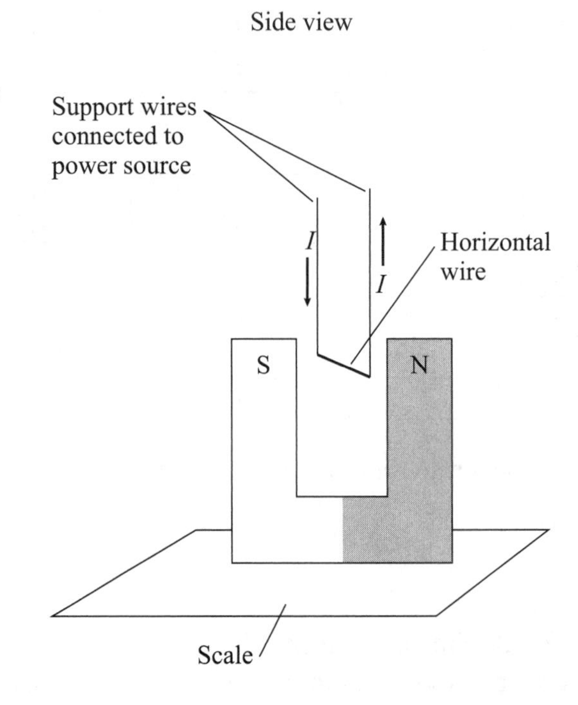

Record the reading of the scale with the magnet on it (no current in the wire).	Explain in words why the reading of the scale changes when the current is turned on.	Write a procedure to predict the reading of the scale when the current is doubled. List your assumptions.	Perform the experiment, record the outcome, and reconcile the outcome with the prediction.
		Procedure: Assumptions:	

17.3.4 Reason A galvanometer is a device that serves as a basis for an ammeter and a voltmeter. The galvanometer consists of a coil hanging between the poles of a horseshoe magnet. The coil is supported by a rod that can turn in a balljoint. A spring opposes its turning. A needle, attached to the rod, changes its orientation as the rod turns. The greater the current flowing through the coil, the greater the torque exerted on it by the magnetic field of the magnet, and the more the needle deflects. Discuss how one can make an ammeter and a voltmeter out of the same galvanometer. Imagine that you have resistors of different resistances.

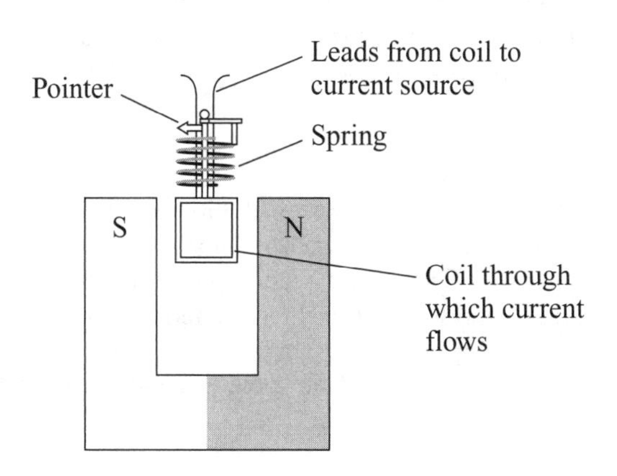

17.4 Quantitative Reasoning

17.4.1 Represent and reason The mass-detecting part of a mass spectrometer is described below in multiple ways. Starting with basic equations, devise a mathematical expression for the particle mass.

Words	Sketch	Physical representation	Mathematical representation
An ion with mass m and charge $+e$ leaves a velocity selector moving at speed v. It then moves in a half circle in a magnetic field that is perpendicular to the plane of its motion. At the end of this trip, it is detected. The radius of the circle can be used to determine the mass of the ion.	Top view Velocity selector $\vec{B}_{\text{out of paper}}$ $2r$ $\vec{v}$ Detector	Radial $\vec{F}_m$	

17.4.2 Represent and reason The speed of blood flow in an artery can be measured using our knowledge of a magnetic field and an electric field. The process is described below in words. Represent the process in other ways.

Description in words Positive and negative ions move with blood in an artery through a magnetic field that is perpendicular to the blood's velocity. The magnetic force causes some positive ions to accumulate on one wall of the artery and negative ions on the other wall. This charge separation causes an electric field that opposes further charge separation. Derive an expression for the speed of the blood in terms of the electric and magnetic fields.		
Draw a sketch.	Write a mathematical representation.	Draw a physical representation.

17.4.3 Regular problem What happens to a cosmic-ray proton flying into the Earth's atmosphere at a speed of about 10^7 m/s? The magnitude of the Earth's $\vec{B}$ field is approximately 5×10^{-5} T. The mass m of a proton is approximately 10^{-27} kg. Consider three cases: The proton enters the Earth's atmosphere parallel to the $\vec{B}$ field, perpendicular to the field, and at a 30° angle.

17.4.4 Equation Jeopardy Two processes are represented mathematically below, using Newton's second law. Fill in the table that follows to describe the processes in other ways.

Mathematical representation	Construct a physical representation.	Sketch the situation.	Write a description of the problem in words.
$(1.6 \times 10^{-19}\,\text{C})(2.0 \times 10^7\,\text{m/s})B = (1.67 \times 10^{-27}\,\text{kg})(2.0 \times 10^7\,\text{m/s})^2/(6000\,\text{m})$			
$0.020\,\text{N} = (0.020\,\text{A})(0.10\,\text{T})(20\,\text{m})(0.50)$			

17.4.5 Evaluate the solution

The problem: You are playing a video air-hockey game in which a hockey puck of mass m with electric charge $+q$ leaves a velocity selector traveling at speed v on a horizontal frictionless surface. You are to write an expression for the magnitude B of a magnetic field and decide on its direction so that it bends the puck in a curving half circle (toward the bottom of the screen) and it hits a target a distance $2R$ below the place the puck left the velocity selector.

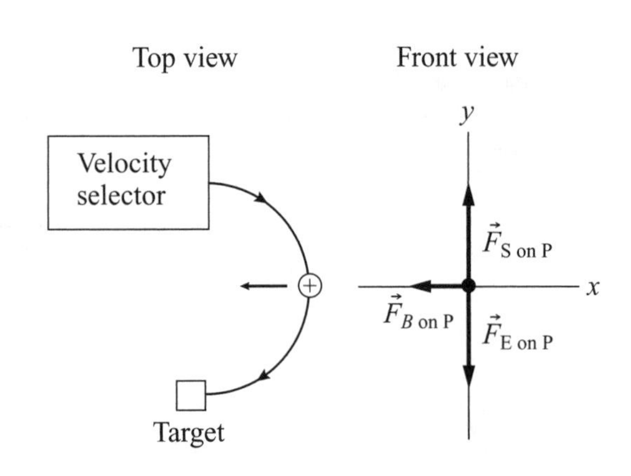

Proposed solution: The situation is pictured at the right, along with a force diagram for the puck as seen in the plane of its motion at the instant the puck reached the halfway point around the half circle. The acceleration direction is indicated.
Mathematical representation:

$$\Sigma F_{\text{rad}} = mv^2/R \quad \text{or} \quad qvB \sin 90° = mv^2/R$$

Consequently, $B = mv^2/qR$ and the field points to the left.

a. Identify any missing elements or errors in the solution.

b. Provide a corrected solution if there are errors.

17.4.6 Evaluate the solution

The problem: You wish to impress your friends with your mystical powers and decide to build a small object that you can cause to levitate at the dinner table. How will you do this?

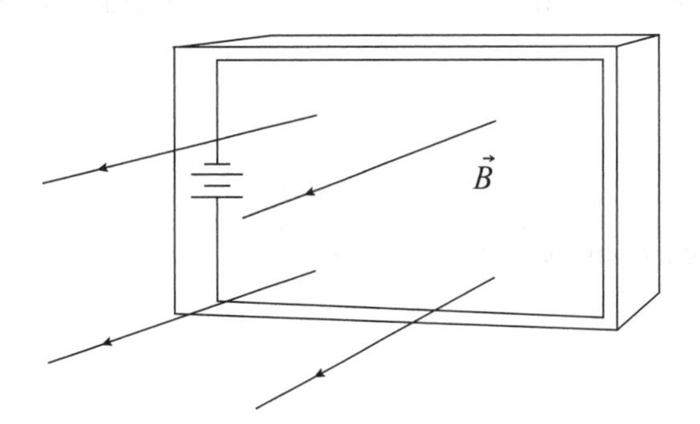

Proposed solution: You decide to place a 0.20-cm-high, 1.0-cm-long rectangular loop of resistance 0.30 Ω inside a 2.0-g box. A tiny 1.5-V battery sends a current through the coil (after a hidden switch is turned on). When you place your hands on each side of the box (with a 2.0-T field caused by the magnets you hide between your fingers), the force of the magnetic field causes the top and bottom parts of the current-carrying wire in the loop to rise up into the air.

Mathematical representation:

$$\Sigma F_y = 2ILB = mg$$

Note that the magnetic force is exerted on the two long horizontal sections of the loop wire. We check to see if the two sides of the equation are equal:

The left side:

$$2(1.5\ \text{V}/0.30\ \Omega)(1.0\ \text{cm})(2.0\ \text{T}) = 20\ \text{N}$$

The right side:

$$(2.0\ \text{g})(10\ \text{m/s}^2) = 20\ \text{N}$$

a. Identify any missing elements or errors in the solution.

b. Provide a corrected solution or missing elements if you find errors.

17.4.7 Evaluate You work in the complaint office of an auto manufacturer. A customer makes a complaint. Fill in the table that follows to describe the process in other ways and decide how you, an employee in the complaint office, will respond.

Description of the argument in words	A customer says that while driving through Earth's magnetic field, a potential difference developed from one end of her car to the other. The car discharged, causing her to lose control and run off the highway into a ditch. Decide the action you should take. Be sure to use your physics knowledge to help complete the recommendation—people have much more confidence in decisions based on physics, don't they?
Sketch the situation.	Construct a physical representation.
Write a mathematical representation.	Evaluate the argument and provide a recommendation.

17.4.8 Evaluate A neighborhood group asks your advice concerning the hazard of a new electric line that is to go into your neighborhood to provide power for streetlights. The neighbors have heard that magnetic fields caused by power lines may cause cancer to people living near the lines. Use your knowledge of physics to try to help you make a decision about whether the proposed electric line in your neighborhood will cause health hazards. Indicate any assumptions you made.

17.4.9 Reason Suppose that the rail for a train had an electric current traveling through it and that the train had a large coil such as that shown on the right. Can the magnetic field produced by the rail exert a magnetic force on the train that helps support it—like magnetic levitation? Explain. *Hint:* The magnetic field strength decreases with distance from the current-carrying wire.

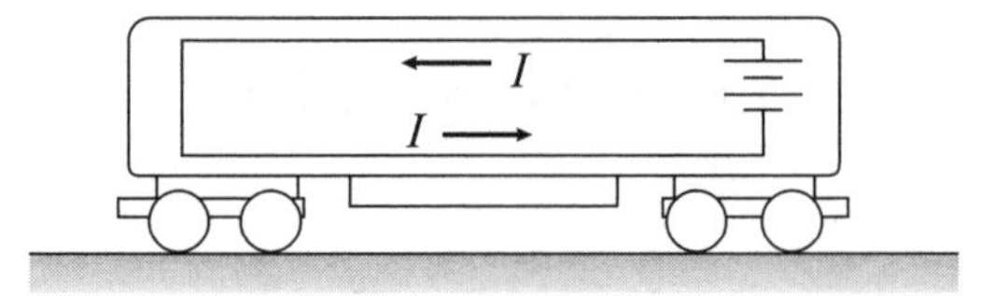

18 Electromagnetic Induction

18.1 Qualitative Concept Building and Testing

18.1.1 Observe and find a pattern The table that follows describes six experiments involving a galvanometer, a bar magnet, and a coil. Perform the experiments and record the outcomes.

Experiment	Illustration	Outcome
a. Hold the magnet motionless in front of the coil, with any orientation.	N S	
b. Hold the magnet perpendicular to the coil with the N pole facing the coil. Move the magnet quickly toward the coil. Then pull it away quickly.	N S I N S I	

(*continued*)

Experiment	Illustration	Outcome
c. Repeat experiment 2, but this time with the S pole facing the coil.	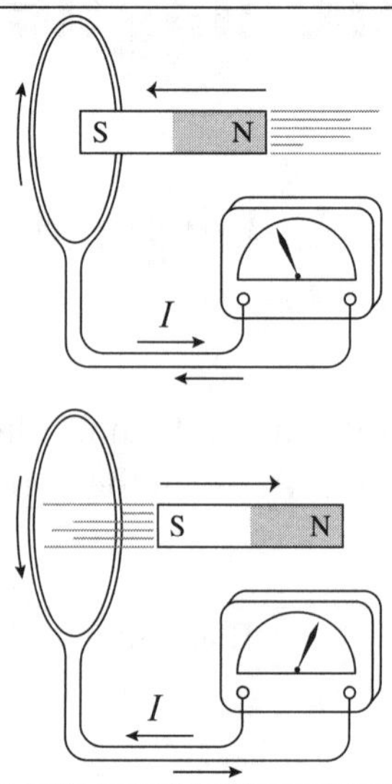	
d. Align the magnet in the same plane as the coil, and move either pole toward or away from the coil.	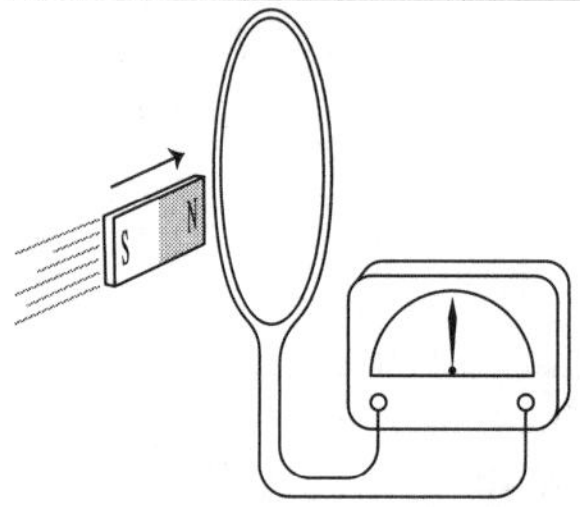	
e. Hold the magnet in front of the coil and rotate it 90° as shown. (The magnet starts out perpendicular to the coil and ends up parallel to it.)	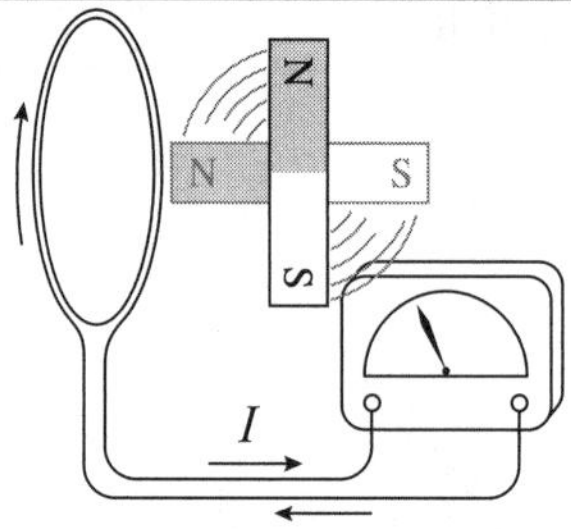	
f. Position the magnet as in experiment 2, but this time grasp the sides of the coil and collapse the coil quickly. Then pull it back open.	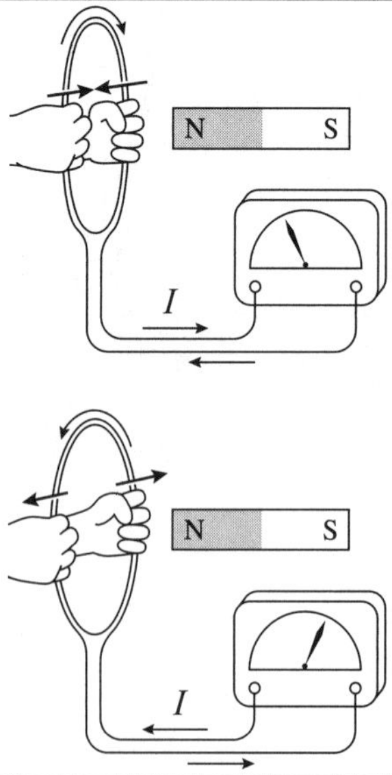	

Devise a rule that summarizes when a current is induced in a coil.

18.1.2 Test your idea Four experiments using a galvanometer, a switch, and a coil are described in the table that follows. Use the rule or rules you devised in Activity 18.1.1 to predict if there should be an induced current in the coil that is not connected to a battery shown in the following illustrations, as detected by the galvanometers.

a. Fill in the table that follows.

Experiment	Illustration	Predict the outcome.	Perform the experiment and record the outcome.
The current in the left coil increases just after the switch is closed.	A B Switch Galvanometer		
The current in the left coil increases just after the switch is closed. The coils are perpendicular.	A B Switch Galvanometer		
The current in the left coil decreases just after the switch is opened.	A B Switch Galvanometer		
The switch in the left coil is closed (steady current) as the right coil moves toward and above the left coil.	A B $\vec{v}$ Galvanometer		

b. If necessary, revise the rule you developed in Activity 18.1.1.

18.1.3 Test your idea The switch in the left coil pictured in the illustration at right is closed (there is a steady current in the left coil), and both coils move right at the same velocity.

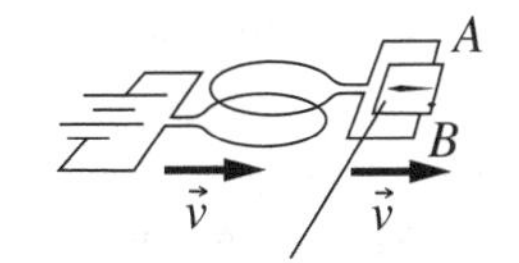

a. Predict whether the galvanometer will register an induced current as the two coils are moving. Explain your prediction.

b. Perform the experiment and record the outcome. If the outcome contradicts your prediction in part a, discuss whether you consistently used the rule developed in earlier activities to make the prediction.

18.1.4 Test an idea David says that the size of the magnet determines whether a current can be induced in a coil. You want to convince him that his idea is not correct. One way to do it is to design an experiment whose outcome might contradict a prediction based on David's idea.

a. Fill in the table that follows.

Describe an experiment to test David's idea.	Make a prediction of its outcome based on David's idea.	Perform the experiment and record the outcome.

b. Discuss whether David will be convinced by your results.

18.1.5 Observe and find a pattern The table that follows describes five new experiments using a galvanometer, bar magnets, and a coil. Perform the experiments and compare the outcomes to the described outcomes. The outcomes of the experiments are included.

Experiment	Illustration	Outcome
a. Position a magnet perpendicular to the coil and move it slowly toward the coil. Repeat the experiment, moving the magnet quickly.	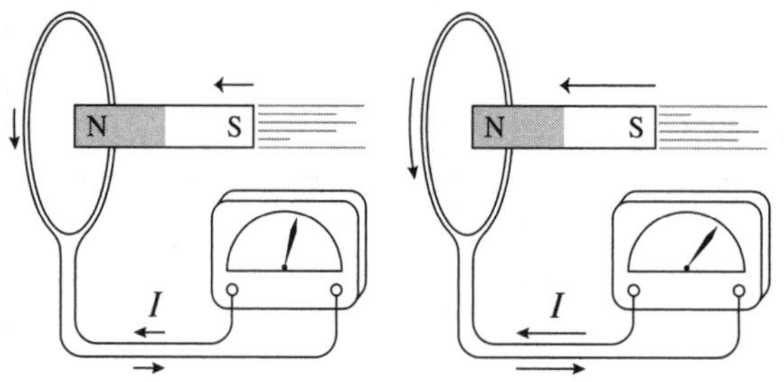 N S N S I I	The quicker the magnet's motion, the stronger the induced current.
b. Position a smaller magnet perpendicular to the coil and move it slowly toward the coil. Repeat the experiment using a bigger magnet.	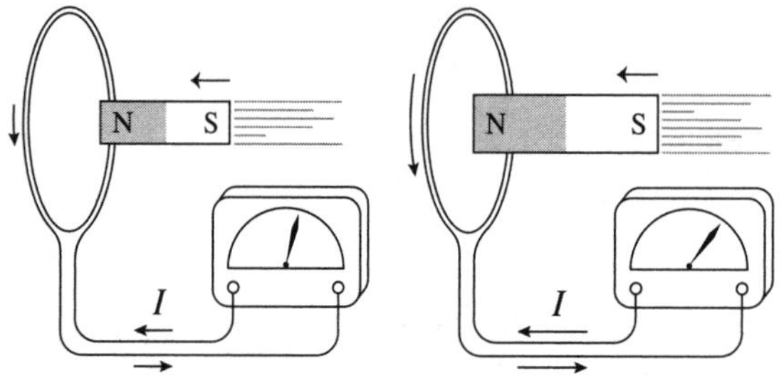 N S N S I I	The bigger magnet induces a stronger current than the smaller magnet when they move at the same speed with respect to the coil.
c. Move a magnet perpendicular to the coil. Then move it so that it makes an angle with the plane of the coil. Keep the speed the same.	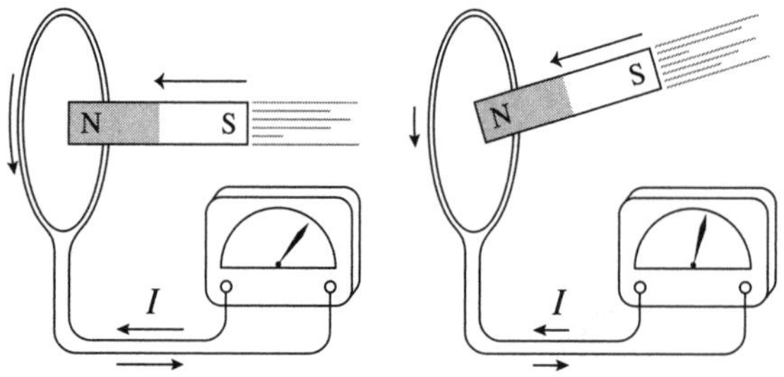 N S N S I I	When the magnet moves perpendicular to the coil, the strongest current is induced.

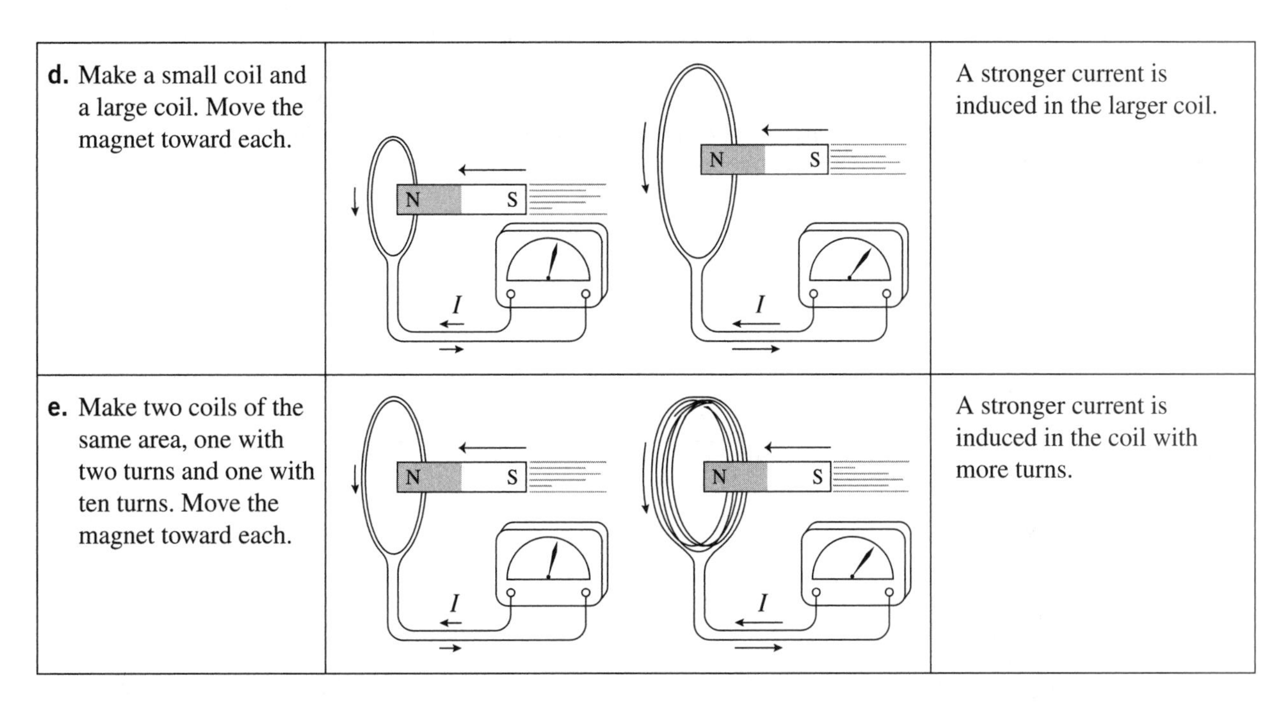

d. Make a small coil and a large coil. Move the magnet toward each.		A stronger current is induced in the larger coil.
e. Make two coils of the same area, one with two turns and one with ten turns. Move the magnet toward each.		A stronger current is induced in the coil with more turns.

Devise a rule that relates the *magnitude* of the induced current to various properties of the magnet, its motion, and properties of the coil.

18.1.6 Observe and find a pattern The table repeats three earlier experiments that used a galvanometer, a bar magnet, and a coil and in which a current was induced. The direction of the induced current is shown in the illustrations.

a. Fill in the table that follows.

Experiment	Draw $\vec{B}_{ext}$ field vectors, and $\Delta\vec{B}_{ext}$ vectors through the coil caused by the moving magnet. Indicate whether the external magnetic flux through the coil is decreasing or increasing. Draw $\vec{B}_{ind}$ field vectors due to the induced current.

(continued)

Experiment	Draw $\vec{B}_{ext}$ field vectors, and $\Delta\vec{B}_{ext}$ vectors through the coil caused by the moving magnet. Indicate whether the external magnetic flux through the coil is decreasing or increasing. Draw $\vec{B}_{ind}$ field vectors due to the induced current.
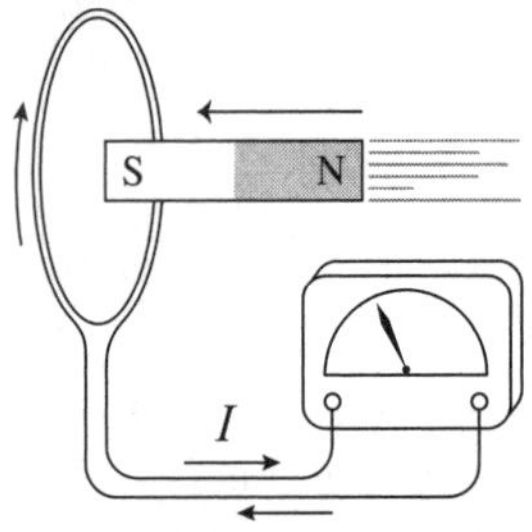 S N I	
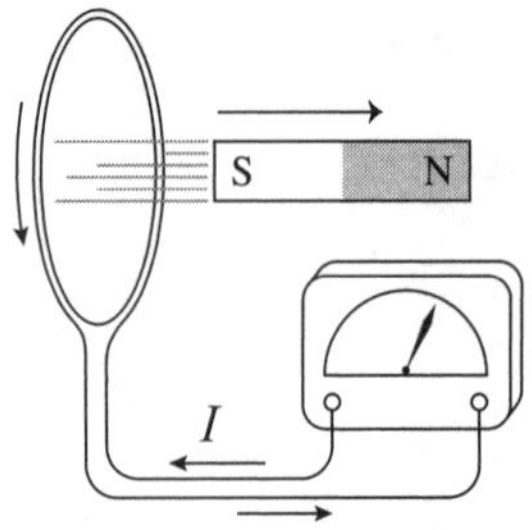 S N I	

b. Use the data in the table to devise a rule relating the direction of the induced current in the coil and the change of external magnetic flux through it. *Hint:* (1) Draw the $\Delta\vec{B}_{ext}$ field vectors of the bar magnet and make a note of whether the flux due to this magnet is increasing or decreasing through the coil. (2) Then draw $\vec{B}_{ind}$ vectors as a result of the induced electric current. (3) Compare the direction of $\vec{B}_{ind}$ vectors to the $\Delta\vec{B}_{ext}$ field vectors of the bar magnet when the flux through the coil increases and (4) when the flux decreases.

c. How does the direction of the induced current in a coil relate to the change of external magnetic flux through it?

18.2 Conceptual Reasoning

18.2.1 Reason For each situation shown in the table that follows, use the rules devised and tested in Section 18.1 to predict if a current is induced through the resistor attached to the loop. If a current is induced, indicate the direction of that induced current.

Experiment	Predict if a current is induced; explain your prediction.	If you predict that a current is induced, what is the direction of the current?
a. The loop is perpendicular to the page. 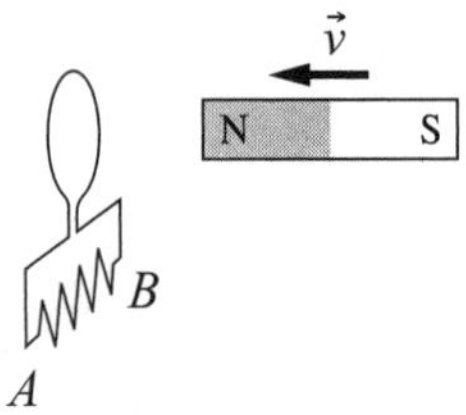 $\vec{v}$ N S B A		

b. The loop is perpendicular to the page and the magnet turns 90°. 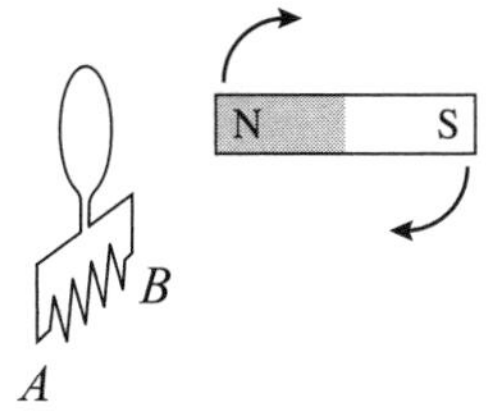 N S B A		
c. The loop is in the plane of the page. 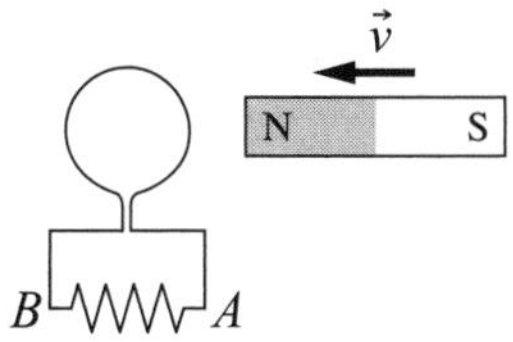 $\vec{v}$ N S B A		
d. The loop, perpendicular to the page, is pulled upward so that it collapses. 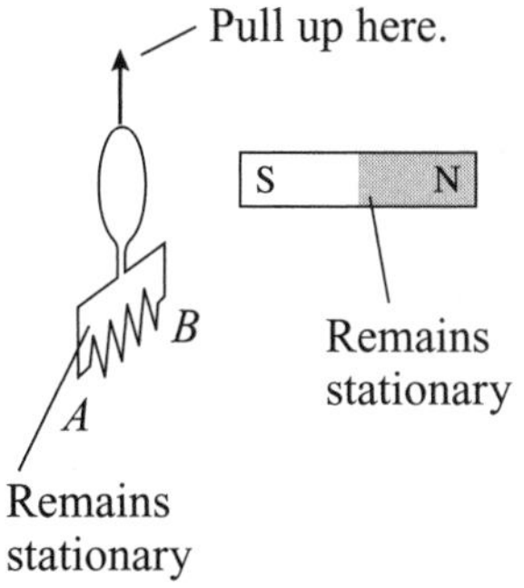 Pull up here. S N B A Remains stationary Remains stationary		
e. The switch in the left circuit is closed, and the current increases abruptly. 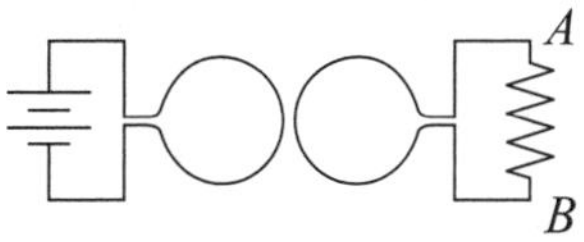 A B		
f. There is a steady current in the left circuit. 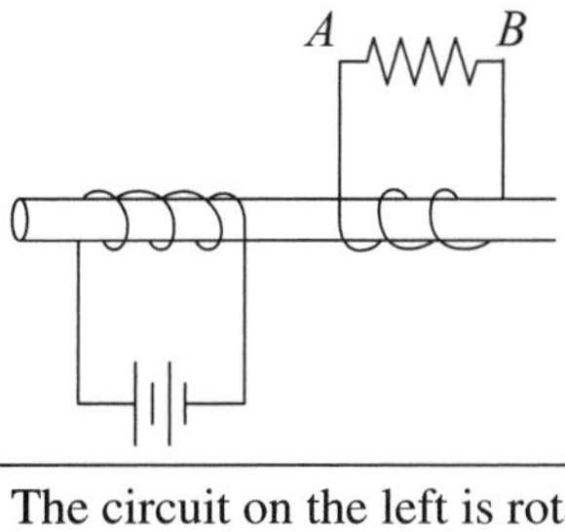 A B		
g. The circuit on the left is rotated 90°. 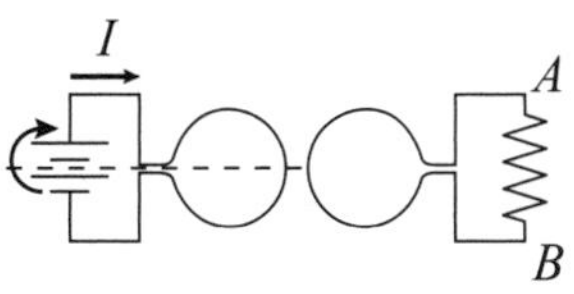 I A B		

(continued)

Experiment	Predict if a current is induced; explain your prediction.	If you predict that a current is induced, what is the direction of the current?
f. The switch in the left circuit is opened, and the current decreases abruptly. *A* *B*		

18.2.2 Represent and reason The rectangular loop with a resistor is pulled at constant velocity through a uniform external magnetic field that points into the paper in the regions shown in the illustration with the crosses (×).

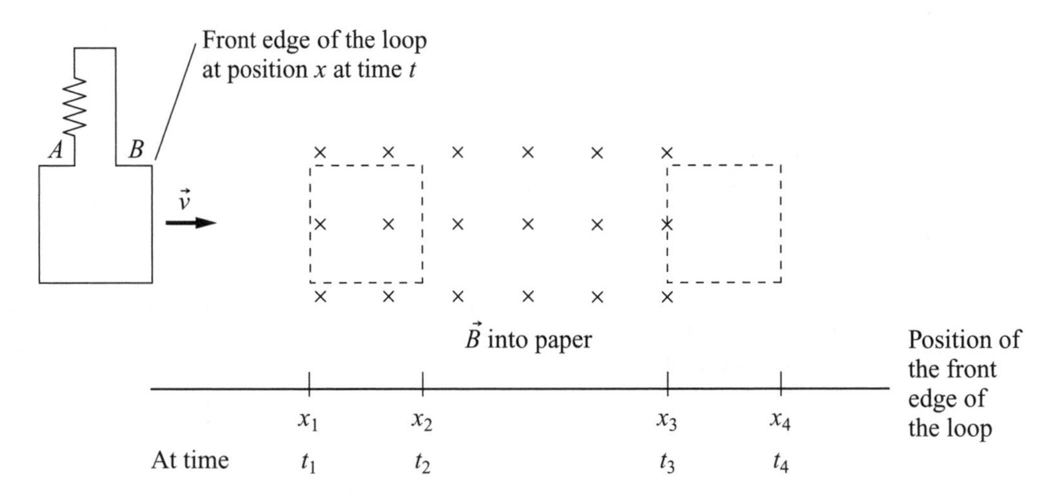

Complete the table that follows to determine qualitatively the shape of the induced current-versus-time graph.

a. Draw a qualitative flux-versus-time graph for the process (positive in and negative out).	Flux Φ — Time t; t_1, t_2, t_3, t_4
b. Draw a qualitative induced magnetic field-versus-time graph for the process.	Induced magnetic field B_{in} — Time t; t_1, t_2, t_3, t_4

c. Draw a qualitative induced current-versus-time graph for the process.	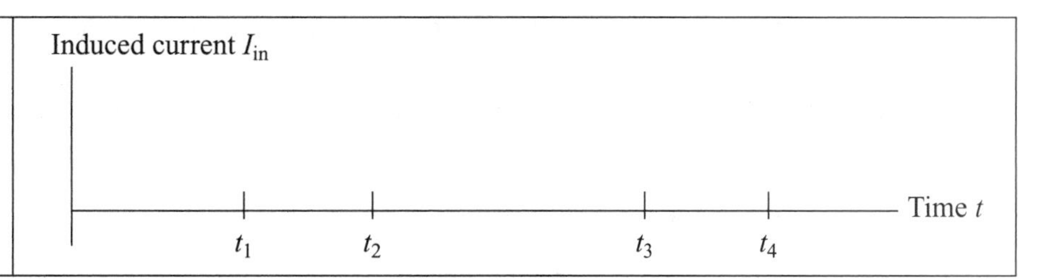

18.2.3 Evaluate the solution

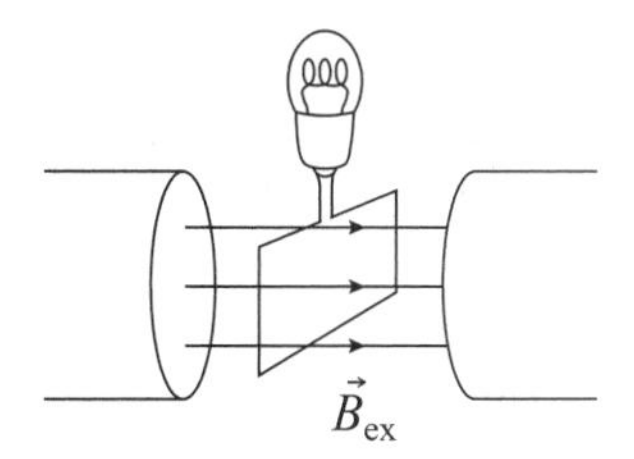

The problem: The magnetic field through the square coil shown in the illustration is at first steady and large (there are many turns in the coil, but only one is shown). The field then decreases to zero in about 1.0 s. A bulb connected to the ends of the coil indicates an induced current in the circuit. When are you most likely to observe light from the bulb?

Proposed solution: A steady light will come from the bulb when the field is steady and large. The brightness of the light will decrease as the field decreases. There is no light when the magnetic field becomes zero.

a. Identify any errors in the proposed solution.

b. Provide a corrected solution if there are errors.

18.3 Quantitative Concept Building and Testing

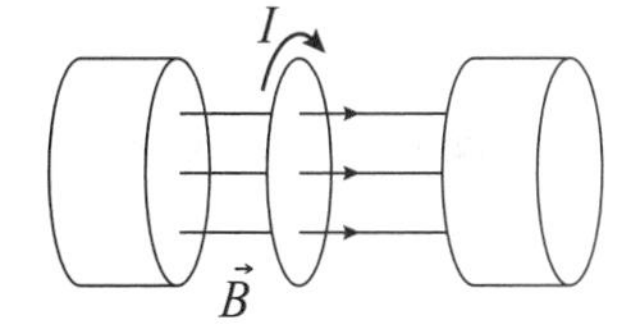

18.3.1 Observe and explain In the table that follows, the results of four experiments are shown in which a changing magnetic field produced by an electromagnet passes through a loop, as illustrated to the right. This changing $\vec{B}$ field causes a changing flux Φ through the loop and an induced current I_{ind} around the loop of resistance R. The product $I_{ind}R$ is also plotted as a function of time.

a. Draw a fourth graph in each table column that shows the induced emf ε_{ind} in the loop.

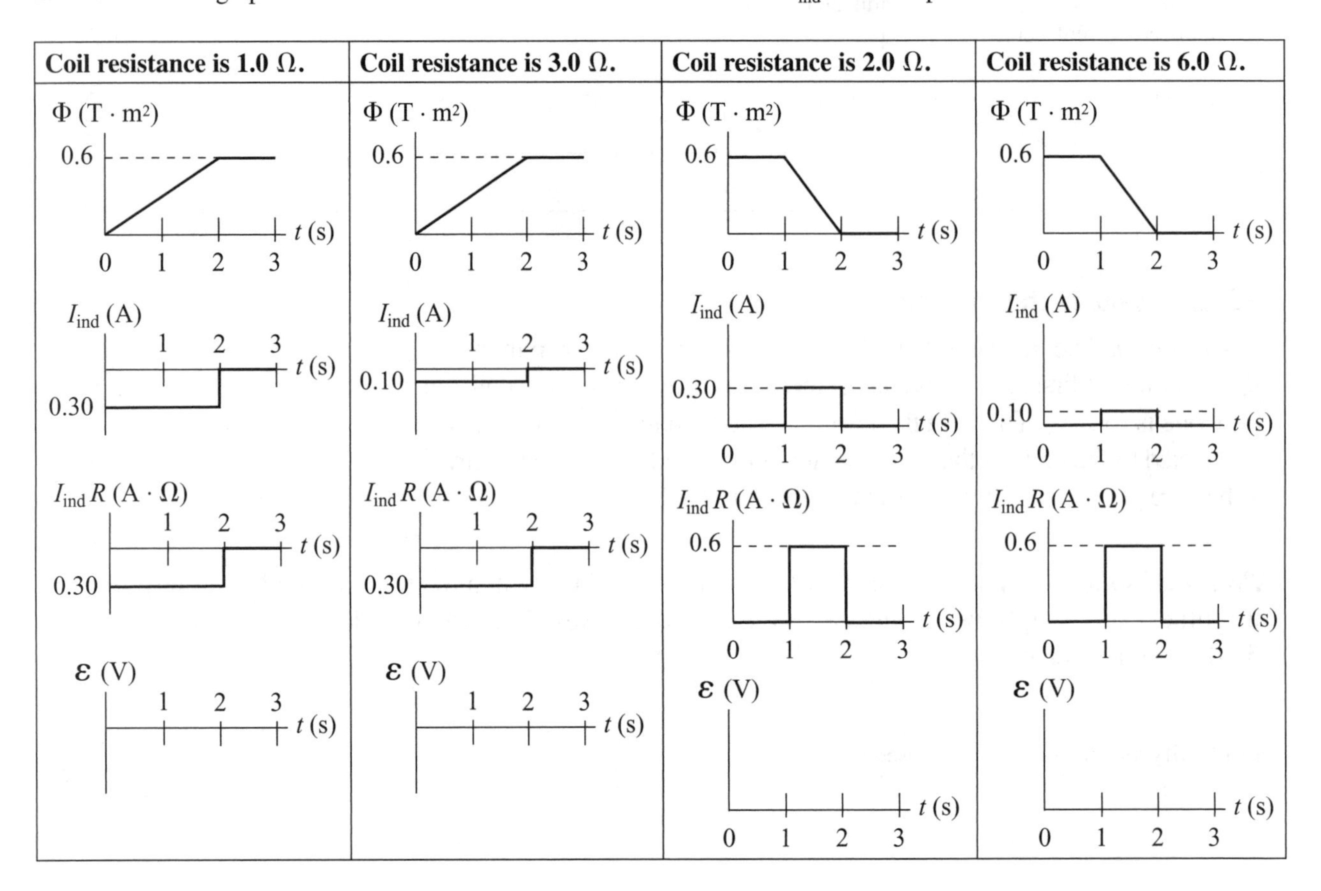

b. Devise a relationship between $\Delta\Phi/\Delta t$ and ε_{ind}. Do not forget the sign!

18.3.2 Observe and explain The analysis of the following experiment will help you devise an expression for the potential difference produced in a loop moving into, through, and out of a magnetic field. This is called *motional emf.*

Experiment	Analysis	Analysis
A metal bar of length L moves at constant velocity through a magnetic field that points into the paper (the crosses).	**a.** Indicate on the bar how the magnetic force exerted by the field on charges in the bar redistributes charges in the bar.	**b.** This charge redistribution, which occurs quickly, produces an electric field inside the bar that prevents further charge redistribution. Draw the $\vec{E}$ field lines.
$\vec{v}$	$\vec{v}$	$\vec{v}$

Analysis	Analysis	Analysis
c. Apply Newton's second law for a charge in the middle of the bar—now in an $\vec{E}$ field and a $\vec{B}$ field.	**d.** Use the expression that relates electric field E_y and the potential difference over a distance $\Delta V/L$ with the previous results to determine an expression for the potential difference (emf) across the ends of the bar.	**e.** Below, a rectangular metal conductor is depicted entering the magnetic field, residing completely in that field, and leaving the field described above. Use the results of parts a and b to draw on the left and right vertical parts of the rectangle the charge distributions due to the force exerted by the magnetic field on the electrically charged particles in the metal. *Note:* The magnetic field only exerts forces on charges in the metal parts that are in the field, not on parts outside the field. Will there be electric current in the conductor? If so, indicate the direction.
f. Use the results of part d to write an expression for the potential difference induced around the rectangular metal conductor while entering the field.	**g.** Use the results of part d to write an expression for the potential difference induced around the rectangular metal conductor when completely in the field. Note that there are now charge distributions that cancel each other.	**h.** Use the results of part d to write an expression for the potential difference induced around the rectangular metal conductor while leaving the field.

18.3.3 Observe and explain Repeat the previous activity, only this time use the ideas of flux and induced emf. When you are finished, check to see if the results are consistent with those in Activity 18.3.2.

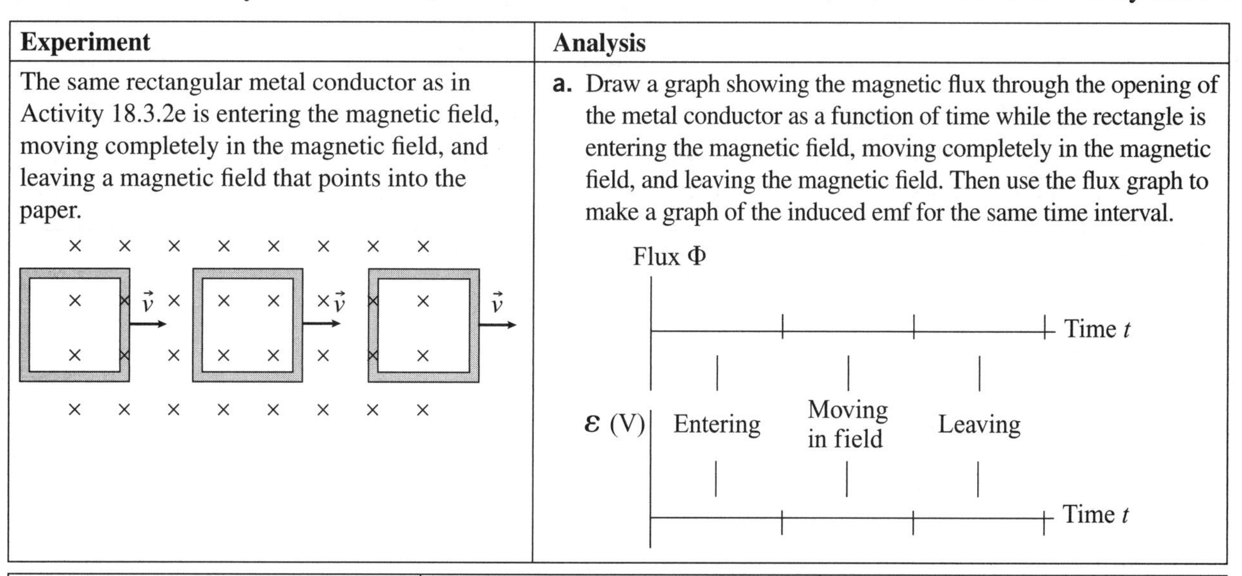

Experiment	Analysis
The same rectangular metal conductor as in Activity 18.3.2e is entering the magnetic field, moving completely in the magnetic field, and leaving a magnetic field that points into the paper.	**a.** Draw a graph showing the magnetic flux through the opening of the metal conductor as a function of time while the rectangle is entering the magnetic field, moving completely in the magnetic field, and leaving the magnetic field. Then use the flux graph to make a graph of the induced emf for the same time interval.

Analysis	Analysis	Analysis
b. Write an expression for the changing flux as the rectangle enters the field. Then use this expression to determine the emf while the rectangle is entering the field.	**c.** Write an expression for the changing flux as the rectangle is completely in the field. Then use this expression to determine the emf while the rectangle is in the field.	**d.** Compare the expressions in parts b and c with the expressions determined in 18.3.2f and g. (We are skipping the calculation for when the rectangle is leaving the field—it's a little more messy.)

18.4 Quantitative Reasoning

18.4.1 Reason The magnitude of the magnetic field in each situation described below is 0.50 T. For each situation in this table, write an expression for the magnetic flux through the loop of radius r.

Situation	Write an expression for the flux.	Situation	Write an expression for the flux.
Loop and $\vec{B}$ in the plane of the paper 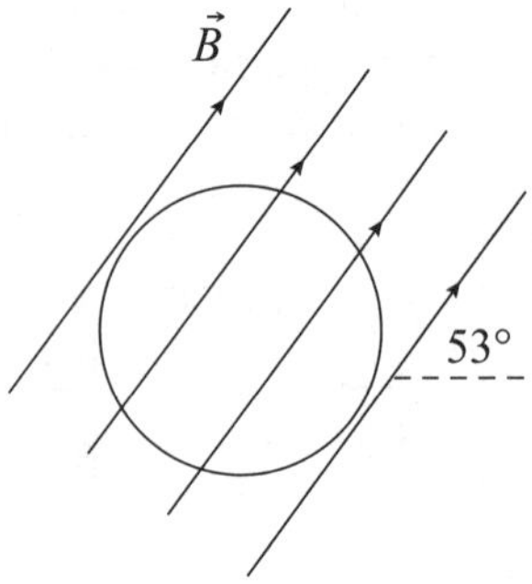 $\vec{B}$ 53°		Loop perpendicular to the paper and $\vec{B}$ in the plane of the paper 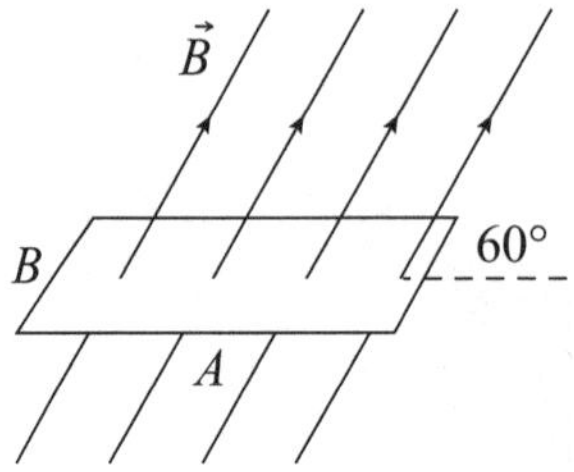 $\vec{B}$ B 60° A	
Loop in the plane of the paper $\vec{B}$ out of paper 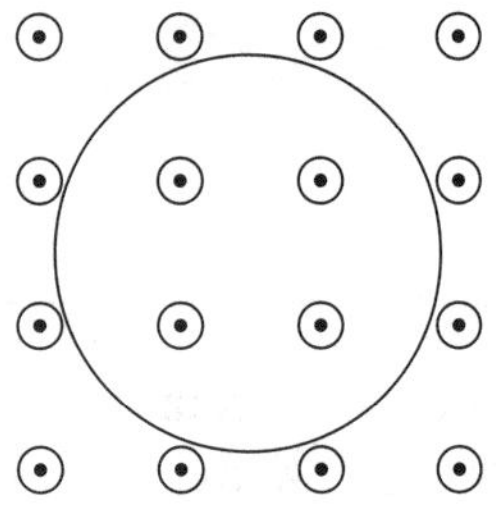		Square loop of side A in the plane of the paper. $\vec{B}$ into the paper 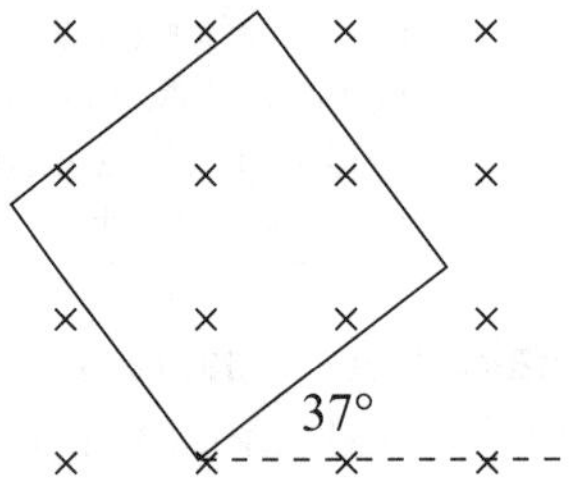 37°	

18.4.2 Represent and reason Four situations are shown in which the external flux through a loop is plotted as a function of time. In the table that follows, draw another graph that shows the induced emf in the loop as a function of time.

Situation 1	Situation 2
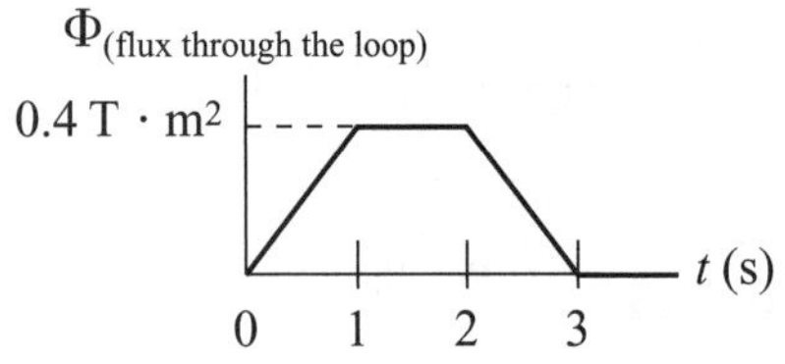 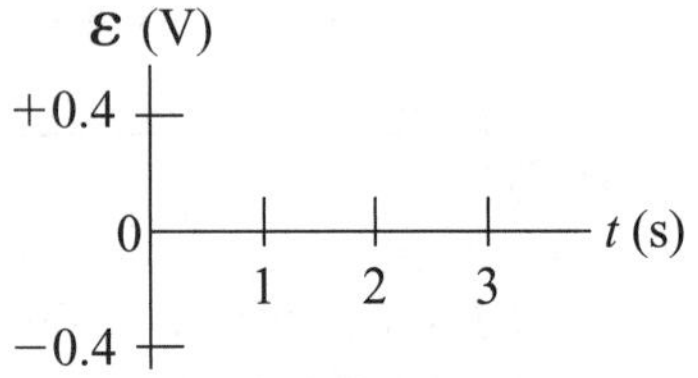 Φ(flux through the loop) 0.4 T · m² 0 1 2 3 t (s) $\mathcal{E}$ (V) +0.4 0 1 2 3 t (s) −0.4	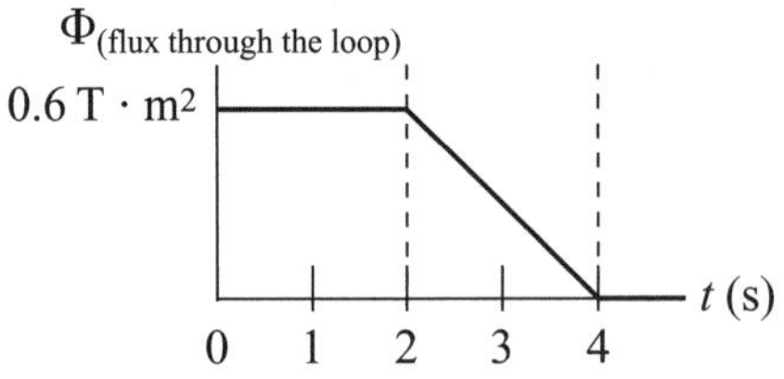 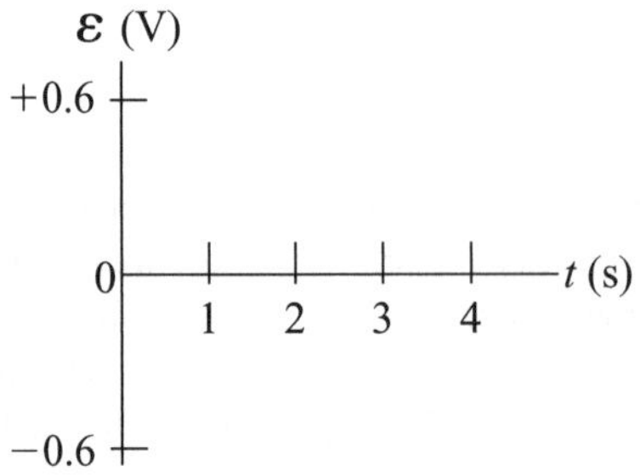 Φ(flux through the loop) 0.6 T · m² 0 1 2 3 4 t (s) $\mathcal{E}$ (V) +0.6 0 1 2 3 4 t (s) −0.6

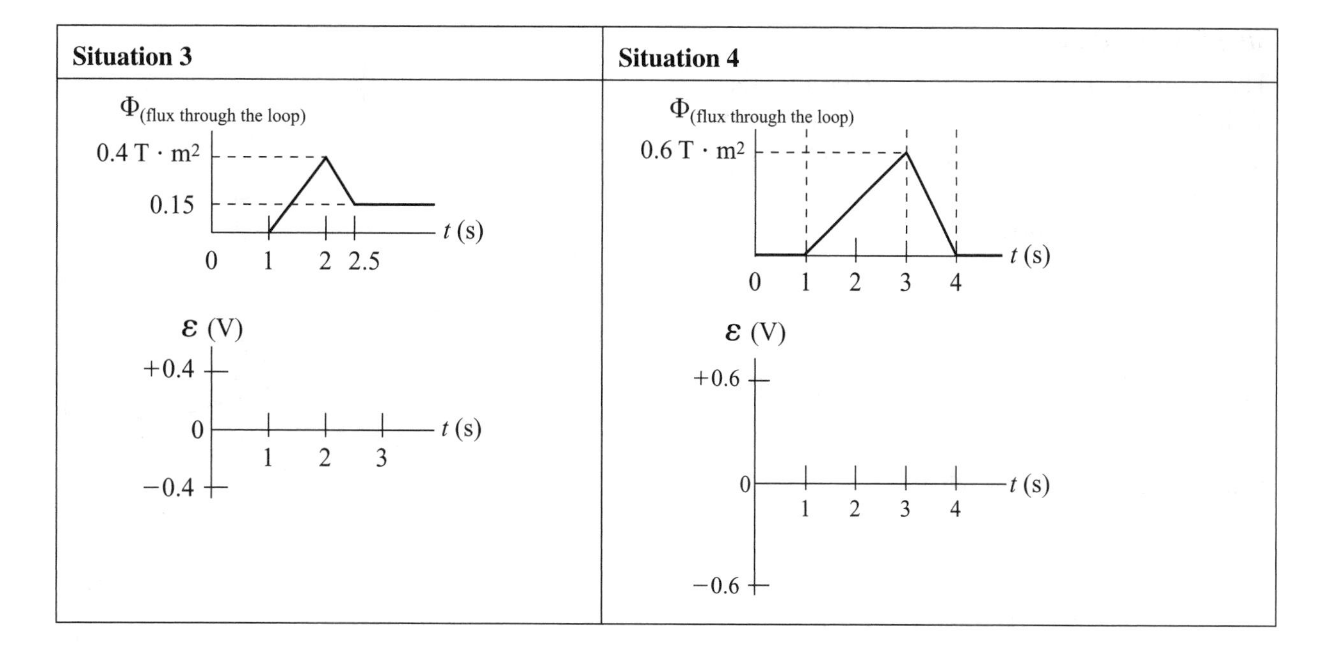

18.4.3 Equation Jeopardy Write a problem in words and construct a sketch for a phenomenon involving electromagnetic induction that is described by each equation below (there is more than one possibility). Provide all the details for this phenomenon.

Mathematical description	Write a description in words of a problem that is consistent with the equation.	Sketch a situation the problem might describe.
$\varepsilon = -20[\pi(0.10\text{ m})^2]\cos 37° \times (0 - 0.40\text{ T})/(2.0\text{ s} - 0)$		
$\varepsilon = -4(0.40\text{ T})\cos 0° \times [0 - (0.10\text{ m} \times 0.20\text{ m})]/(2.0\text{ s} - 0)$		

18.4.4 Evaluate the solution

The problem: A single 0.10-m × 0.10-m square loop is between the poles of a large electromagnet. The surface of the loop makes a 53° angle with respect to the magnetic field. The magnetic field varies as shown in the illustration. Determine the induced emf produced by the loop.

Proposed solution:

Sketch and translate

See the drawing at the right.

Simplify and diagram

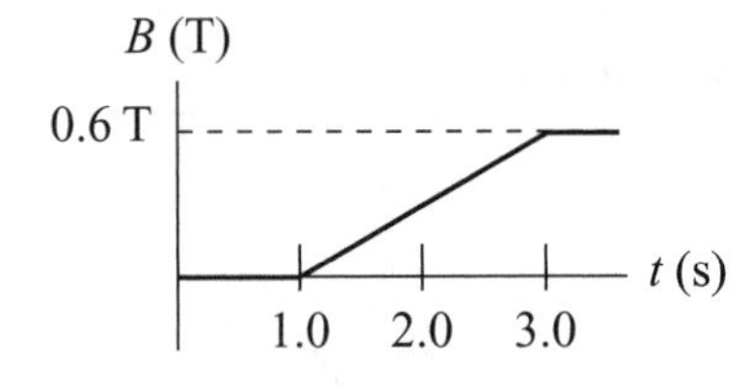

The magnetic flux Φ through the loop increases as the magnetic field increases (see the graph). The emf ε has the same shape, only with a negative sign (the lower graph line).

Represent mathematically and solve

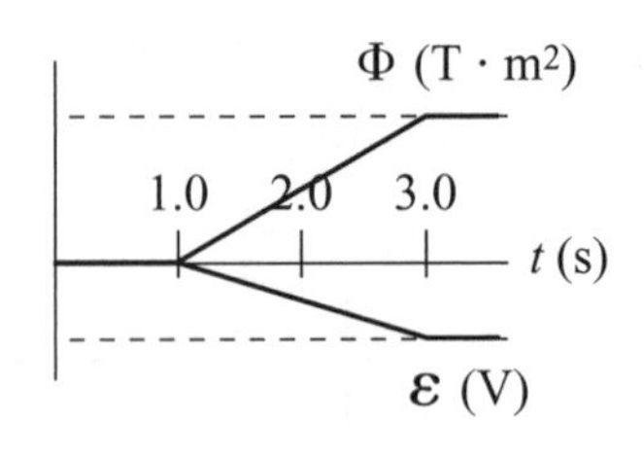

$$\varepsilon_{\text{ind av}} = N\Delta(BA\cos\theta)/\Delta t = NA\cos 53^\circ[(B_f - B_i)/(t_f - t_i)]$$
$$= (1)(0.10\text{ m})^2\{0.60[(0.60\text{ T} - 0)/(3.0\text{ s} - 0)]\}$$
$$= 1.2 \times 10^{-3}\text{ T}\cdot\text{m}^2/\text{s} = 1.2 \times 10^{-3}\text{ V}$$

a. Identify any errors in the solution.

b. Provide a corrected solution if there are errors.

18.4.5 Evaluate the solution

The problem: A single square loop is between the poles of a large electromagnet with its surface perpendicular to the external magnetic field. The magnetic field decreases steadily to zero. Determine the direction of the induced current in the loop.

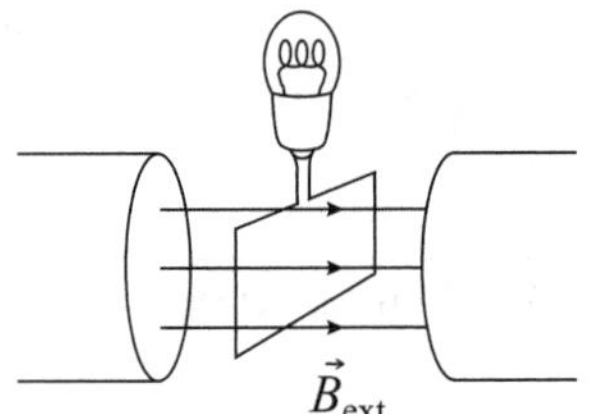

Proposed solution: The induced magnetic field (shown with dashed arrows) must point left to oppose the external magnetic field (the solid arrows). A clockwise induced current (the curved arrow) will produce this induced magnetic field.

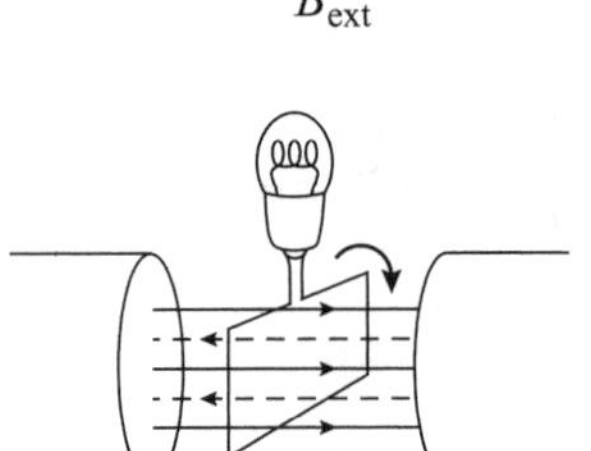

a. Identify any errors or missing elements in the proposed solution.

b. Provide a corrected solution if there are errors.

18.4.6 Pose a problem Design a problem that involves a graphical representation of magnetic flux-versus-time. You can provide the graph and ask for some other information based on the graph or provide other information and construct the graph as part of the problem solution.

18.4.7 Pose a problem Design a problem that involves a graphical representation of induced emf-versus-time. You can provide the graph and ask for other information based on the graph or provide other information and construct the graph as part of the problem solution.

18.4.8 Regular problem A horizontal bar is pulled at a constant velocity through a downward-pointing magnetic field. The bar slides on two horizontal, frictionless metal rails moving away from a resistor connected between their ends. Derive an expression for the induced current through the resistor of resistance R in terms of any or all quantities that you choose to include in a sketch of this system. Be sure to identify the quantities in the sketch.

a. Draw top-view sketches showing the rails and the bar location at an initial time and at a later time. Include symbols for quantities involved in the problem.

b. Describe the assumptions you are making.

c. Construct labeled graphs for the process below.

Plot the flux through the area surrounded by the rails, moving bar, and the resistor at the end of the rails as a function of time.

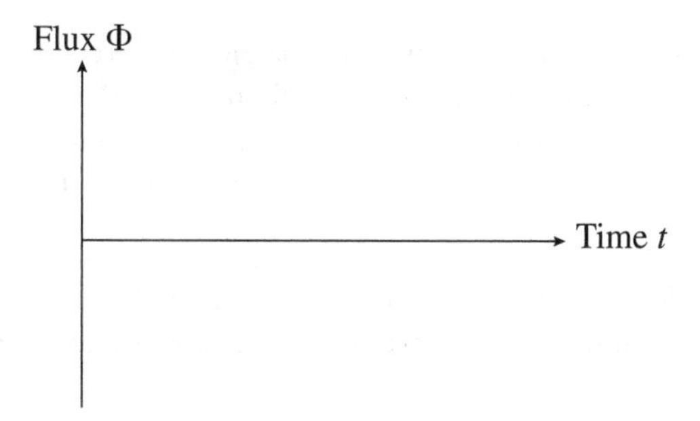

Consistent emf-versus-time graph

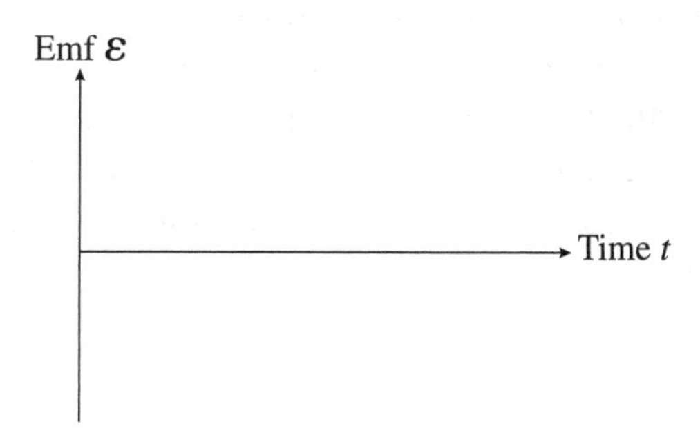

d. Represent flux and emf mathematically.

e. Combine the mathematical representation with Ohm's law to get the desired expression for the current.

18.4.9 Regular problem Loop *KLMN* is made of metal rods, where rod *KL* slides at a constant speed on the side rods *NK* and *ML* in the direction indicated by the arrow. A constant external magnetic field either points down into the loop (the crosses) or up out of the loop (the dots). For each situation, use two different methods to determine the direction (not the magnitude) of the electric current in loop *KLMN*.

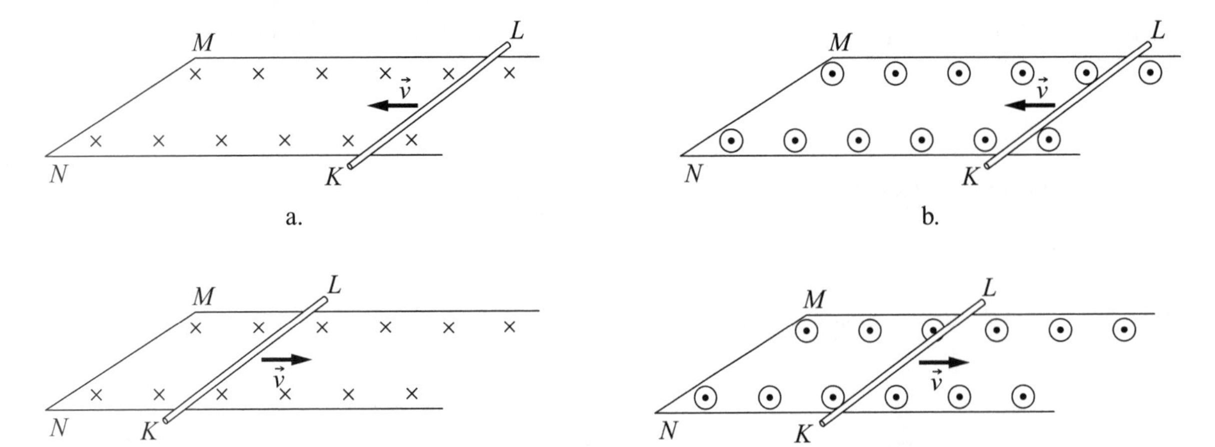

a. Use the magnetic force on free charges in the rod to find charge separation on the rod, the electric force that this charge separation produces that prevents additional charge separation, and the relationship between the electric field caused by the charge separation and electric potential difference to determine which side of the moving rod is at higher electric potential and which way electric current flows around the loop *KLMN*. Draw the direction of the current in the figure above.

b. Next use Lenz's law and right-hand rule for the magnetic field to determine the direction of the induced current. Draw the direction of the current determined with this method in the figure above using a different-colored pencil.

c. Do these two approaches agree? Explain.

18.4.10 Regular problem The Dreamworld's Tower of Terror at the Gold Coast in Australia is the fastest, tallest thrill ride in the Southern Hemisphere. Its L-shaped track has a smooth curve between the horizontal section of the track and the vertical section. The 15-passenger cart goes from 0 to 100 mph in 7 s on the horizontal section of the track and then coasts up to the top of the track, like a ball thrown up into the air. The cart then falls back down and is stopped by eddy-current braking on the horizontal track where it started.

a. Describe how this braking might work.

b. Estimate the distance traveled while the rollercoaster cart gets up to speed.

c. Estimate the acceleration in *g*'s while the rollercoaster cart starts.

d. Estimate the maximum height reached by the cart.

18.4.11 Pose a problem An airplane flies at a speed of 900 km/h. The distance between the tips of its wings is about 12 m. The vertical component of Earth's magnetic field is about 5×10^{-5} T. Pose a problem using this information.

19 Vibrations

19.1 Qualitative Concept Building and Testing

19.1.1 Observe and find a pattern Try the following simple experiments and describe common patterns concerning the behavior of the block.

a. Fill in the table that follows.

Experiment	Record your observations.
Tie a string to a small, heavy block and let the block hang freely. Now pull the block to the side and release it.	
Hang a heavy block from a spring, pull the block down, and release it.	

b. Identify patterns common to both experiments.

19.1.2 Explain In Activity 19.1.1 you found that both the block on a string and the block on a spring had repeatable motion, either back and forth or up and down. The blocks moved about the place where they resided when not vibrating—that is, about the *equilibrium position.* Explain why each block returns to this equilibrium position, first moving in one direction and then a short time later in the opposite direction. To help your thinking, draw force diagrams for the block when on each side of the equilibrium position.

19.1.3 Test your idea Clamp the top of a spring to a ring stand. Attach a horizontal metal bar to the bottom of the spring. Suppose you rotate the bar about 90° to the side perpendicular to the spring, so that it twists the spring. Predict what happens to the bar when you release it. Be sure to identify the equilibrium position for the bar and the reason it moves as it does.

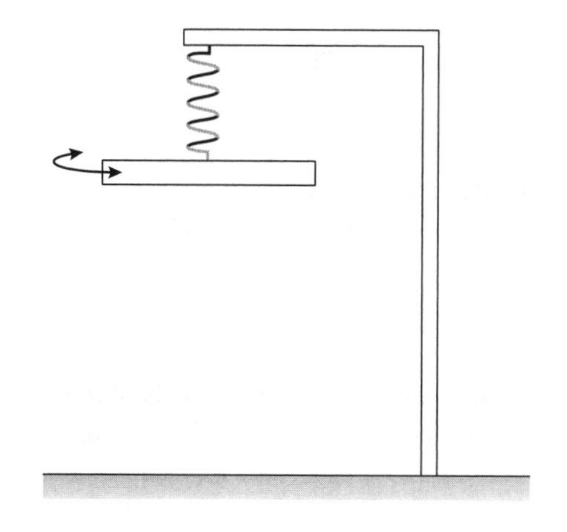

19.2 Conceptual Reasoning

19.2.1 Represent and reason The cart in the figure is attached to a special spring that can stretch and compress equally well. The spring is very light. The cart and spring rest on a low-friction horizontal surface. The cart is pulled to position I and then released. It moves to position V, where it then reverses direction and returns again to position I. It repeats the motion. Represent with motion diagrams and force diagrams the cart's motion between the points indicated in the table that follows.

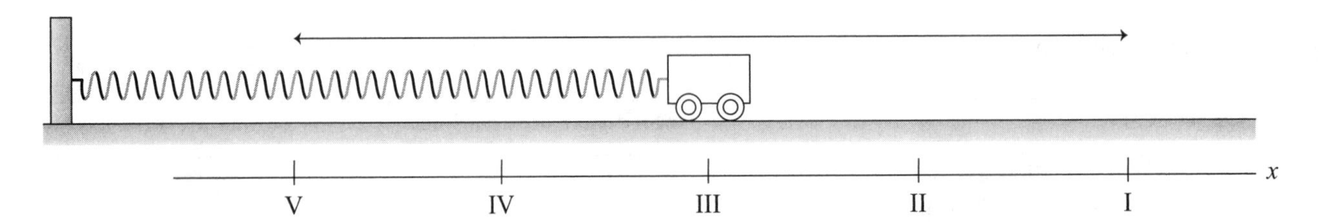

Draw a motion diagram for motion between points I–III.	Draw a motion diagram for motion between points III–V.	Draw a motion diagram for motion between points V–III.	Draw a motion diagram for motion between points III–I.
Draw a force diagram for point I, cart moving left.	Draw a force diagram for point III, cart moving left.	Draw a force diagram for point V, cart moving left.	Draw a force diagram for point II, cart moving left.
Draw a force diagram for point I, cart moving right.	Draw a force diagram for point III, cart moving right.	Draw a force diagram for point V, cart moving right.	Draw a force diagram for point II, cart moving right.

a. Do the force diagrams depend on whether the cart was moving left or right? Explain.

b. Are the force descriptions consistent with the motion description? For example, is the net horizontal force in the same direction as the acceleration? Give several specific examples.

c. At each position, compare the direction of the net force exerted by the spring on the cart and the cart's displacement from equilibrium when at that position.

19.2.2 Represent and reason

a. Construct five qualitative work–energy bar charts for the cart–spring system described in Activity 19.2.1 at the points described in the table that follows.

Construct a work–energy bar chart for point V.	Construct a work–energy bar chart for point IV.	Construct a work–energy bar chart for point III.
K U_s Other + 0 –	K U_s Other + 0 –	K U_s Other + 0 –
Construct a work–energy bar chart for point II.	**Construct a work–energy bar chart for point I.**	
K U_s Other + 0 –	K U_s Other + 0 –	

b. Do the charts depend on whether the cart is moving left when at a particular position or moving right? Explain.

c. How would the charts change if the surface had considerable friction? Explain.

19.2.3 Reason and explain Summarize the results of Activities 19.2.1–19.2.2 to describe and explain the motion of the cart. The description should include your observations, and the explanations should include reasoning based on force and energy analyses for the observed phenomena.

19.2.4 Represent and reason You have a small bob on a long string (a pendulum). The pendulum bob swings back and forth, as shown in the figure. At each of the marked points in the figure, the coordinate system consists of an axis in the radial direction (r axis) and a perpendicular axis in the tangential direction (t axis). Disregard air resistance.

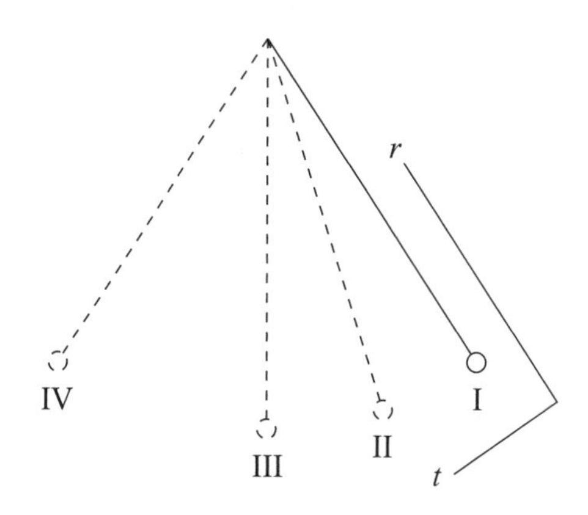

a. Complete the table that follows for positions shown in the figure.

Use the graphical velocity–subtraction method to estimate the bob's acceleration direction.	Position I	Position II	Position III	Position IV
Draw a force diagram for the bob.	Position I	Position II	Position III	Position IV
Draw the r component of the net force $\vec{F}_{net}$. Does the acceleration have a component in the radial direction that points in the same direction as the r component of the net force?	Position I	Position II	Position III	Position IV
Draw the t component of the net force $\vec{F}_{net}$. Does the acceleration have a component in the tangential direction that points in the same direction as the t component of the net force?	Position I	Position II	Position III	Position IV
Construct an energy bar chart.	Position I 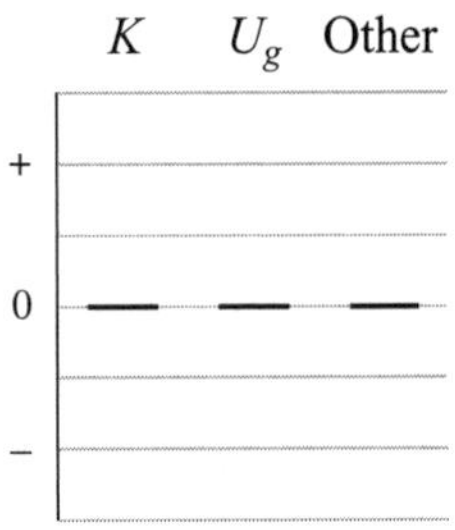 K U_g Other + 0 –	Position II 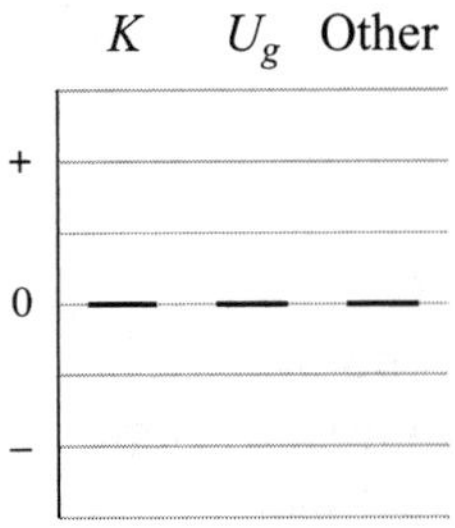 K U_g Other + 0 –	Position III 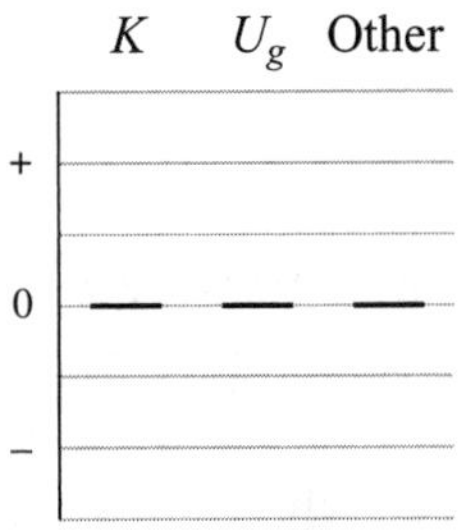 K U_g Other + 0 –	Position IV K U_g Other + 0 –

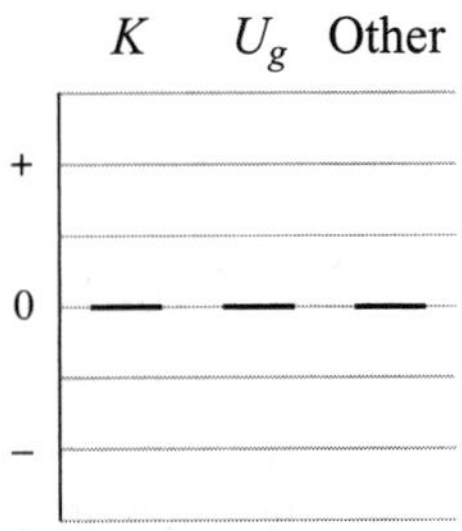

b. Is there a relationship between the t component of the net force and the displacement of the bob from the equilibrium position? Explain.

c. Compare the patterns of the net force and acceleration for the vibrating pendulum bob to the net force and acceleration of the vibrating cart in Activity 19.2.1.

19.3 Quantitative Concept Building and Testing

19.3.1 Observe and find a pattern Suppose that when the cart in Activity 19.2.1 was vibrating at the end of a spring, you used a motion detector to record the cart's motion. Graphs of position-versus-time, velocity-versus-time, and acceleration-versus-time are shown below.

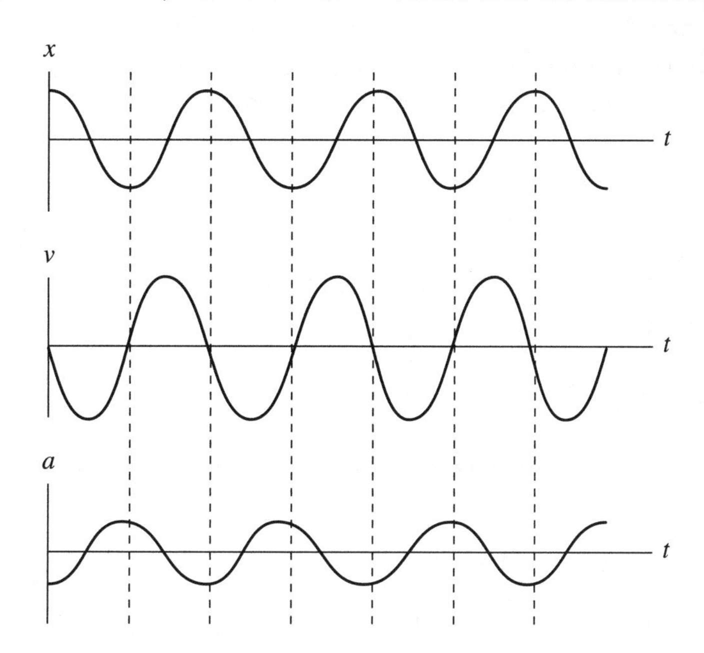

a. Examine the graphs carefully and answer the questions in the table that follows.

Recall that $v = \Delta x/\Delta t$. Is the shape of the velocity-versus-time graph consistent with this mathematical definition and with the position-versus-time graph? Compare the slope of $x(t)$ with the value of v at the maximum, minimum, and zero points. Explain.	Are these three graphs consistent with the force and motion diagrams in Activity 19.2.1? Explain.
Recall that $a = \Delta v/\Delta t$. Is the shape of the acceleration-versus-time graph consistent with this definition and with the velocity-versus-time graph? Compare the slopes of $v(t)$ with the value of a at the maximum, minimum, and zero points. Explain.	Are the direction and magnitude of the acceleration consistent with the direction and magnitude of the restoring force? Explain.

(continued)

Describe the relationship between the position-versus-time graph and the acceleration-versus-time graph. Explain why they mirror each other. Think about Newton's second law and the expression for the force that the spring exerts on the cart ($F_{\text{S on C}x} = -kx$).	Decide what mathematical function can be used to describe the position of the cart as a function of time.

b. Period is a physical quantity that characterizes the time interval for one complete vibration. Indicate in each of the three graphs at the beginning of this activity the period of the vibration.

19.3.2 Reason Assume that position, velocity, and acceleration change with time, as you saw in Activity 19.3.1. Write mathematical expressions for $x(t)$, $v(t)$, and $a(t)$ as cosine or sine functions of time. (In your expressions, try to use the quantities amplitude A and period T.) How do you know if the expressions you wrote make sense? *Hint:* Think about whether a sine or cosine function will work for the position-versus-time graph in Activity 19.3.1. Examine special cases (such as $t = 0$) and the relationships between the functions.

19.3.3 Observe and explain We would like to derive an approximate expression for the period of the cart's vibration at the end of the spring shown in Activity 19.2.1. Begin the derivation by answering two questions (a–b).

a. What physical quantities might affect the period?

b. Describe experiments that you could perform to decide if these quantities do in fact affect the period.

Now we move to the actual step-by-step derivation (c–g).

c. What is the total distance, in terms of the amplitude A, traveled by the cart during one complete cycle of vibration?

d. Assume that the cart's average speed v_{ave} during that cycle of vibration is half the cart's maximum speed $v_{\max}$. Write an expression for the period T of the cart in terms of A and $v_{\max}$.

e. Now use the fact that the maximum elastic energy of the spring equals the maximum kinetic energy of the cart to rewrite the expression for T in terms of the cart's mass m and the spring constant k.

f. A more rigorous method of derivation using calculus gives us a different expression for the period: $T = 2\pi(m/k)^{\frac{1}{2}}$. What is the difference between this expression and the one you provided in the previous part?

g. List all of the assumptions that you made to derive the expression for the period.

19.3.4 Test your ideas Assemble a set of springs, a set of bobs of different mass, and a ruler. Design an experiment to test each relationship proposed in the left column of the following table (some relationships may be incorrect). Fill in the table.

The period of vibration of a cart–spring system (as discussed in Activity 19.2.1) or of a pendulum bob depends on the amplitude of vibration.	Describe the experiment and include a sketch.	List the controlled variables (i.e., what you keep constant).	Describe the procedure and the predictions that you make based on the relationship you will test.	Perform the experiment and record the outcome.
The period of vibration of a pendulum bob depends on the mass of the bob.	Describe the experiment and include a sketch.	List the controlled variables (i.e., what you keep constant).	Describe the procedure and the predictions that you make based on the relationship you will test.	Perform the experiment and record the outcome.
The period of vibration of a system consisting of a bob attached to a vertical spring depends on the mass of the bob.	Describe the experiment and include a sketch.	List the controlled variables (i.e., what you keep constant).	Describe the procedure and the predictions that you make based on the relationship you will test.	Perform the experiment and record the outcome.

19.3.5 Observe and find a pattern You have a pendulum consisting of a long string with a metal bob hanging at one end. You can vary the mass m of the metal bob, the length L of the string, and the amplitude A of its back-and-forth vibration. You vary one of these three quantities at a time, measure the period of the pendulum, and then calculate the frequency (if needed). The data are shown in the table.

Bob mass (kg)	**String length (m)**	**Amplitude (m)**	**Period (s)**
1	1	0.05	2.00
2	1	0.05	2.00
3	1	0.05	2.00
1	2	0.05	2.80
1	3	0.05	3.45
1	4	0.05	4.00
1	1	0.07	2.00
1	1	0.10	2.00

a. Based on this data, decide what physical quantities affect the period. Decide which quantities do not affect the period. Explain.

b. An expression for the period of a *simple pendulum* (a small object vibrating with small amplitude on a long string) derived using calculus is:

$$T = 2\pi\sqrt{\frac{L}{g}}$$

where g is the acceleration due to gravity. Use the data in the table to decide whether the pendulum in the experiment can be considered a simple pendulum. Explain your decision.

19.3.6 Test your ideas Assemble a pendulum with a long string and a small metal bob. Use the relationship between the period and the length of the string for a simple pendulum to predict the period of its vibrations. Write the predicted value; take experimental uncertainties into account. Then perform the experiment and test your prediction. Record the experimental value and compare it to the prediction. Can your pendulum be considered a simple pendulum?

19.4 Quantitative Reasoning

19.4.1 Represent and reason A 2.0-kg cart attached to a horizontal spring vibrates on a low-friction track (similar to the situation shown in Activity 19.2.1). The cart's displacement-versus-time is described by:

$$x = (0.20\text{ m})\sin\left[\left(\frac{2\pi}{2.0\text{ s}}\right)t\right]$$

The positive direction of the x axis is to the right. Determine the cart's position at $t = 0$, $t = T/4$, $t = T/2$, and $t = (3T)/4$.

$t = 0$: _______ $t = T/4$: _______ $t = T/2$: _______ $t = (3T)/4$: _______

19.4.2 Represent and reason A 2.0-kg cart attached to a spring undergoes simple harmonic motion on a low-friction surface. Its displacement-versus-time is described by:

$$x = (0.20\,\text{m})\sin\left[\left(\frac{2\pi}{2.0\,\text{s}}\right)t\right]$$

Complete the table that follows.

Determine the period and the amplitude of the motion.
Determine the spring constant.
Determine the maximum elastic potential energy of the system.
Determine the maximum speed of the cart.
Write an expression for the velocity as a function of time.

19.4.3 Represent and reason A 2.0-kg cart attached to a spring undergoes simple harmonic motion so that its displacement-versus-time is described by:

$$x = (0.20\,\text{m})\sin\left[\left(\frac{2\pi}{2.0\,\text{s}}\right)t\right]$$

a. Draw a motion diagram for one cycle of the cart's motion.

b. Draw a force diagram for the cart at the times indicated in the table that follows.

$t = 0$	$t = T/4$	$t = T/2$	$t = 3T/4$	$t = T$

19.4.4 Represent and reason A 2.0-kg cart attached to a spring undergoes simple harmonic motion so that its displacement-versus-time is described by:

$$x = (0.20\,\text{m})\sin\left[\left(\frac{2\pi}{2.0\,\text{s}}\right)t\right]$$

Complete the table that follows. Make sure the graphs are consistent with the motion diagram and the force diagrams in Activity 19.4.3.

Construct a position-versus-time graph.	x vs. t axes (0, 1, 2)
Construct a velocity-versus-time graph.	a vs. t axes (0, 1, 2)
Construct an acceleration-versus-time graph.	v vs. t axes (0, 1, 2)

19.4.5 Represent and reason A 2.0-kg cart attached to a spring undergoes simple harmonic motion so that its displacement-versus-time is described by:

$$x = (0.20\text{ m})\sin\left[\left(\frac{2\pi}{2.0\text{ s}}\right)t\right]$$

Construct qualitative energy bar charts for the cart–spring system at the times indicated in the table that follows.

$t = 0$	$t = T/4$	$t = T/2$
K U_s Other; +, 0, −	K U_s Other; +, 0, −	K U_s Other; +, 0, −
$t = 3T/4$	$t = T$	
K U_s Other; +, 0, −	K U_s Other; +, 0, −	

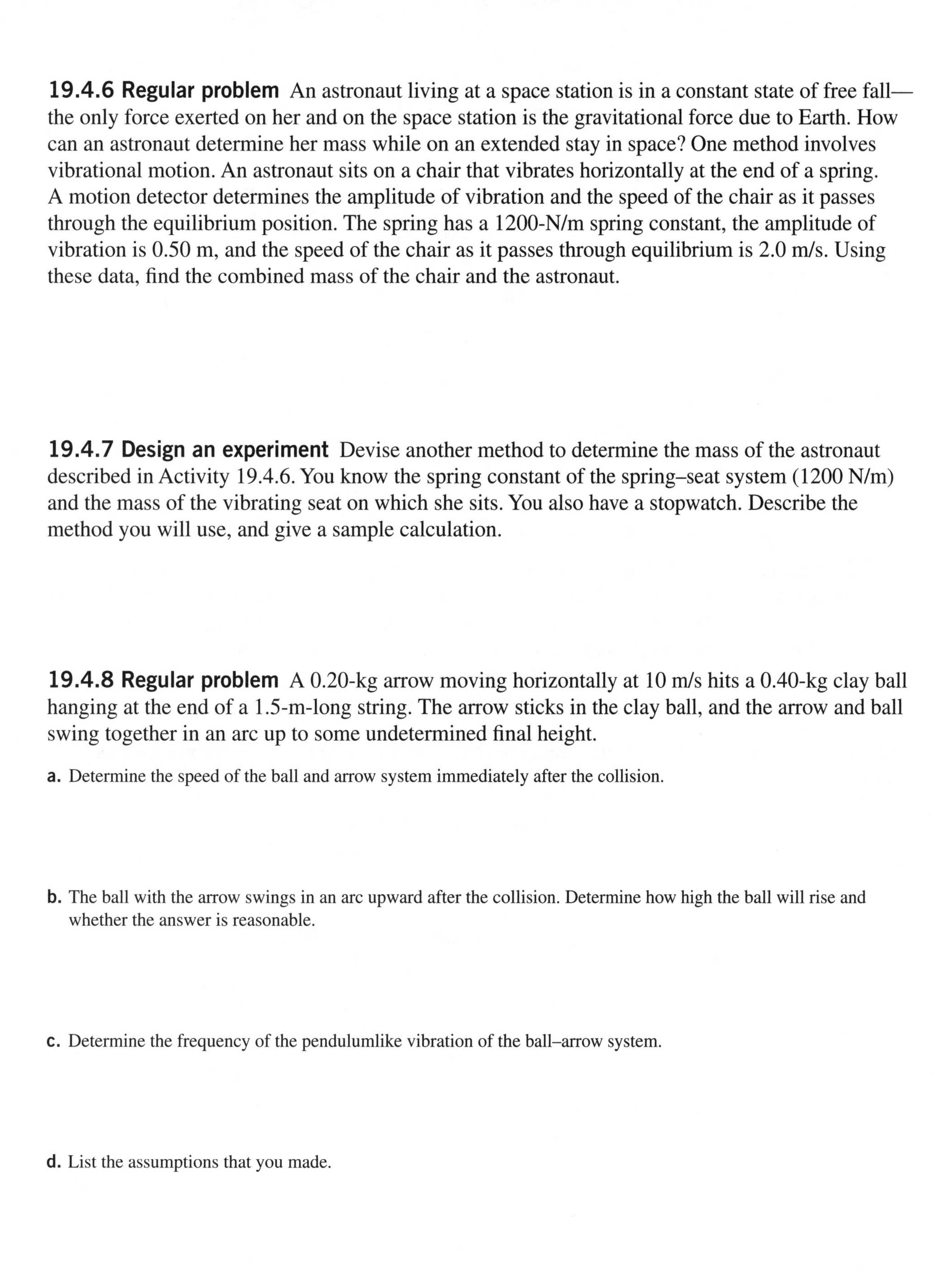

19.4.6 Regular problem An astronaut living at a space station is in a constant state of free fall—the only force exerted on her and on the space station is the gravitational force due to Earth. How can an astronaut determine her mass while on an extended stay in space? One method involves vibrational motion. An astronaut sits on a chair that vibrates horizontally at the end of a spring. A motion detector determines the amplitude of vibration and the speed of the chair as it passes through the equilibrium position. The spring has a 1200-N/m spring constant, the amplitude of vibration is 0.50 m, and the speed of the chair as it passes through equilibrium is 2.0 m/s. Using these data, find the combined mass of the chair and the astronaut.

19.4.7 Design an experiment Devise another method to determine the mass of the astronaut described in Activity 19.4.6. You know the spring constant of the spring–seat system (1200 N/m) and the mass of the vibrating seat on which she sits. You also have a stopwatch. Describe the method you will use, and give a sample calculation.

19.4.8 Regular problem A 0.20-kg arrow moving horizontally at 10 m/s hits a 0.40-kg clay ball hanging at the end of a 1.5-m-long string. The arrow sticks in the clay ball, and the arrow and ball swing together in an arc up to some undetermined final height.

a. Determine the speed of the ball and arrow system immediately after the collision.

b. The ball with the arrow swings in an arc upward after the collision. Determine how high the ball will rise and whether the answer is reasonable.

c. Determine the frequency of the pendulumlike vibration of the ball–arrow system.

d. List the assumptions that you made.

19.4.9 Equation Jeopardy Mathematical expressions describe two situations involving vibrational motion. Fill in the table that follows. Provide all the details for these situations. *Note:* There is more than one possible solution for each problem.

Mathematical description	Sketch a situation the equation(s) might describe.	Write in words a problem for which the equation(s) is a solution.
$(1/2)(20{,}000\ \text{N/m})(0.20\ \text{m})^2 = (1/2)(100\ \text{kg})v^2 + (1/2)(20{,}000\ \text{N/m})(0.10\ \text{m})^2$		
$0.20\ \text{Hz} = (1/2\pi)[k/(100\ \text{kg})]^{1/2}$ $(1/2)k(0.40\ \text{m})^2 = (1/2)(100\ \text{kg})v_{\text{max}}{}^2$		

19.4.10 Evaluate the solution *The problem:* You are helping design a stopping system for Soapbox Derby race cars. The proposed stopper is a padded cushion that catches the car at the end of the race. The far end of the cushion, opposite the car, is attached to a spring that compresses at collision. The car's mass with driver is 60 kg, and it is traveling at 12 m/s when it first contacts the stopper; the mass of the stopper is 20 kg. The car is to stop in 2.0 m after contacting the stopper. What is the spring constant of the spring needed for this system?

Proposed solution: The initial kinetic energy of the car is converted into elastic potential energy of the compressed spring when the car stops. Thus,

$$(1/2)mv^2 = (1/2)kA^2$$

or

$$\begin{aligned} k &= m(v^2/A^2) \\ &= (60\ \text{kg})[(12\ \text{m/s})^2/(2.0\ \text{m})^2] = 2160\ \text{kg/s}^2 \end{aligned}$$

a. Identify any errors in the solution to this problem.

b. Provide a corrected solution if there are errors.

19.4.11 Design an experiment You have a stop watch and a metal ball attached to a 1.0-m-long string. Design an experiment using this equipment to measure the acceleration of free fall g. Fill in the table that follows to help you.

<table>
<tr><td>Describe the experiment in words.</td><td rowspan="2">List the physical quantities that you will measure and the quantities that you will calculate.

To be measured:

To be calculated:</td></tr>
<tr><td>Draw a labeled sketch of the apparatus.</td></tr>
<tr><td>Describe the mathematical procedure you will use to determine g.</td><td rowspan="3">List sources of experimental uncertainties and ways to minimize them.

Uncertainties:

Ways to minimize:</td></tr>
<tr><td>List additional assumptions.</td></tr>
<tr><td>Perform the experiment and record the outcome. Does your result make sense?</td></tr>
</table>

19.4.12 Design an experiment Design an experiment to determine if a human arm can be treated as a simple pendulum when it swings back and forth during a walk. Fill in the table that follows.

<table>
<tr><td>Describe the experiment in words.</td><td rowspan="2">List the physical quantities that you will measure and the quantities you will calculate.

To be measured:

To be calculated:</td></tr>
<tr><td>Draw a labeled sketch of the apparatus.</td></tr>
<tr><td>Describe the mathematical procedure you will use.</td><td rowspan="3">List sources of experimental uncertainties and ways to minimize them.

Uncertainties:

Ways to minimize:</td></tr>
<tr><td>List additional assumptions.</td></tr>
<tr><td>Describe how you will make a judgment about whether an arm can be treated as a simple pendulum.</td></tr>
</table>

19.4.13 Pose a problem You have a rubber band, a 100-g object, a stopwatch, and a meterstick. Pose an experimental problem that you can solve using this equipment. Describe how you would solve the problem.

20 Mechanical Waves

20.1 Qualitative Concept Building and Testing

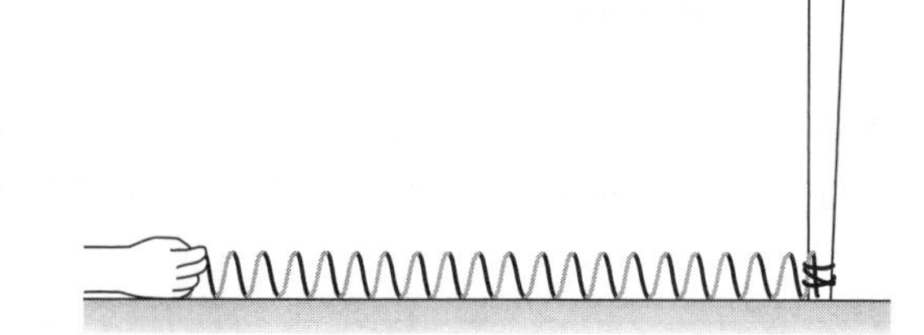

20.1.1 Observe and find a pattern Fasten one end of a metal Slinky® toy to the leg of a chair resting on a hard floor—or alternatively, to a clamp, which in turn is fastened to the end of a lab bench so the Slinky lies on the smooth lab bench surface. Grasp the free end of the Slinky and stretch it so that the Slinky is about 3- to 4-m long. Do not lift it off the smooth surface. Fill in the table that follows.

a. Keeping the Slinky on the smooth surface and stretched along a straight line, give the end of the Slinky in your hand a quick push along its axis. Describe what you observe.	
b. Sketch the Slinky at one instant of time during the propagation of the disturbance you created in part a.	
c. Indicate in words or draw how an individual Slinky ring in the middle of the Slinky moves with respect to the Slinky as the disturbance passes.	
d. If you repeat the procedure described in part a, but this time push more abruptly, does the disturbance move faster along the Slinky? How do you know?	
e. If you push less abruptly than in part a, does the speed of the disturbance change? How do you know?	

Slinky® is a registered trademark of Poof-Slinky Inc.

20.1.2 Observe and find a pattern Keep the Slinky toy from Activity 20.1.1 on a smooth, hard surface and fastened securely at one end. Again grasp the free end of the Slinky and stretch it so that the Slinky is about 3- to 4-m long. Fill in the table that follows.

a. Give the end of the Slinky in your hand an abrupt sideways shake, perpendicular to the Slinky (see the top-view illustration), all the while keeping it on the smooth surface. Describe what you observe. Top view	
b. Sketch the Slinky at one instant of time during the propagation of the disturbance you created in part a.	
c. Indicate in words or draw how an individual Slinky ring in the middle of the Slinky moves with respect to the Slinky as the disturbance passes.	
d. If you repeat the procedure described in part a, but this time make a larger abrupt sideways shake, does the disturbance move faster along the Slinky?	
e. If you make a smaller abrupt sideways shake than in part a, does the speed of the disturbance change? How do you know?	

20.1.3 Observe and find a pattern Suppose you stand in the water a meter or two from one end of a swimming pool.

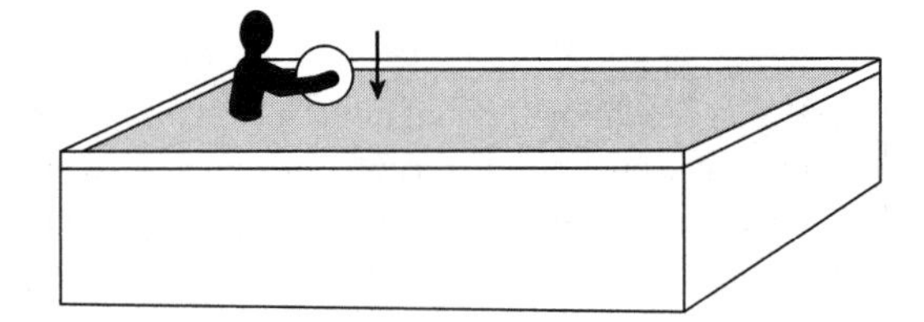

a. You push down hard once on a large beach ball floating on the water in front of you. What do you observe?

b. How can you estimate the speed at which the disturbance moves away from the ball?

c. Now you push down on the ball a few times (do not push too many times). Does the speed at which the crests of the waves move away from you depend on how frequently you push the ball up and down? Explain.

d. Draw a top view of the wave-crest pattern that you would see in the water at one instant of time. How does the distance between adjacent crests differ if you bob the ball up and down more frequently or less frequently? Explain.

20.1.4 Reason and explain Small balls of mass m are connected with small springs of spring constant k, as shown in the figure. The balls and springs rest on a smooth, frictionless surface. Imagine that you vibrate one end of this chain of balls and springs back and forth, parallel to the axis of the chain, causing a wave disturbance that moves along the spring–mass chain at some speed v.

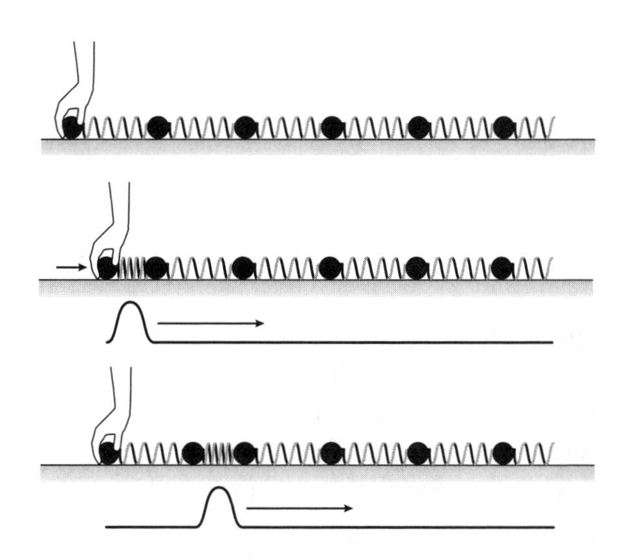

a. Do you think the speed of the waves along the chain depends on the spring constant k? If so, do you think the speed is greater or less for greater spring constants? Explain.

b. Do you think the speed of the waves along the chain depends on the mass m of the balls? If so, do you think the speed is greater or less for greater mass? Explain.

c. By analogy, identify two properties of stretched strings (for example, a violin, guitar, or piano string) that might affect the speed of waves on the strings.

20.1.5 Design an experiment Design an experiment to determine if a transverse pulse or a longitudinal pulse moves with greater speed along a Slinky.

Sketch the experimental setup.	**Explain in words how you will measure the speed for each kind of pulse.**	**List quantities that you will measure and quantities that you will calculate. List your assumptions and experimental uncertainties.**	**Perform the experiment; record the results and describe your conclusion.**
		To be measured: To be calculated: Assumptions: Uncertainties:	

20.1.6 Observe and explain Each of the two waves depicted in the table represents a disturbance of a medium at one instant of time. The disturbance travels away from the source of the disturbance. For each wave, construct a graph that shows along the vertical axis how the medium is disturbed at that instant of time at different positions x along the horizontal axis. Be sure to label the vertical axes with the names of the quantities that you are plotting on those axes.

Wave on string	**Sound wave in air**
Picture of the disturbance at one instant in time	Picture of the disturbance at one instant in time
Graphical representation of the disturbance y vs. x	Graphical representation of the disturbance ρ vs. x

20.1.7 Observe and explain Consider a beach ball bobbing up and down in the center of a swimming pool. Imagine that the ball remains at the same position at the center of the pool. The illustration shows several consecutive wave crests at one instant of time. Suppose that observer *A* is moving toward the source and observer *B* is moving away from the source. Does *A* or *B* observe higher frequency water wave vibrations, or do they observe the same frequency vibrations? Explain.

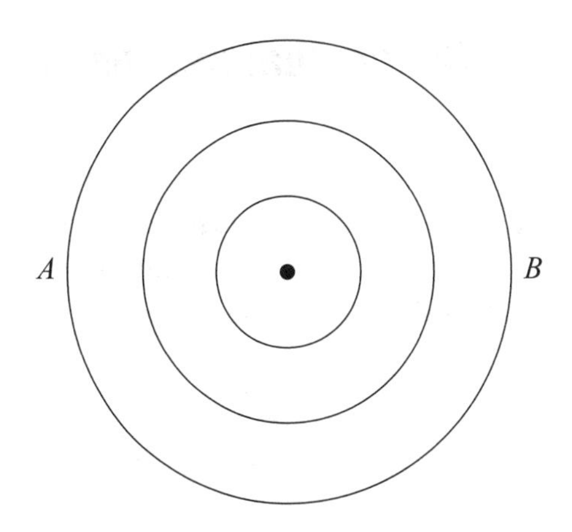

20.1.8 Observe and explain Four wave pulses produced by a large beach ball bobbing up and down in a pool are shown in the figure as the ball moves to the right. Wave-crest 1 of large radius was created when the ball was at position 1, and the wave-crest 4 of small radius was created when it was at position 4. Explain why observer *B*, standing stationary in the water in the direction the source moves, feels higher frequency water wave pulses than observer *A*, standing stationary behind the moving wave source.

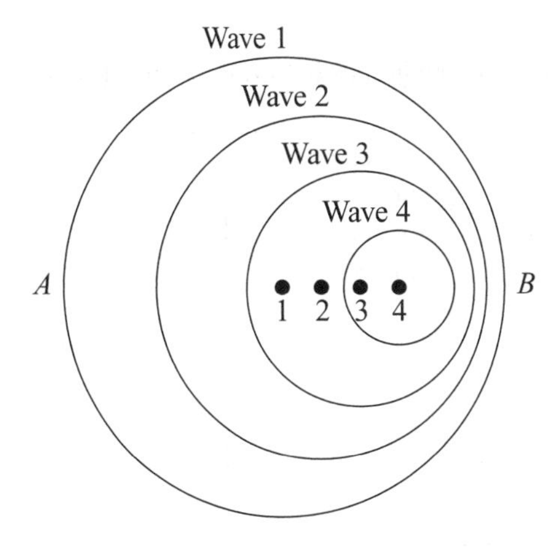

20.1.9 Describe and explain Look for a pattern in the two experiments described next.

Experiment I: You stand beside a train track; a train blowing its whistle moves toward you, passes you, and then moves away. The sound from the whistle changes from a higher pitch (higher frequency) as it moves toward you to a lower pitch as it moves away.

Experiment II: Your professor swings a ball tied to a rope in a horizontal circular path. A whistle inside the ball makes a higher-pitched sound as the ball moves toward you and a lower-pitched sound as it moves away. Devise a qualitative explanation for these observations using the ideas you developed in Activity 20.1.8.

20.2 Conceptual Reasoning

20.2.1 Represent and reason A longitudinal wave of amplitude 3.0 cm, frequency 2.0 Hz, and speed 3.0 m/s travels on an infinitely long Slinky. Displacement y is the distance that some part of the Slinky is displaced from its equilibrium position.

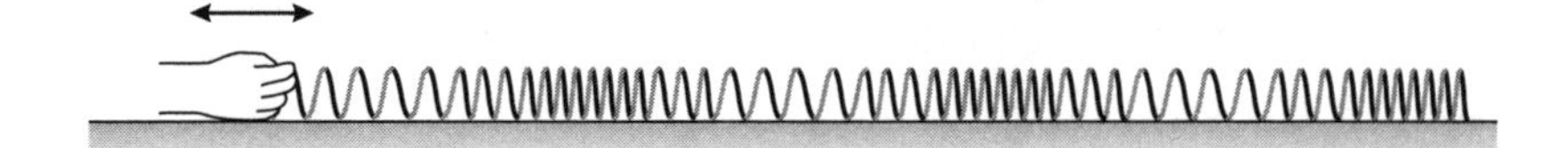

a. How far apart are the two nearest points on the Slinky that at one particular time both have the maximum displacements from their equilibrium positions? Explain your reasoning.

b. Complete the graphs below. Position x is a point along the axis of the Slinky. Be sure to put scales on the graphs.

Construct a displacement-versus-time graph for one coil of the Slinky. Show the period T of the wave on the graph.	y / t	Construct a displacement-versus-position graph for a segment of the infinitely long Slinky. Show the wavelength of the wave on the graph.	y / x

20.2.2 Reason The frequency f of a wave equals $1/T$.

a. Explain why this makes sense.

b. Suppose that there are 10 vibrations in 5 s. What is the frequency of such a wave, and what is its period?

c. If the wave travels at speed 4.0 m/s, how far will it travel during one period?

d. Show that $\lambda = v/f$.

20.2.3 Reason The speed of a wave depends on properties of the medium through which the wave travels. The speed v can also be determined in a different way—if you know the wavelength λ of the wave and the period T of the vibration, then $v = \lambda/T$.

a. Explain why this equation makes sense.

b. Is the equation $v = \lambda/T$ an operational definition or a cause-effect relationship for the wave speed? Explain.

20.2.4 Reason The graph describes the varying air pressure against a microphone as a sound wave passes. The readings are given with respect to the atmospheric pressure when the sound is not present. The negative pressure means that the pressure at a particular time is less than atmospheric pressure. *Note:* Sound travels at about 340 m/s. Note that 1 Pa (pascal) $= 1\ \text{N/m}^2$.

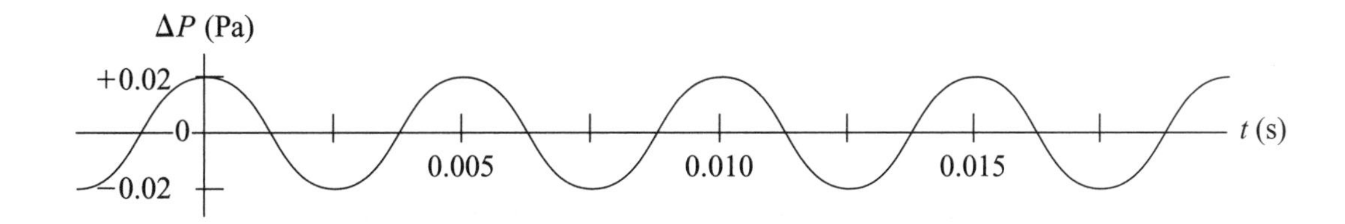

a. Determine the amplitude of the wave.

b. Determine the frequency of the wave.

c. Determine the wavelength of the wave.

20.2.5 Reason The graph describes the varying air pressure at different positions in space at one particular time due to a sound wave. Again, remember that sound travels at about 340 m/s.

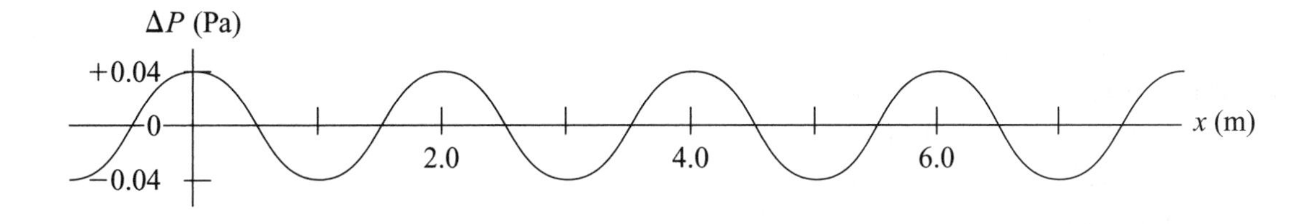

a. Determine the amplitude of the wave.

b. Determine the wavelength of the wave.

c. Determine the frequency of the wave.

20.2.6 Represent and reason Two waves are shown in the illustration that represent pressure variation-versus-time; they could, for example, be the pressure variations caused by two different sound waves at a microphone.

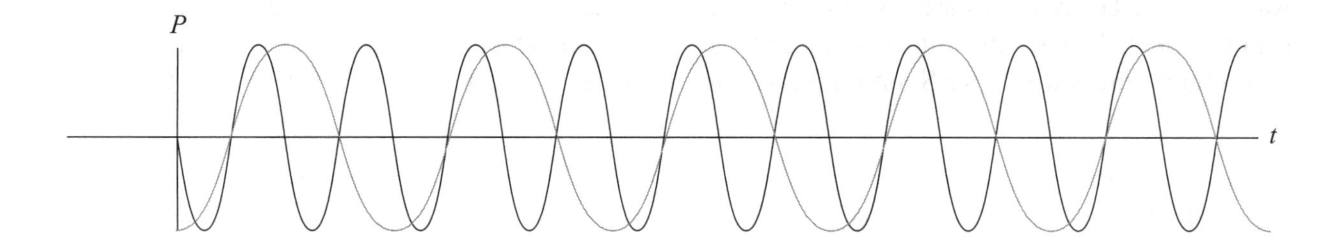

Construct the complex wave that would be produced if both waves were present at the same location during that same time interval.

20.3 Quantitative Concept Building and Testing

20.3.1 Describe and explain A periodic wave disturbance at one particular time (call it $t = 0$) is represented by the graph. In a way the graph is a snapshot of the wave. Your physics major friend claims that the equation that follows describes this periodic wave disturbance at different positions x at different times t:

$$y = A \cos 2\pi(t/T - x/\lambda)$$

where y is the disturbance at time t of the medium at position x (the distance from the source). Answer the following questions to try to disprove or support her claim. Note that A is the amplitude of the wave, T is its period and λ is its wavelength.

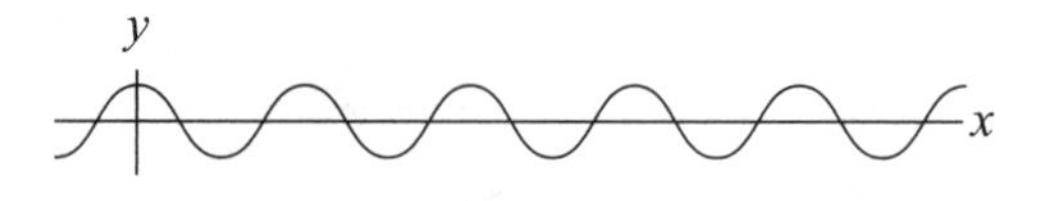

a. At $t = 0$ and $x = 0$, this particular wave disturbance has a value $y = A \cos 0 = A$, which matches what we see in the figure. At the same time ($t = 0$), what will the value y of the disturbance be at one wavelength forward? What will it be at two wavelengths forward? Three wavelengths forward? Does the equation give you the desired value? Explain.

b. At $t = 0$, what will the value y of the wave disturbance be at positions $x = \lambda/2, 3\lambda/2, 5\lambda/2$, and so forth? Does the equation give you the desired values? Explain.

c. At $t = 0$, what will the value y of the wave disturbance be at positions $x = \lambda/4, 3\lambda/4, 5\lambda/4$, and so forth? Does the equation give you the desired value? Explain.

d. At $t = T$, what will the value of y be at positions $x = 0, \lambda/4, \lambda/2, 3\lambda/4$, and λ? Does the equation give you the desired values? Explain.

e. Does the mathematical description seem appropriate based on your analysis? Explain.

20.3.2 Reason The equations that follow describe the variation of pressure at different positions and times (relative to atmospheric pressure) caused by sound waves. Fill in the table that follows. *Note:* The speed of sound in air is 340 m/s.

Equation $\Delta P = (2.0\ \mathrm{N/m^2})\cos 2\pi[t/(0.010\ \mathrm{s}) - x/(3.4\ \mathrm{m})]$			
Identify the amplitude of the pressure variation.	Identify the period for one vibration.	Determine the frequency of the sound.	Identify the wavelength of the sound.
Equation $\Delta P = (4.0\ \mathrm{N/m^2})\cos 2\pi[t/(0.0010\ \mathrm{s}) - x/(0.34\ \mathrm{m})]$			
Identify the amplitude of the pressure variation.	Identify the period for one vibration.	Determine the frequency of the sound.	Identify the wavelength of the sound.

20.3.3 Observe and explain Imagine that you have three long springs. If you measure the speed of wave pulses along the springs, you would accumulate the data given in the table.

Spring number	Force exerted on the end of the spring (tension, N)	Amplitude (cm)	Frequency (Hz)	Mass/length (kg/m)	Speed (m/s)
1	4.0	10	2	0.16	5.0
1	8.0	10	2	0.16	7.1
1	16.0	10	2	0.16	10.0
1	4.0	10	2	0.16	5.0
1	4.0	20	2	0.16	5.0
1	4.0	30	2	0.16	5.0
1	4.0	10	2	0.16	5.0
1	4.0	10	3	0.16	5.0
1	4.0	10	4	0.16	5.0
1	4.0	10	2	0.16	5.0
2	4.0	10	2	0.080	7.1
3	4.0	10	2	0.040	10.0

Come up with an expression that can be used to determine the speed of the wave as a function of different properties of the springs.

20.3.4 Describe

a. Write an equation that describes the graph line shown at the right. Note that v is the speed of a wave on a string, F is the magnitude of the force exerted on the end of the string (equal to the tension in the string), and μ is the mass per unit length of the string. Is this equation consistent with the data in Activity 20.3.3?

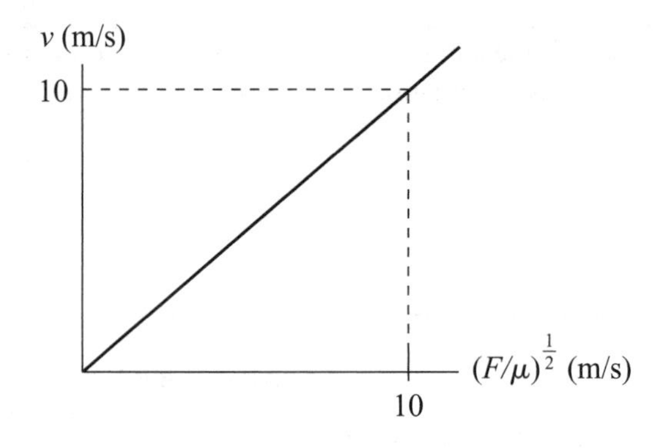

b. Determine the speed of a wave pulse on a 0.080-kg/m string when a person pulls exerting a 2.0-N force on the end of the spring.

20.3.5 Observe and find a pattern Tie one end of a rope (or one end of a long, tightly wound spring) securely to a post. Hold the other end in your hand and vibrate it up and down at different frequencies. At most frequencies, the rope responds little to your efforts. However, at special frequencies, big amplitude vibrations occur (the figure that follows shows four of these vibrations). If you videotape the process and view the video frame by frame, you will find that the period of the second vibration is half the period of the first, the period of the third is one-third of the period of the first, and so forth.

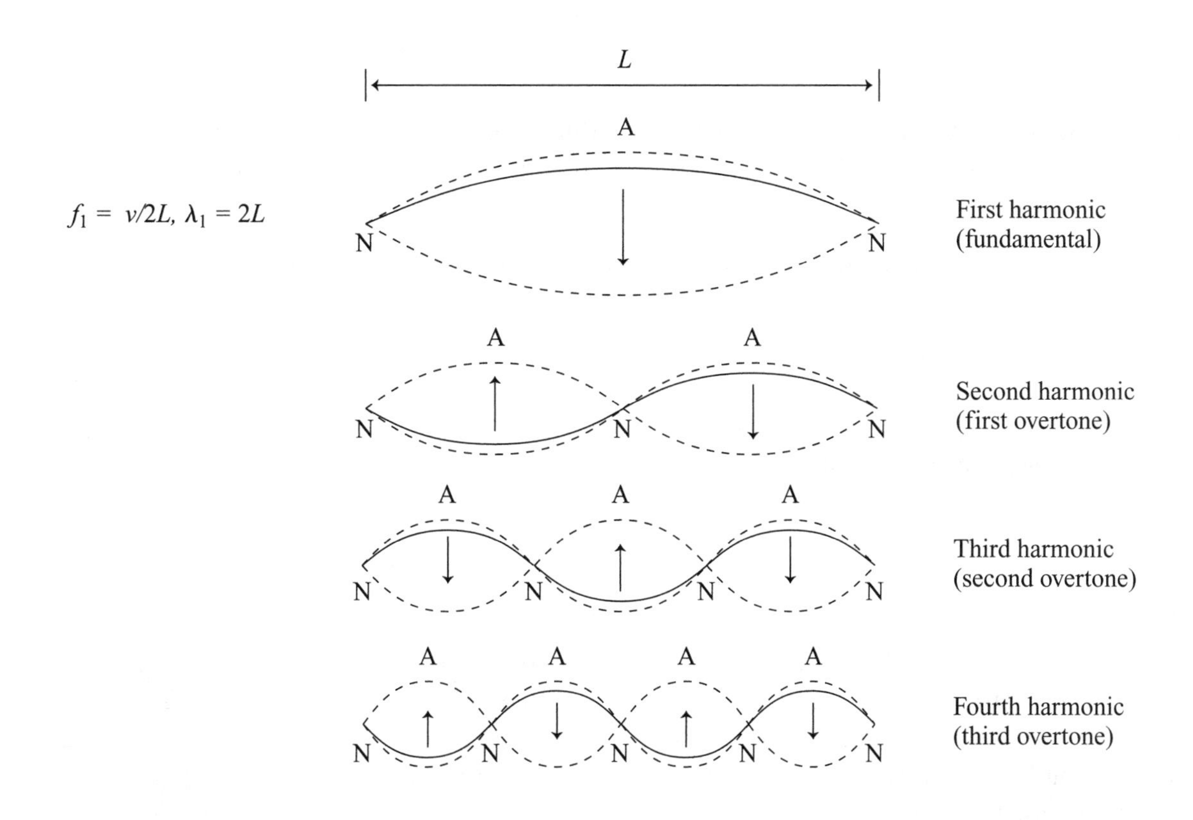

a. Write expressions for the frequency of the other standing-wave vibrations; there are many more possible than are shown in the figure.

b. Write expressions for the wavelengths of the observed and of other standing-wave vibrations.

20.3.6 Test your idea Pluck the A string on a violin (or another stringed instrument). Now tune an electronic oscillator attached to a sound speaker, and you will find that this string vibrates at 440 Hz (called concert A). The string is 0.33 m long.

a. Find the speed of a pulse on the string.

b. Use the information in part a and what you learned in Activity 20.3.5 to predict the frequency of the string's vibration if you press your finger against the fingerboard, thus effectively changing the length of the string to 0.22 m. You then pluck it again.

c. Check your prediction by plucking the string and matching the sound to that of the oscillator. Did your prediction match the outcome? (You can try the same type of experiment with other stringed instruments.)

20.3.7 Explain Wind instruments such as trumpets, flutes, clarinets, and the pipes in organs consist of columns of air inside tubes. They also have opening and closing valves (or slides, in the case of a trombone).

a. Explain how the valves and slides allow a musician to change the frequency of sound that these instruments produce.

b. Blow a whistle and try to change the frequency of sound it produces. How did you do it? Is the reason consistent with the explanation you provided in part a?

20.3.8 Test your idea We express the fundamental frequencies of vibration of a tube open at both ends (called an open tube) the same as the expression for a string $(f_n = nv/2L)$, where v is the speed of sound in the air inside the tube, L is the length of the tube, and n is an integer $(1, 2, 3\ldots)$. You have a whirly tube that is 0.86 m long and is open at both ends. When swung in the air, it produces a sound of a particular frequency. If swung faster, the frequency is higher. You can get about three distinct frequency sounds from a whirly tube.

a. Use the expression for the standing-wave frequencies in tubes to predict the frequencies of the whirly tube.

b. Check your predictions by whirling the tube while simultaneously tuning an electronic oscillator connected to a sound speaker to get matching frequencies. Compare what you measure with your predictions.

c. Do your predictions agree (with minor difference) with the measured value? If not, discuss the observed difference.

20.3.9 Reason and explain

a. We found in Activity 20.1.7 that the observed frequency is higher if the observer moves toward the stationary source and is lower if the observer moves away from the stationary source. Use the equation given earlier and sign conventions to show that these changes in frequency are consistent with the equation.

b. We observed and predicted in Activity 20.1.8 that the observed frequency is higher if the source moves toward the stationary observer and is lower if the source moves away from the stationary observer. Show that these changes in frequency are consistent with the equation.

20.4 Quantitative Reasoning

20.4.1 Represent and reason A traveling wave is represented in different ways in the table that follows.

Description in words	A tuning fork of frequency 100 Hz produces sound in the air. The sound travels at speed 340 m/s and causes a maximum pressure variation of 0.020 N/m^2. The wave is represented along one direction in other ways.
***Graph:* Pressure variation-versus-time for one position**	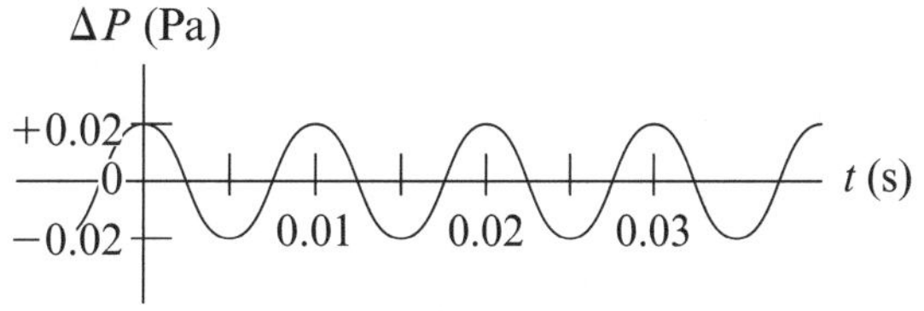 ΔP (Pa) +0.02 0 −0.02 0.01 0.02 0.03 t (s)
Mathematics:	$T = 0.01$ s
***Graph:* Pressure variation-versus-position graph at one instant in time**	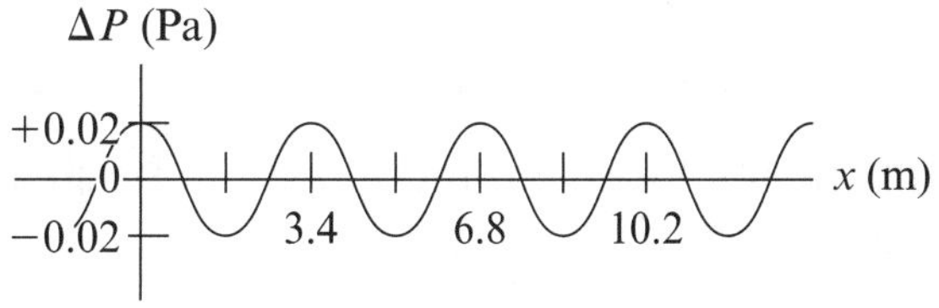 ΔP (Pa) +0.02 0 −0.02 3.4 6.8 10.2 x (m)
Mathematics:	$\lambda = 3.4$ m
***Mathematics:* Equation of displacement as a function of position and time**	$\Delta P = (0.020\ \text{N/m}^2)\cos 2\pi[t/(0.010\ \text{s}) - x/(3.4\ \text{m})]$

Are the representations consistent with each other? Explain.

20.4.2 Represent and reason Fill in the table that follows to describe in multiple ways a traveling wave on a very long string. The period of the wave is 0.20 s, the amplitude is 6.0 cm, and the wavelength is 6.0 cm. Make sure you use consistent units.

Construct a displacement-versus-time graph for one point on the string.		Determine the wave's frequency and speed.	
Construct a displacement-versus-position graph for one particular time.		Write an equation for the displacement as a function of position and time.	

20.4.3 Represent and reason Two ropes are the same length. The speed of a pulse on rope 1 is 1.4 times the speed on rope 2.

Write an expression for the ratio of the speeds (v_1/v_2) in terms of the ratios of the rope tensions (F_1/F_2) and of the rope masses (m_1/m_2). Use no numbers yet.	If the forces pulling on the ends of the rope (rope tensions) are the same, determine the ratio of their masses for the speed ratio given in the problem statement.	If the masses of the ropes are the same, determine the ratio of the forces pulling on the ends of the ropes—for the speed ratio given in the problem statement.

20.4.4 Represent and reason Read the descriptions of the situations described below and answer the questions that follow.

Situation I: A 0.50-m-long string vibrates in three segments, with a frequency of 240 Hz.

Situation II: A 0.68-m pipe is open at both ends. The speed of sound is 340 m/s.

a. What is the fundamental frequency of the string?

b. What is the speed of a wave on this string?

c. What is the fundamental frequency of the pipe?

d. What is the fundamental frequency of the pipe when one end is closed?

20.4.5 Evaluate the solution

A friend proposes a solution for the following problem.

The problem: A violin A string is 0.33 m long and has mass 0.30×10^{-3} kg. It vibrates at a fundamental frequency of 440 Hz (concert A). What is the tension in the string?

Proposed solution: Speed depends on the tension and string mass $(v = [T/m]^{\frac{1}{2}})$. Thus,

$$T = v^2 m = (340 \text{ m/s})^2 (0.30 \text{ g}) = 34{,}680 \text{ N}$$

a. Evaluate the solution and identify any errors.

b. Provide a corrected solution if you find errors.

20.4.6 Evaluate the solution

The problem: A shepherd blows on the end of a bone pipe (it is considered closed at one end and open at the other) that is 0.30 m long. She can play the first harmonic by blowing gently and higher harmonics by blowing harder. Determine the frequencies of these first three harmonics.

Proposed solution: The speed of sound in the solid bone material is about 3000 m/s and in air is about 340 m/s. Thus, the first three harmonic frequencies are:

$$f_1 = v/2L = (3000 \text{ m/s})/[2(0.30 \text{ m})] = 5000 \text{ Hz}$$
$$f_2 = 2v/2L = 2(3000 \text{ m/s})/[2(0.30 \text{ m})] = 10{,}000 \text{ Hz}$$
$$f_3 = 3v/2L = 3(3000 \text{ m/s})/[2(0.30 \text{ m})] = 15{,}000 \text{ Hz}$$

a. Identify any errors in the solution to the problem.

b. Provide a corrected solution if there are errors.

20.4.7 Regular problem You form a jug-and-bottle band where the musicians blow across the tops of the bottles (and the jug) to initiate sounds. Your band wants to play songs with notes ranging in frequency from 120 Hz to 300 Hz.

Determine the size of the jug or bottle to play the 120-Hz sound. Explain how you made your choice.		Determine the size of the jug or bottle to play the 300-Hz sound.	

20.4.8 Represent and reason How can you distinguish the sound of a violin and a flute both playing the same note—for example, concert A at 440 Hz? Most sounds are made up of a combination of the fundamental frequency of the vibrating object and a combination of higher harmonics. The quality of the sound depends in part on the number and relative amplitudes of these higher harmonics compared to the amplitude of the fundamental frequency. A rich violin sound may include 20 harmonics of A, and a flute sound may include only a few. The graph above represents the pressure disturbance-versus-position at one time of a sound wave of frequency 100 Hz.

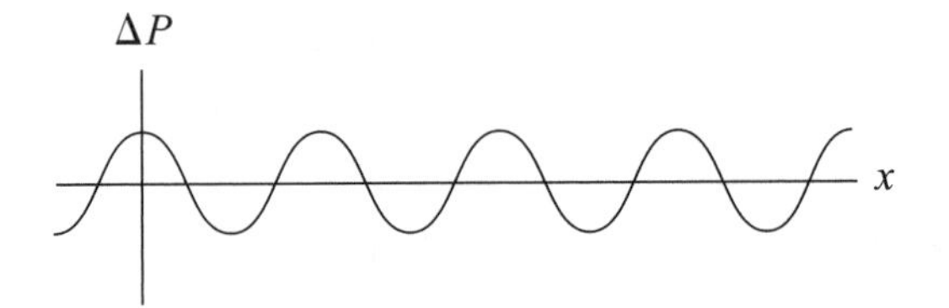

a. Draw another wave of frequency 200 Hz with half the amplitude.

b. Add the two harmonics together to construct a so-called complex wave made of the two harmonic waves.

20.4.9 Regular problem The Doppler effect can be used to determine the speed of red blood cells (as well as baseballs and cars). The Doppler speed detector emits sound at a particular frequency and detects the reflected sound at a different frequency. The difference in the emitted and detected sound frequencies indicates the speed of the object being measured. Assume that sound of frequency 100,000 Hz enters an artery opposite the direction of blood flow, which travels at speed 0.40 m/s. Answer the questions below to see how detecting the frequency of the sound reflected from a red blood cell indicates how fast it is moving.

a. Use the Doppler equation to determine the frequency that the cell would detect as it moves toward the sound source.

b. Suppose that the moving cell emits sound at the same frequency it detected in part a. What frequency does the Doppler detection system measure coming from the cell?

c. Often the Doppler detection system measures a beat frequency. The beat frequency is the magnitude of the difference between the emitted sound and the reflected sound that it received back from the moving blood cell. What beat frequency is observed in the case described above?

21 Reflection and Refraction

21.1 Qualitative Concept Building and Testing

21.1.1 Observe and explain Go to a room that is isolated from all external light sources—natural and artificial. Turn off the internal lights and wait in the dark room for several minutes. Record your observations and propose an explanation.

21.1.2 Observe and explain Place a laser pointer on a horizontal surface (say, a desk) in the center of a room and observe a bright spot on the wall toward which the laser points. Fill in the table that follows.

What path did light follow to reach the wall? You can find it by trial and error—by trying to block the light with a small piece of paper at several locations along its path to the wall.	
What can you say about the path of the light from the laser to the wall? Represent that light path by a long arrow, called a *ray*. A ray is not real; it is just a way to show the direction that light is traveling.	
Why can't you see the beam of light itself but you can see the bright spot on the wall or on a piece of paper that intersects with the beam? Write possible explanations.	
Now sprinkle chalk dust along the line of light propagation; you will see the beam of light in the air. Explain why the chalk dust made it possible to see the light beam.	
Discuss the conditions needed for us to see something.	

21.1.3 Test an idea Place a powered, frosted lightbulb on a table in the center of a dark room and observe that the walls are almost uniformly lighted. A friend draws two ray diagrams to try to explain this observation.

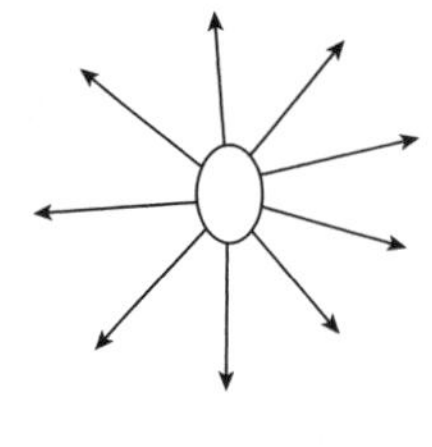
a.

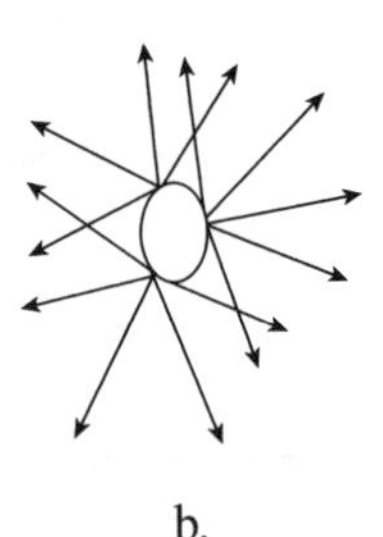
b.

a. Describe the main difference between the two diagrams. Consider how each point of the bulb emits light according to the diagrams.

b. Design an experiment to test which of the diagrams represents the way a lightbulb emits light—does each point emit one ray, or does each point emit rays in all directions? Describe the experiment with a picture and write a prediction based on each diagram.

c. Perform the experiment and decide which diagram did not predict the outcome. Which diagram will you use to represent how each point of the bulb emits light?

21.1.4 Observe and explain Place a powered, frosted lightbulb on a table in the center of a dark room. Take the point of a sharpened pencil and place it *close* to the wall. Record what you observe. Next move the pencil toward the bulb and away from the wall.

a. Represent what you observe with words, a sketch, and a ray diagram.

b. Discuss how your ray diagram uses ideas from Activity 21.1.3.

21.1.5 Observe and explain Place a candle on a table and set a piece of thin cardboard with an opening cut into it on the table between the candle and a nearby wall (see the figure). Move the cardboard closer to the wall and then move it closer to the candle. Observe the changes on the wall.

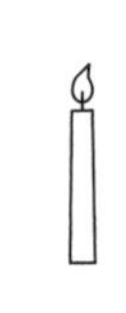
Candle

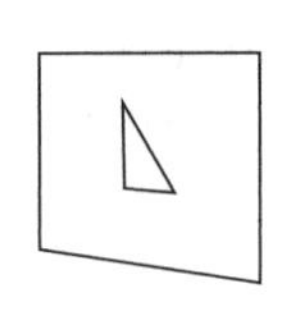
Cardboard

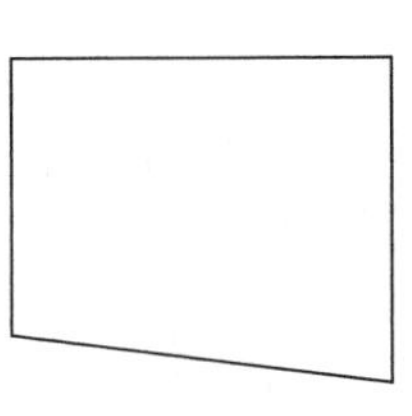
Wall

Positions of the cardboard	Draw ray diagrams to represent what you observe on the wall.
Cardboard close to the wall	
Cardboard in an intermediate position between the candle and the wall	
Cardboard close to the candle	

21.1.6 Test your ideas Imagine that you put a candle on a table and place a piece of thin cardboard between the candle and a nearby wall. Use the figure to draw a ray diagram to predict what you will see on the wall if you make a tiny hole in the cardboard. Then light the candle and turn off the room lights to observe the outcome of the experiment and revise your diagram if necessary.

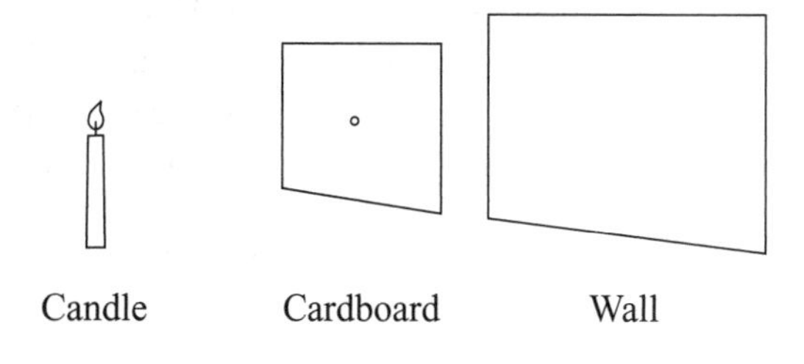

21.1.7 Observe and find a pattern Place a protractor on a tabletop and put the flat edge of the protractor against a mirror that is held upright and perpendicular to the tabletop and protractor. Shine a laser pointer across the protractor so that the beam hits slightly above the center zero point on the protractor. The reflected light returns across the protractor (see the figure). Then change the angle at which the laser ray hits the mirror and see if there is any pattern in the direction of the reflected light. Record your results in the table. Notice the normal line *CO* in the figure—this is a line perpendicular to the surface of the mirror at a point where the incident beam hits the mirror.

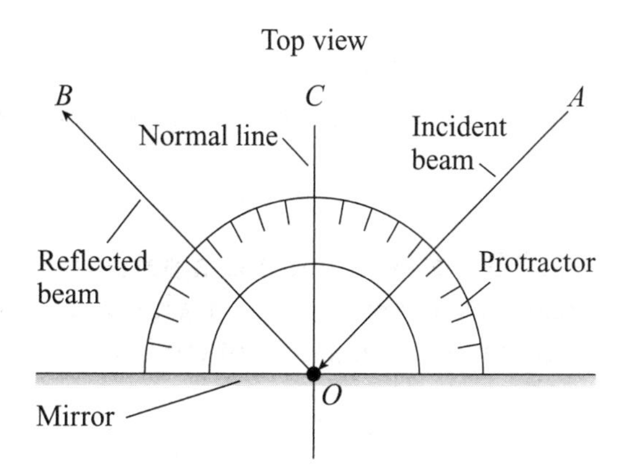

Angle *AOB* between the incident and reflected beams	Angle between the incident beam *AO* and the mirror	Angle between the reflected beam *BO* and the mirror	Angle between the incident beam *AO* and the normal line *CO*	Angle between the reflected beam *BO* and the normal line *CO*
	90°			
	70°			
	60°			
	45°			
	30°			
	10°			

Find a pattern in the data and express the pattern in words and mathematically.

21.1.8 Test your idea Assemble two mirrors on a flat surface so that their faces make a right angle, as shown in the figure. Place a target on the wall or on the other side of the table. Use any of the relationships that you found in Activity 21.1.7 to predict how you need to aim a laser beam to hit mirror 1 so that light reflected from mirror 1 then hits mirror 2 and finally hits the target.

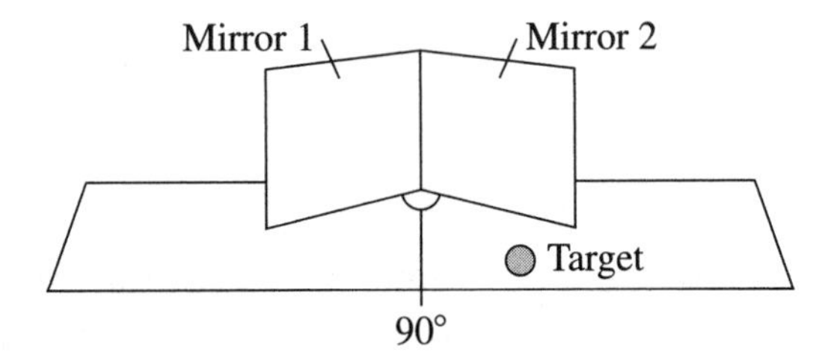

a. Draw a ray diagram to make a prediction.

b. Perform the experiment, record the results, and check to see if your prediction matches the outcome of the experiment.

c. Can you hit the target with the light that hits mirror 1 from a different direction? Explain.

21.1.9 Observe and find a pattern Shine the light from a laser beam on to the surface of a clear plastic container filled with water. The path of the light is shown in the illustration for three different incident angles of the laser beam.

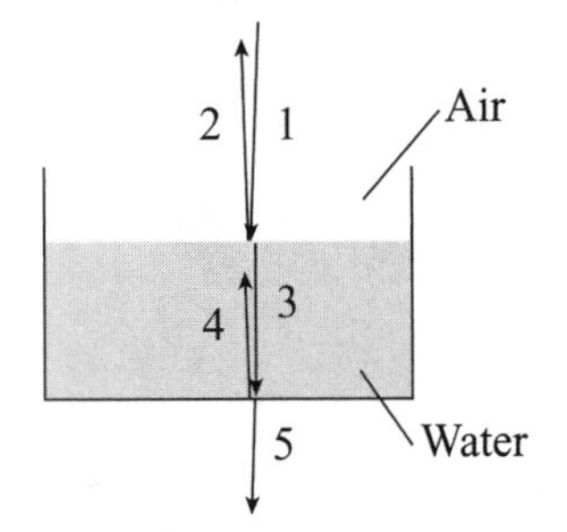

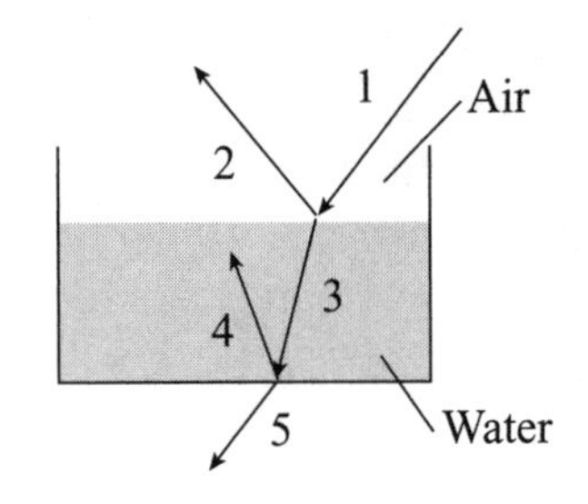

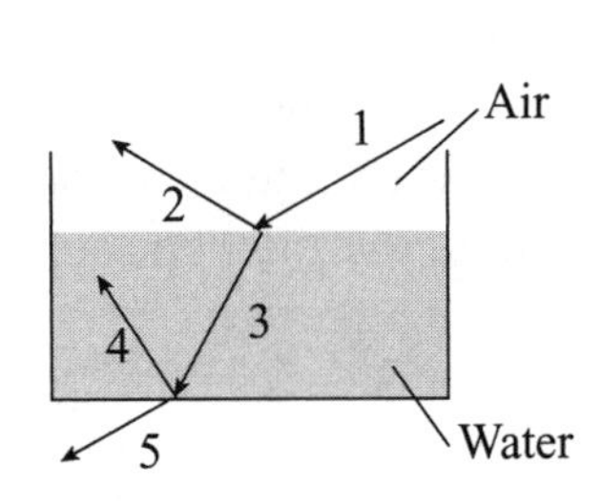

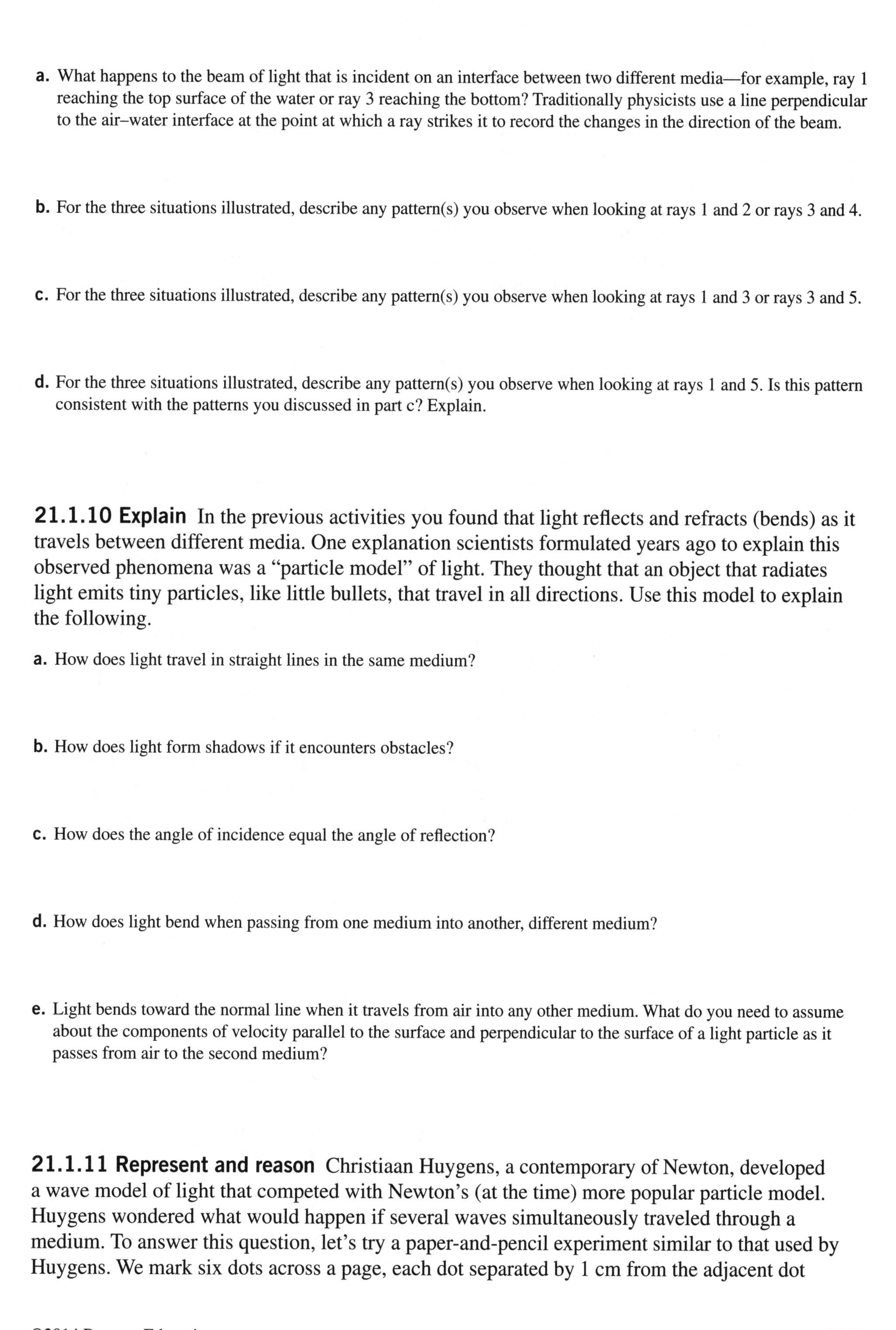

a. What happens to the beam of light that is incident on an interface between two different media—for example, ray 1 reaching the top surface of the water or ray 3 reaching the bottom? Traditionally physicists use a line perpendicular to the air–water interface at the point at which a ray strikes it to record the changes in the direction of the beam.

b. For the three situations illustrated, describe any pattern(s) you observe when looking at rays 1 and 2 or rays 3 and 4.

c. For the three situations illustrated, describe any pattern(s) you observe when looking at rays 1 and 3 or rays 3 and 5.

d. For the three situations illustrated, describe any pattern(s) you observe when looking at rays 1 and 5. Is this pattern consistent with the patterns you discussed in part c? Explain.

21.1.10 Explain In the previous activities you found that light reflects and refracts (bends) as it travels between different media. One explanation scientists formulated years ago to explain this observed phenomena was a "particle model" of light. They thought that an object that radiates light emits tiny particles, like little bullets, that travel in all directions. Use this model to explain the following.

a. How does light travel in straight lines in the same medium?

b. How does light form shadows if it encounters obstacles?

c. How does the angle of incidence equal the angle of reflection?

d. How does light bend when passing from one medium into another, different medium?

e. Light bends toward the normal line when it travels from air into any other medium. What do you need to assume about the components of velocity parallel to the surface and perpendicular to the surface of a light particle as it passes from air to the second medium?

21.1.11 Represent and reason Christiaan Huygens, a contemporary of Newton, developed a wave model of light that competed with Newton's (at the time) more popular particle model. Huygens wondered what would happen if several waves simultaneously traveled through a medium. To answer this question, let's try a paper-and-pencil experiment similar to that used by Huygens. We mark six dots across a page, each dot separated by 1 cm from the adjacent dot (see the illustration). The dots represent points on the crest of a wave moving toward the top of the page. According to Huygens, each dot is the source of a small wave disturbance that moves up the page in the direction the wave is traveling. In the figure the 3-cm-radius half circles, called *wavelets,* represent these disturbances. On the illustration provided, note places above the dots where the net disturbance from the six wavelets is two or more times bigger than the disturbance caused by any one wavelet—places where the wavelets add together to form bigger waves. This is the new crest of the wave. Draw a line on the sketch indicating the location of the new wave crest that was formerly at the position of the dots. Also, draw a ray indicating the direction the wave is traveling. The pattern is even clearer if you make many more dots and wavelets in the same space.

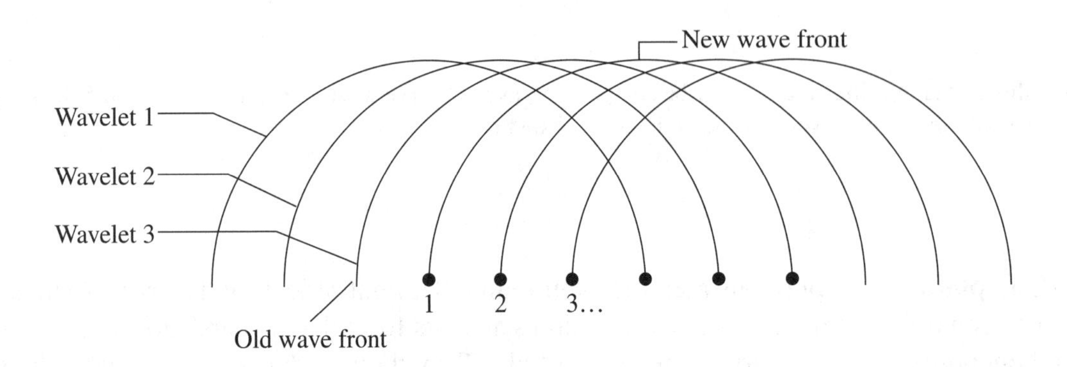

21.1.12 Represent and reason In Activity 21.1.11, semicircular wavelets all traveling at the same speed make up a new wave front. Waves often travel at different speeds in different places. For example, water waves travel slower in shallow water than in deeper water. Sound travels slower in cold air than in warm air. The difference in speed in different regions causes the wave to bend—to change direction. Huygens' principle can be used to understand this better. In the sketch below, the six dots are part of a wave crest (a wave front) moving toward the top of the page. The wave travels slower on the left side than on the right side. Thus, the wavelets originating from the left side have smaller radii than those farther to the right.

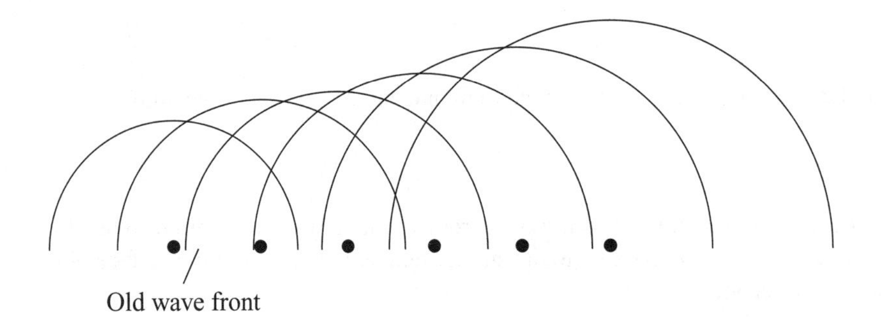

a. On the sketch, draw a new wave front that is produced by the wavelets leaving the positions of the six dots—that is, leaving the old wave front. Also, draw a ray that approximately indicates the wave's path as it moves up the page.

b. Based on this activity, which way do you think that waves tend to bend—toward regions in which they travel slower (the left side of the page) or regions in which they travel faster (the right side of the page)? Explain. (Remember your answer; you'll use this idea later.)

c. Based on your answer to part b and looking at the sketch in Activity 21.1.9, decide if light seems to travels faster or slower in water than in air. Explain your answer.

21.1.13 Represent and reason Imagine a wave whose wave fronts moving in one medium are incident on a boundary with another medium. The wave ray is not perpendicular to the boundary of the two media (see the illustration). The wave travels faster in medium 1 than in medium 2. During a certain time interval, the wavelet that earlier left the right edge of the wave front is just reaching the boundary between medium 1 and medium 2. The wavelet that left the lower left edge of the wave front at the same time travels less distance (i.e., moves more slowly) in medium 2.

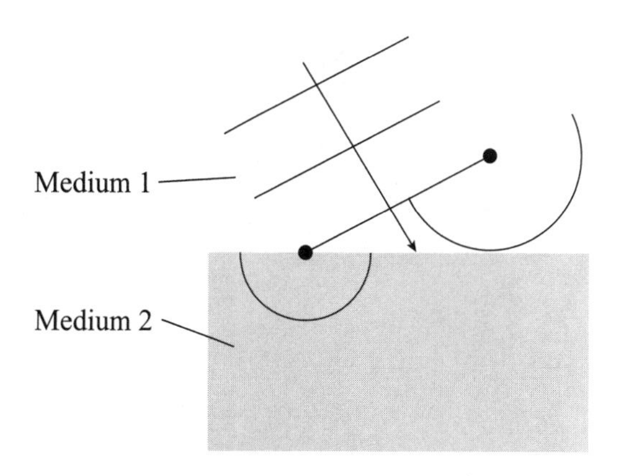

a. The wavelets leaving the middle of the wave front travel part of the time in faster medium 1 and part of the time in slower medium 2. Note that their radii should be longer than the wavelet on the lower-left edge but shorter than the wavelet on the upper-right edge. What is the orientation of the new wave front formed from these wavelets—now completely in medium 2? Draw a ray indicating the direction of the wave in medium 2.

b. Compare your sketch with the illustration in the Activity 21.1.9. Based on your analysis here and on that sketch, decide if light travels faster or slower in water than in air. Explain your answer.

21.2 Conceptual Reasoning

21.2.1 Represent and reason Draw ray diagrams to explain why the shadow of your standing body gets shorter as the Sun rises higher above the horizon. Assume that the Sun is infinitely far away and Earth receives only parallel rays of light.

21.2.2 Represent and reason Draw a ray diagram to determine the angle of the Sun relative to the horizon when the shadow of your body is the same length as your body.

21.2.3 Represent and reason Parts a and b of the figure below depict the path of light beams moving from air into water (the container holding water is made of glass with very thin walls, so that the boundary air/glass and water/glass can be disregarded) or glass and then out into air again. (The reflected beams are not shown.) Note the bending of the light at the interfaces between the two media. In particular, keep track of the light path relative to the normal lines that are perpendicular to the interfaces. Based on the patterns you observe in parts a and b, draw onto the figure for part c a normal line and a ray indicating the light path after it moves from the water to the air. For part d, draw normal lines and rays as the light moves from the air into the glass, through the glass, and out into the air on the other side.

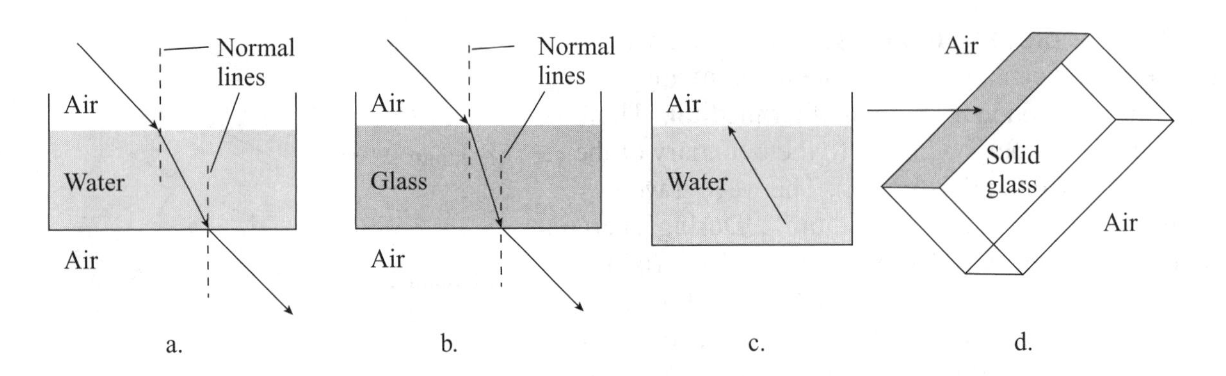

21.2.4 Represent and reason Imagine that you place a closed, empty glass box (with thin glass walls) filled with air underwater and hold it there. A beam of light shines on the top surface of the water, as represented by the ray in the illustration. Draw arrows on the illustration that indicate the beam's path from the top of the water out of the bottom of the container (the walls of the container are made of infinitely thin glass and can be disregarded).

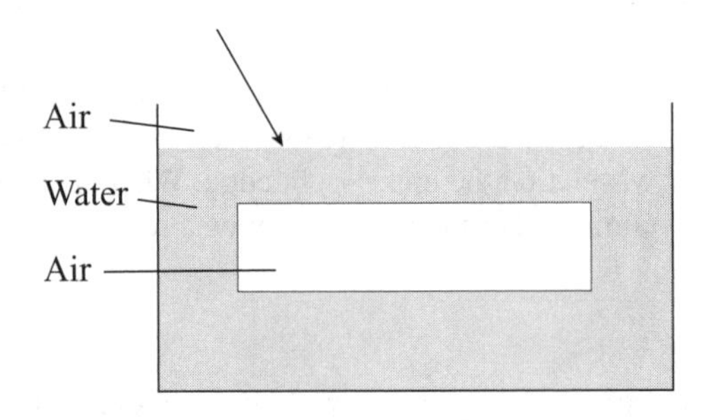

21.2.5 Represent and reason Rays in the illustrations represent beams of light from a laser pointer.

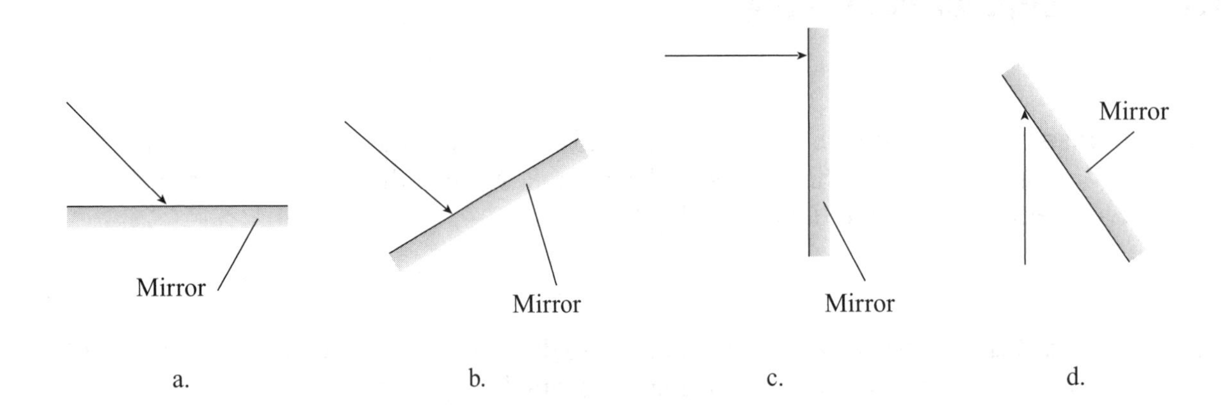

a. Draw on the illustrations reflected rays for each arrangement.

b. Draw on the illustrations two locations for each situation in which you could place a small piece of paper that would be illuminated by the reflected ray.

21.2.6 Explain In Activity 21.1.2 you learned that you only see an object if emitted or reflected light travels from the object to your eye. In Activity 21.1.8 you learned that when a laser beam is reflected off a smooth surface, the incident beam and the reflected beam are at the same angles relative to a line perpendicular to the surface. Imagine that a laser beam hits a wall. If you stand at any place in the room, you see a bright spot on the wall where the laser beam hits it. How can you reconcile these two phenomena? *Hint:* Examine the surface of the wall and compare it to the mirror.

21.3 Quantitative Concept Building and Testing

21.3.1 Observe and find a pattern Fill a fish tank with water and shine the light from a laser pointer at different angles on the top surface of the water. The angle θ_1 of the incident beam relative to the normal line and the angle of the beam that propagates into the water (the so-called refracted ray θ_2) are shown in the diagram and recorded in the table. Various trigonometric functions of those angles are recorded as well.

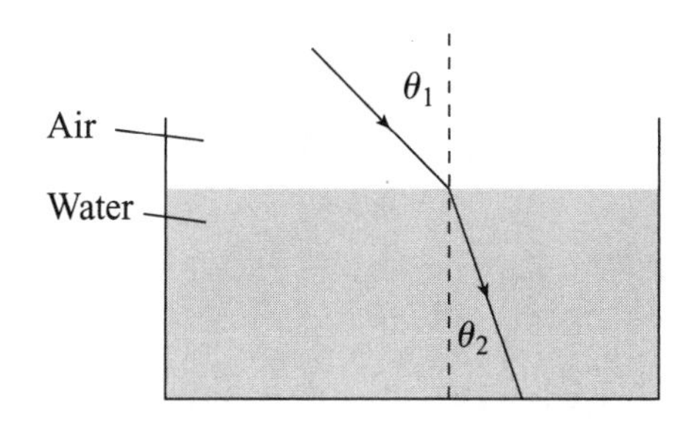

Incident angle θ_1	$\cos\theta_1$	$\sin\theta_1$	Refracted angle θ_2	$\cos\theta_2$	$\sin\theta_2$
20°	0.94	0.34	15°	0.97	0.26
30°	0.87	0.5	22°	0.93	0.37
40°	0.77	0.64	29°	0.87	0.48
50°	0.64	0.77	35°	0.82	0.57
60°	0.50	0.87	41°	0.76	0.65

a. Use any or all of the values given in the table to devise a rule that relates the angle of incidence and the angle of refraction. *Hint:* You might see if the ratio of two quantities has the same value for all angles. If so, use this to help devise a rule.

b. While performing the experiment, you see a dot of light on the ceiling. Explain.

21.3.2 Observe and find a pattern Repeat Activity 21.3.1, only this time imagine that you shine the light through air onto a block of glass with a smooth surface. The table records the angle between the incident light and the normal line and the angle between the refracted light and the normal line.

Incident angle θ_1	Refracted angle θ_2
20°	13°
30°	19°
40°	25°
50°	30°
60°	35°

a. Use the rule relating the angle of incidence and the angle of refraction that you devised in the previous activity, only this time apply it for light propagation from air into glass. Compare and contrast the air–glass refraction with the air–water refraction.

b. Use the wave model of light and the results from part a and Activity 21.3.1 to compare the speeds of light in air, water, and glass. Justify your answer.

21.3.3 Observe and explain Place a pencil in a glass that is half full of water. Observe the shape of the pencil. Draw a picture of what you observe and indicate where your eye is on that picture. Explain using the pattern that you found in Activity 21.3.1.

21.3.4 Test your idea We observed earlier in this chapter that light moving from air to glass or to a liquid bends (refracts) toward a normal line that is perpendicular to that surface. Also, light moving in a glass or liquid into air bends away from a line that is perpendicular to that surface. Use this idea to predict qualitatively what happens to a laser beam in each of the experiments below. *Hint:* Do not forget to draw a normal line at the location at which the light beam hits the border of the two media. Use a solid glass prism and a hollow glass prism to complete the table that follows.

Illustration of the experiment	Use your knowledge of refraction to predict qualitatively the path of the beam.	Perform the experiment and record the results (i.e., the path of the beam).	Discuss whether your prediction was successful or if the relationship needs to be modified.
Solid glass prism in air 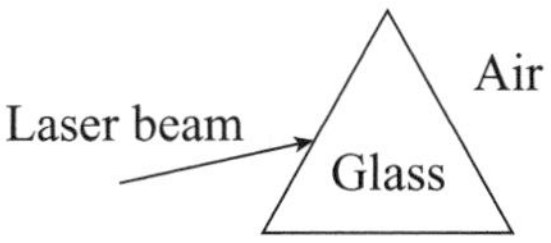 Laser beam / Air / Glass			
Hollow glass prism in water 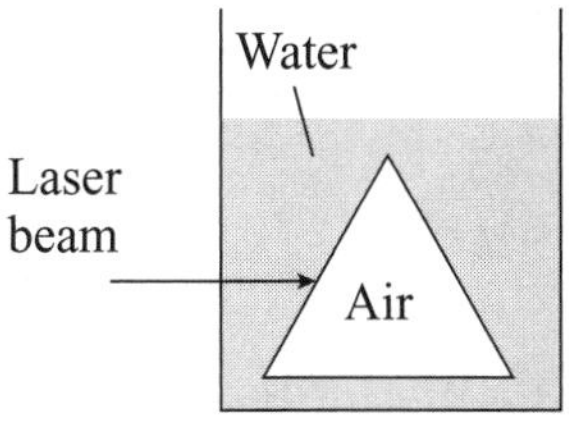 Water / Laser beam / Air			
Solid glass prism in water. Note that the light bends toward the perpendicular line going from water to glass, and vice versa in going from glass to water. 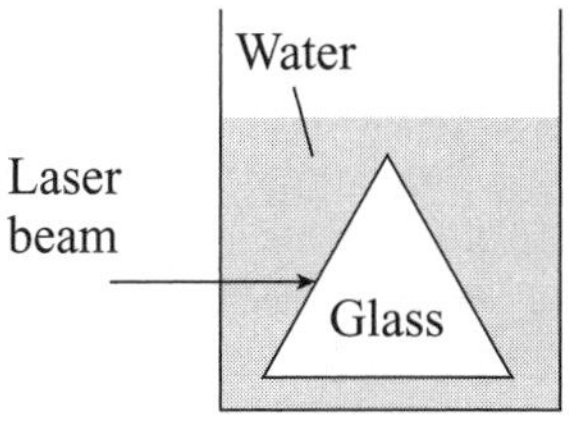 Water / Laser beam / Glass			

21.3.5 Test your ideas Use your knowledge of refraction to predict qualitatively and quantitatively what will happen in the described experiment. Complete the table that follows. Note that the index of refraction of air is 1.0 and of water is 1.33.

Shine a laser beam so that it passes through glass and then refracts into the air above. Vary the angle of incidence on the glass–air interface (e.g., see rays 1 and 2).	Predict what will happen if you gradually increase the angle of incidence. Identify a special angle of incidence where there is no longer a refracted ray. Explain your prediction.	After making the prediction, perform the experiment—did you correctly identify the special angle? What happens at incident angles greater than this special "critical" angle?
Air Glass 2 1		

21.4 Quantitative Reasoning

21.4.1 Represent and reason A beam of light hits a plane mirror perpendicular to the mirror's surface. Determine the angle between the incident and reflected beams if you tilt the mirror 30°. Include a sketch of the initial situation before tilting the mirror and the final situation after tilting it. Draw in labeled rays representing the incident and reflected beams for both orientations.

21.4.2 Represent and reason Two mirrors are placed together at a right angle, with one mirror oriented vertically and the other oriented horizontally. A ray strikes the horizontal mirror at an incident angle of 60° relative to the normal line, reflects from it, and then hits the vertical mirror.

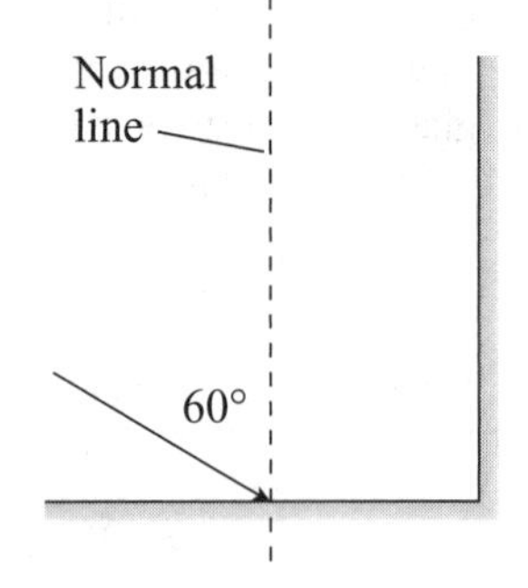

a. Determine the angle of incidence relative to the vertical mirror.

b. Use the law of reflection and the drawing to show that the ray leaves the vertical mirror parallel to its original direction.

21.4.3 Represent and reason Imagine that you shine a laser beam at a glass plate, as shown in the illustration. Use your knowledge of reflection and refraction to predict the path of the beam. Draw rays to indicate the path of the beam. (*Hint:* Before you draw the rays, decide on the direction of the normal line at the point at which the laser light first hits the glass.)

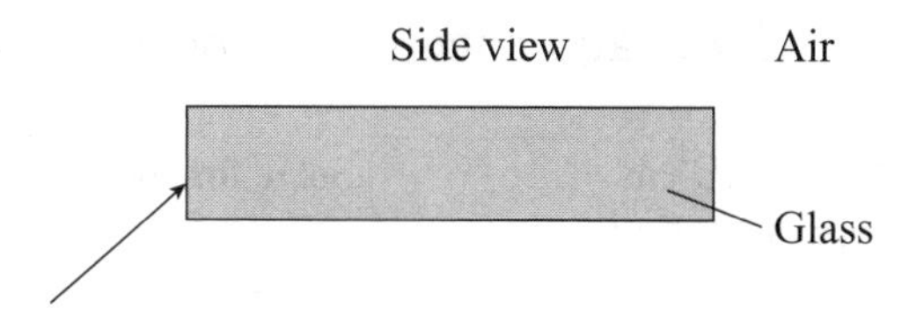

21.4.4 Represent and reason A laser beam shines up through a piece of glass of refractive index 1.56 and reaches the glass–air interface at the top, as shown in the figure.

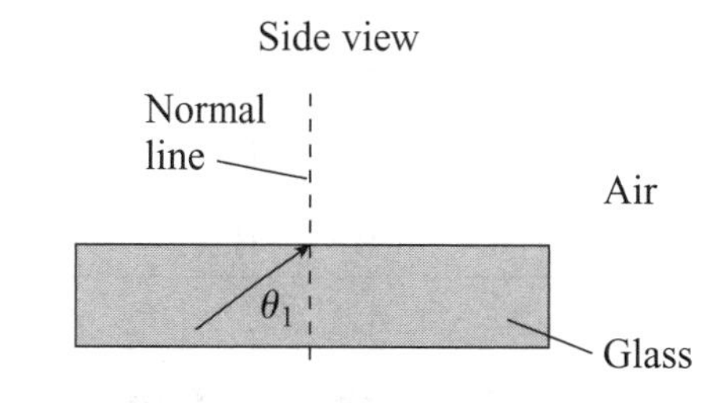

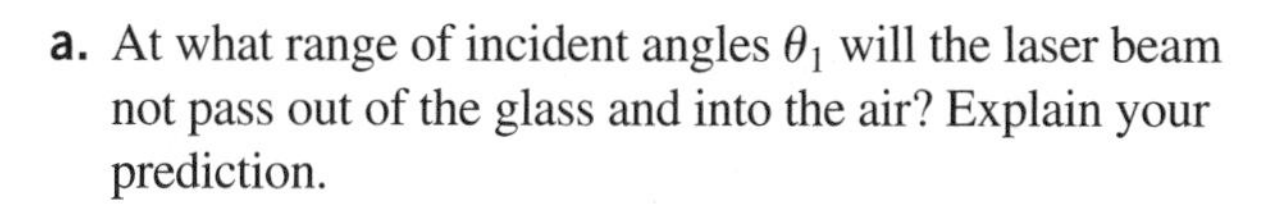

a. At what range of incident angles θ_1 will the laser beam not pass out of the glass and into the air? Explain your prediction.

b. Use your prediction in part a to explain how a glass rod (actually a thin glass fiber) can become a pipe that transmits light signals without light losses out the side walls of the fiber. Indicate any assumptions you made.

21.4.5 Pose a problem You have a block of light crown glass and a laser pointer. The index of refraction of light crown glass is 1.517. Pose a problem for which you need to use the knowledge of refraction and of total internal reflection to solve.

21.4.6 Represent and reason Light enters a right-angle prism as shown in the illustration and experiences total internal reflection. Draw the path of the light on the illustration. What is the minimum refractive index for the prism for total internal reflection to occur?

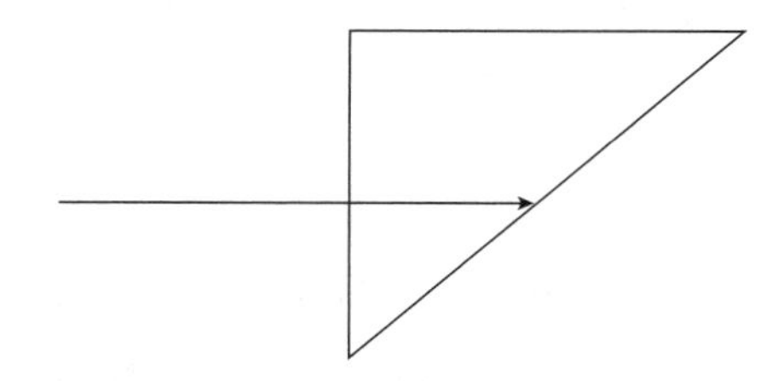

21.4.7 Equation Jeopardy Complete the table that follows.

Mathematical representation	Solve for the unknown(s).	Sketch the situation that the equation describes.	Write a word description of the process.
$1.00 \sin 53° = n_2 \sin 41°$			
$1.00 \sin 53° = 1.56 \sin \theta_2$ and $1.56 \sin \theta_2 = 1.00 \sin \theta_3$			

21.4.8 Evaluate the solution

The problem: The eyes of a person standing at the edge of a 1.2-m-deep swimming pool are 1.6 m above the surface of the water. The person sees a silver dollar at the bottom of the pool at a 37° angle below the horizontal. Determine the horizontal distance d from the person to the dollar.

Proposed solution:

Sketch and translate

The situation is sketched at the right.

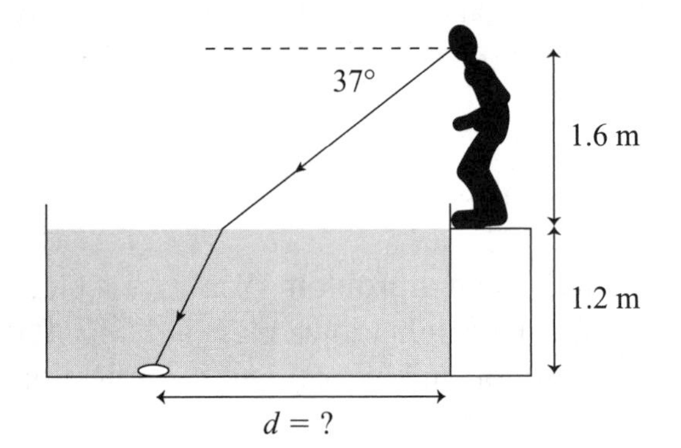

Simplify and diagram

The water surface is smooth, and the index of refraction of water is 1.33.

A ray from the eye to the coin is shown in the figure at the right.

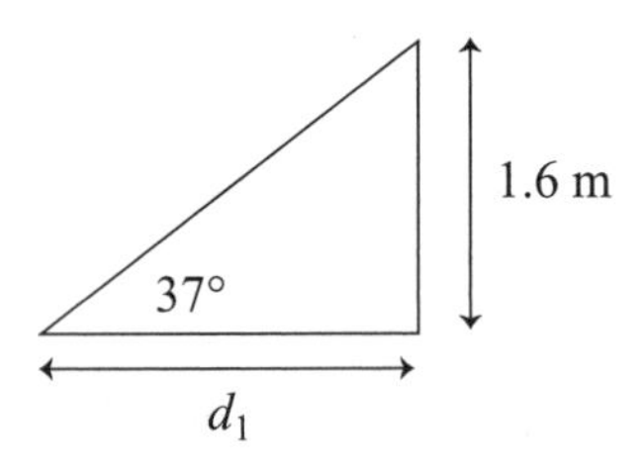

Represent mathematically, solve, and evaluate

The horizontal distance d_1 from the eye to the water surface where the ray enters is determined using trigonometry (see the triangle in the figure at the right).

$$\sin 37° = (1.6 \text{ m})/d_1 \quad \text{or} \quad d_1 = 2.67 \text{ m}$$

Apply Snell's law to find the angle of the ray while in the water:

$$1.00 \sin 37° = 1.33 \sin \theta_{\text{in water}} \quad \text{or} \quad \theta_{\text{in water}} = 26.9°$$

Finally, we can use the triangle in the water to determine the extra horizontal distance d_2 in the water:

$$\sin 26.9° = d_2/(1.2\text{ m}) \qquad \text{or} \qquad d_2 = 0.54\text{ m}$$

The total horizontal distance from the edge of the pool under the person's feet is:

$$d = d_1 + d_2 = 2.67\text{ m} + 0.54\text{ m} = 3.2\text{ m}$$

a. Identify any errors in the student solution.

b. Provide a corrected solution if you find errors.

21.4.9 Reason and explain Different colors of light have different indexes of refraction when passing though water droplets (see the table).

Color	Index of refraction
Red	1.613
Yellow	1.621
Green	1.628
Blue	1.636
Violet	1.661

How can this information qualitatively explain why we see a rainbow while it rains? Think of where the Sun is located with respect to an observer when the observer sees a rainbow. What assumption about the shape of the water droplets in the air can we make?

21.4.10 Design an experiment You have a semicircular piece of transparent material. Design two independent experiments to find the index of refraction of this material.

a. Fill in the table that follows.

Experiment 1			
Describe the experiment; draw a ray diagram to depict the experimental setup.	**Describe the procedure to determine *n*.**	**List experimental uncertainties.**	**Perform the experiment, record the outcomes, and calculate *n*.**
Experiment 2			
Describe the experiment; draw a ray diagram to depict the experimental setup.	**Describe the procedure to determine *n*.**	**List experimental uncertainties.**	**Perform the experiment, record the outcomes, and calculate *n*.**

b. Did the two experiments give you the same value of *n*? Explain the discrepancies.

21.4.11 Observe and explain

Use a glass microscope slide and carefully submerge it in vegetable oil. Complete the table that follows.

Experiment	Describe your observations and suggest an explanation.	Design and perform an experiment to compare the indexes of refraction of oil and glass.	Write an explanation of why we cannot see objects that have the same index of refraction as the medium in which light travels.
As you submerge the slide into oil, the submerged part of the slide disappears.			

22 Geometrical Optics

22.1 Qualitative Concept Building and Testing

22.1.1 Observe and explain Three friends stand behind a candle that is positioned 20 cm in front of a plane mirror. They observe the image of the candle, and each of them points a ruler in the direction of the image of the tip of the candle they see in the mirror. The dashed lines in the illustration indicate the orientations of their rulers.

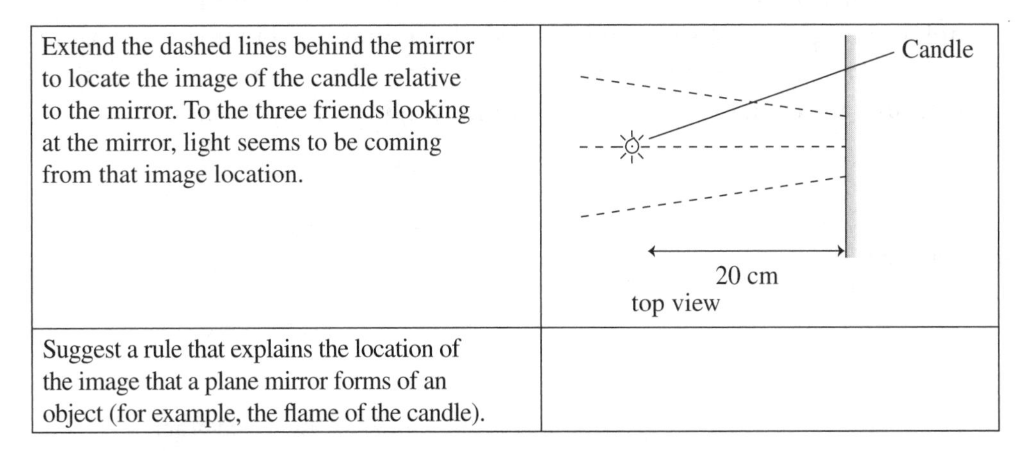

Extend the dashed lines behind the mirror to locate the image of the candle relative to the mirror. To the three friends looking at the mirror, light seems to be coming from that image location.	Candle 20 cm top view
Suggest a rule that explains the location of the image that a plane mirror forms of an object (for example, the flame of the candle).	

22.1.2 Reason and explain Use two arbitrary rays to explain why the image of the candle in Activity 22.1.1 is at the same distance behind the mirror as the candle is in front of the mirror. Complete the table that follows to help you.

Two rays beginning from a point on the candle and moving toward the mirror	1 2	The same rays after reflection from the mirror. Do the reflected rays shown above ever meet? If not, how does the mirror form an image?	1 2
Explain why we see an image of the candle behind the mirror.		Extend the reflected rays back behind the mirror to find the image of the candle. Use geometry to prove that the image is the same distance behind the mirror as the candle is in front.	

22.1.3 Test your idea Ari and Samantha are investigating the location of the image of a candle produced by a plane mirror. Ari says that the image is on the surface of the mirror. Test his idea by designing an experiment whose outcome contradicts the prediction based on Ari's idea.

Describe an experiment to test Ari's idea.	Predict the outcome of the experiment based on Ari's idea.	Perform the experiment and record the outcome.	Discuss whether the experiment disproves the idea that the image is on the surface of the mirror.

22.1.4 Observe and explain Shine a light from a laser pointer on a plane mirror and observe the reflected ray. Then shine it on a curved mirror, which is a section of a sphere of radius R (concave). For two examples, the light reflects as shown in the illustration. Explain the behavior in terms of the law of reflection. *Note:* The dashed line R perpendicular to the curved mirror's surface in the illustration passes through the center O of the sphere from which the mirror was cut.

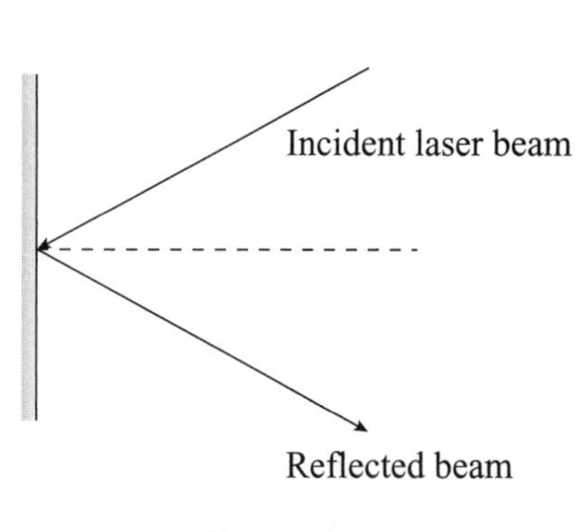

Plane mirror

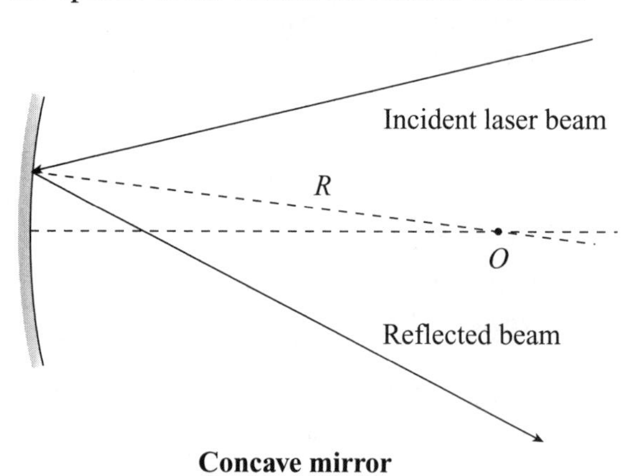

Concave mirror

22.1.5 Observe and explain Shine the light from a laser pointer on a curved mirror, which is a section of a sphere of radius R (convex). The light reflects as shown in the illustration. Explain the behavior in terms of the law of reflection.

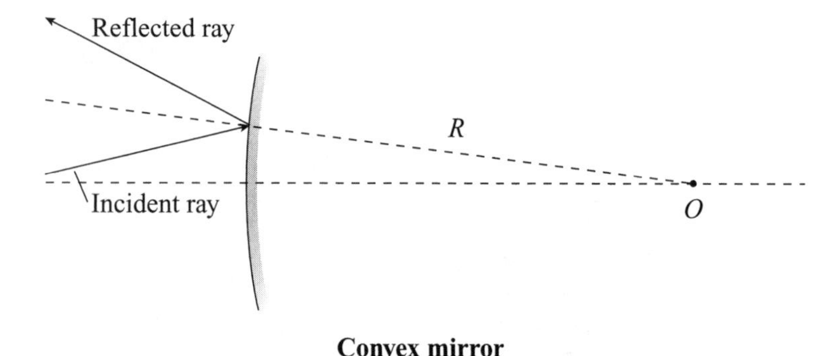

Convex mirror

22.1.6 Observe and explain Assemble a concave mirror and three handheld lasers. Point the beams of the lasers parallel to the main axis of the mirror (a horizontal axis through the center of the mirror). After reflection, the rays all pass through the same point exactly in the middle between the mirror and the center of the sphere from which the mirror was cut. This point is called the *focal point*—the point through which rays parallel to the axis of the concave mirror pass after reflection from the mirror. Use the law of reflection to explain this observation.

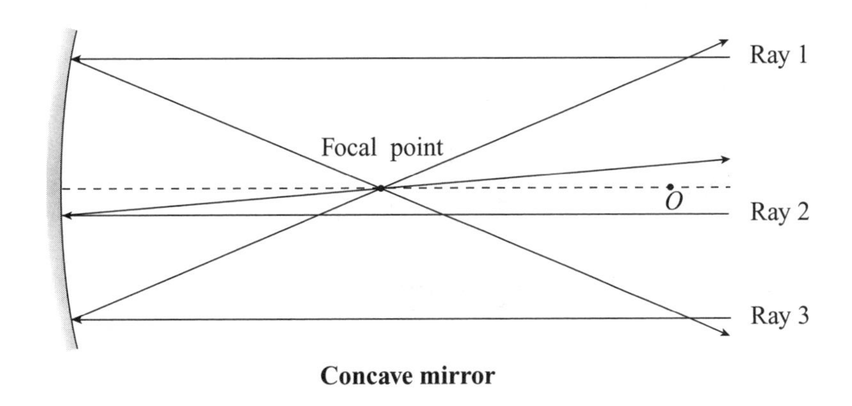

Concave mirror

22.1.7 Test your idea Assemble a concave mirror and two handheld lasers. Fill in the table that follows. *Hint:* Reviewing Activity 22.1.6 might help you make your prediction.

Draw a ray diagram to predict what will happen if you aim the beams of the lasers so that they both pass through the focal point before hitting the mirror.	Perform the experiment and record the outcome. Discuss whether the prediction was confirmed.	Compare and contrast this experiment with the experiment in Activity 22.1.6.

22.1.8 Observe and explain Assemble a concave mirror and three handheld lasers. Aim the laser beams toward the mirror parallel to each other but not parallel to the main axis of the mirror. One of the rays passes through the center of the sphere from which the mirror was cut (point O). This ray reflects back along the same path. After reflection, the other two rays pass through the point where the center-passing ray crosses the focal plane of the mirror.

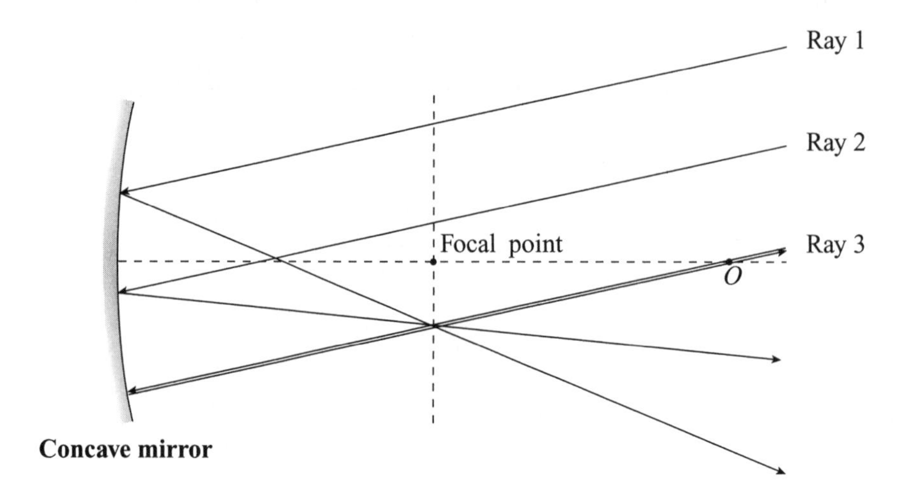

Use the law of reflection to explain this observation.

22.1.9 Test your idea

a. Predict using the law of reflection where the rays parallel to the main axis of a spherical convex mirror of radius R will meet after reflection or, alternatively, a point from which all of the reflected rays seem to originate.

b. Use several laser beams to perform the experiment. Record the outcome.

c. Does a convex mirror have a focal point? If so, where?

22.1.10 Observe and explain Shine three parallel beams of light from laser pointers on a thin convex lens made of glass (its surfaces are segments of a sphere). Refracted beams cross at a point on the main axis—called the *focal point.* Explain qualitatively the path of each ray using the law of refraction—what happens at each glass–air interface to cause the net refraction of the rays shown in the illustration?

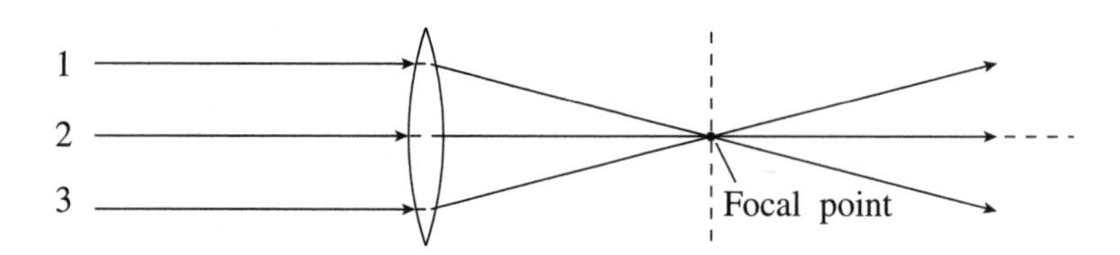

Notice that if you aim the rays from the right side toward the left side of the lens (opposite the way shown in the sketch), after refraction they will pass through another focal point on the left side of the lens.

22.1.11 Observe and explain Shine three parallel beams of laser light onto a thin concave lens made of glass (its surfaces are segments of a sphere, as shown).

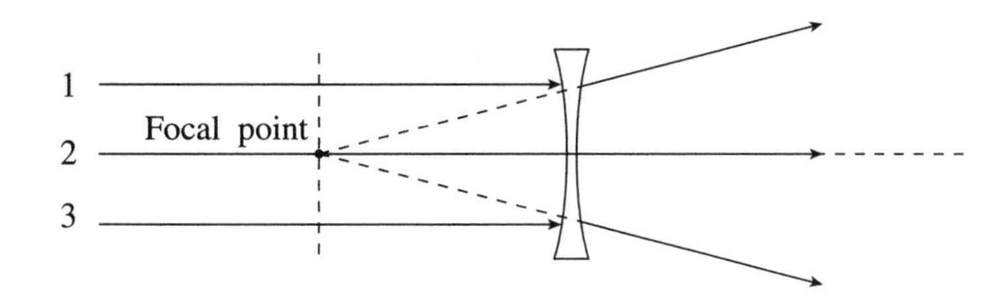

After passing through the lens, the rays diverge and appear to come from a focal point on the main axis. Explain qualitatively the path of each ray using the law of refraction.

22.2 Conceptual Reasoning

22.2.1 Represent and reason A candle burns in front of a plane mirror, as shown in the illustrations. Consider the flame to be a pointlike source of light. For each case, locate the flame's image by drawing any two rays on the illustration. (Rays can extend in any direction that strikes the mirror.)

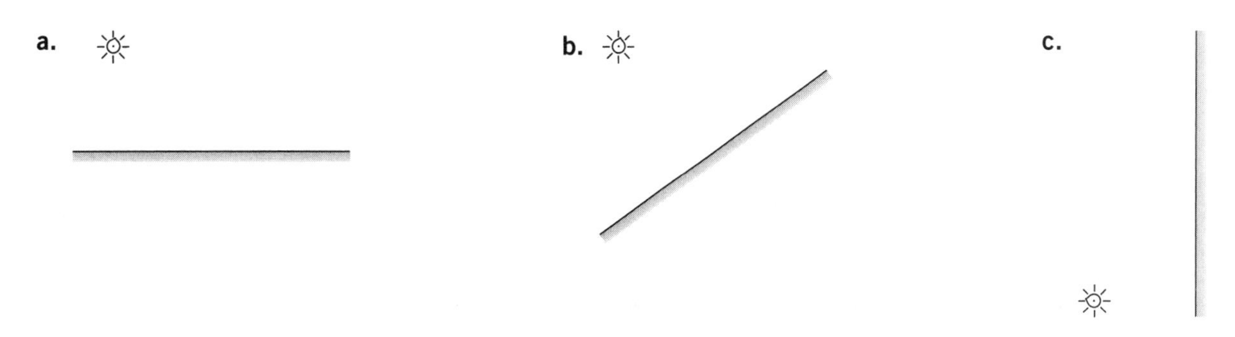

Then devise a rule that can be used to locate plane-mirror images without using rays.

22.2.2 Represent and reason Draw ray diagrams to answer the following questions: How does the size of a plane-mirror image change (increase or decrease) when the object is moved away from a plane mirror? How does the size change when the object is moved toward the plane mirror?

22.2.3 Represent and reason Imagine that you stand in front of a plane mirror to look at your image.

a. Draw a ray diagram to determine the minimum size of a mirror in which you can see your entire body.

b. Where should you put the top of the mirror relative to the top of your head?

22.2.4 Represent and reason Draw a ray diagram showing the path of each of the rays described below after reflection from a concave mirror. Remember that you can locate the image of an object formed by a concave mirror by using any two of these four rays:

a. a ray that is parallel to the main axis of the mirror,	**b.** a ray that passes though the focal point of the mirror,	**c.** a ray that passes through the center of the sphere from which the mirror was cut,	**d.** a ray that is parallel to the ray in part c.

22.2.5 Represent and reason A small, shining object is placed above the main axis of a concave mirror at a distance $s > R$ from the mirror. Two rays are used to find the image of the top of the object. Explain the path of each ray in the illustration.

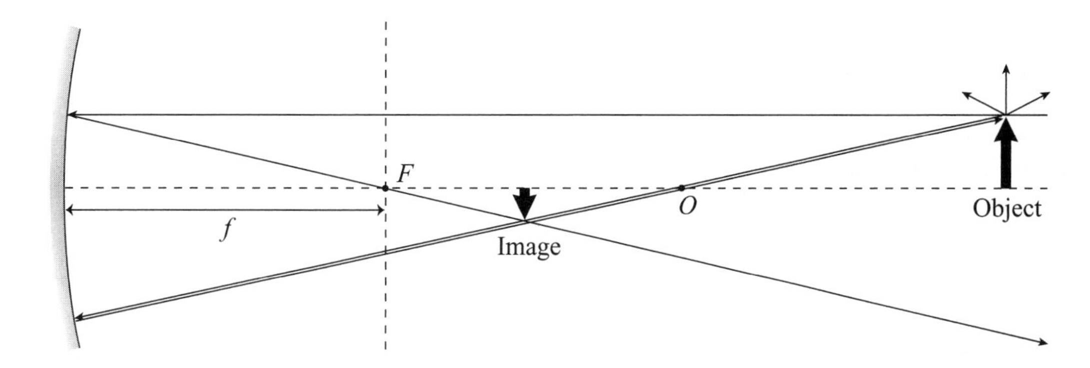

What assumptions were made in the diagram? Point O is the center of curvature of the mirror—that is, the center of the sphere from which the mirror was cut. The focal point is indicated by F, and the focal length is f.

22.2.6 Observe and explain Place a candle or small lightbulb in front of a concave mirror. Use a small white card between the light and the mirror to find a place where a sharp image of the candlelight is formed. Explain the results using a ray diagram.

22.2.7 Represent and reason A ray diagram helps us understand how to find the image of an object produced by a convex mirror. Explain the path of each ray and how we know where the image is located.

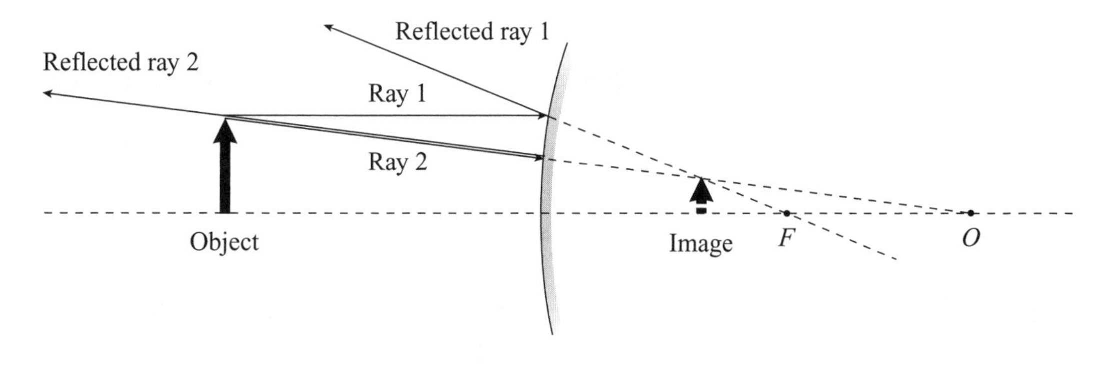

Why is the image drawn above with a dashed line? What assumptions were made in the diagram?

22.2.8 Represent and reason Fill in the table that follows.

Experiment	Sketch	Description
Place a small shining object at a distance s ($s > R$) from a concave mirror. Draw two rays to represent the situation—one parallel to the main axis and one going through the focal point. Draw the reflected rays and find the location of the image of the object.		Describe the image using adjectives: upright (inverted), real (virtual), enlarged (reduced).
Place a small shining object at a distance s ($f < s < R$) from a concave mirror. Draw two rays to represent the situation—one parallel to the main axis and one going through the focal point. Draw the reflected rays and find the location of the image of the object.		Describe the image using adjectives: upright (inverted), real (virtual), enlarged (reduced).
Place a small shining object at a distance s ($s < f$) from a concave mirror. Draw two rays to represent the situation—one parallel to the main axis and one reflected from the center of the mirror. Draw the reflected rays and find the location of the image of the object.		Describe the image using adjectives: upright (inverted), real (virtual), enlarged (reduced).

22.2.9 Represent and reason Use ray diagrams to determine how the size of an image changes (increases or decreases) when the object is moved away from a concave mirror.

22.2.10 Evaluate the reasoning Your friend Brian thinks that if you cover the bottom half of a concave mirror, the real image of a candle produced by the mirror will be cut in half. Do you agree or disagree? If you disagree, how can you convince Brian of your opinion?

22.2.11 Represent and reason Fill in the table that follows.

Place a small shining object at a distance s ($R/2 < s < R$) from a convex mirror. Use a ray diagram to find the location of the image. Describe the image using adjectives: upright/inverted, real/virtual, enlarged/reduced.	Place a small shining object at a distance s ($s < R/2$) from a convex mirror. Use a ray diagram to find the location of the image. Describe the image using adjectives: upright/inverted, real/virtual, enlarged/reduced.

22.2.12 Represent and reason Draw in any two rays to construct images of the objects shown as arrows in the figures below. Confirm by drawing in a third ray. Describe the images using adjectives: upright/inverted, real/virtual, enlarged/reduced.

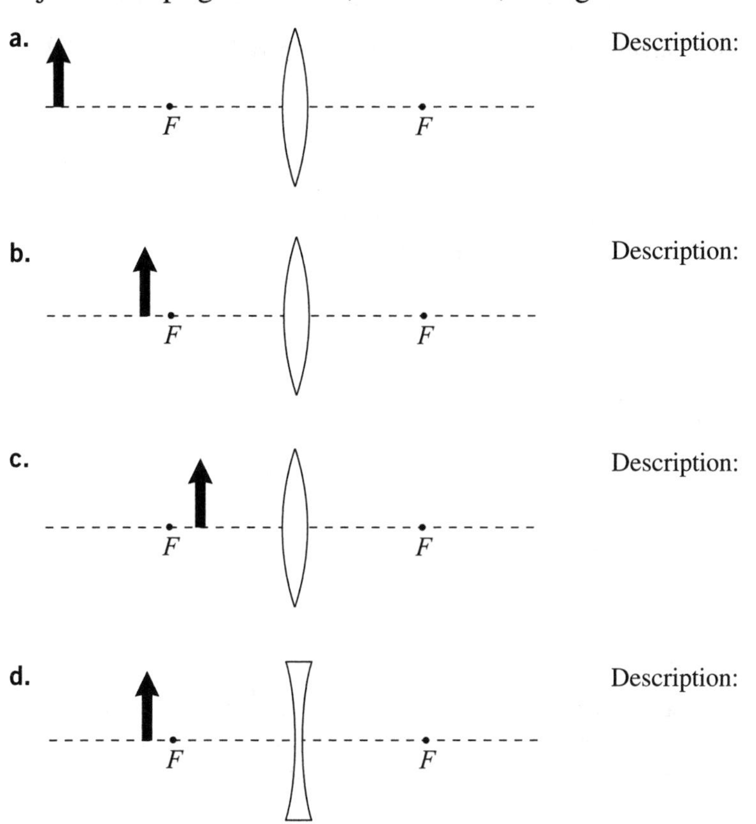

Description:

Description:

Description:

Description:

22.2.13 Evaluate the reasoning Your friend Ritesh says that it's appropriate to call a convex lens a converging lens and a concave lens a diverging lens. How would you convince him that his classification is not always correct?

22.3 Quantitative Concept Building and Testing

22.3.1 Test the mirror equation Assemble a concave mirror and a candle. Do not light the candle yet.

a. Place the candle behind the focal point away from the mirror. Do not light the candle yet. Fill in the table that follows. *Note:* Use a paper screen between the candle and the mirror to check your prediction.

Measure the distance s between the candle and the mirror. Draw a ray diagram and use the mirror equation to predict the distance s'.	**Light the candle, perform the experiment, and measure s'. Record the difference between your prediction and the measured value of s'.**

b. Explain the difference between the predicted and measured values of s' (think of experimental uncertainties and assumptions that we made deriving the mirror equation).

22.3.2 Test the lens equation You have a convex lens. Use a source of light that is far away (you can use the sun on a good day or a window in the classroom if it is far enough away—15–20 m) to find a focal length of the lens.

a. Fill in the table that follows.

Draw a ray diagram for the light source and the lens.	**Perform the experiment and find the focal length of the lens.**

b. Assemble the same lens and a candle. Place the candle behind the focal point and away from the lens, measure the distance between the candle and the lens, and use the lens equation to predict the location of the image of the candle. Use a paper screen to locate the image and hence to check your prediction. Fill in the table that follows.

Measure the distance s.	Predict the distance s'.	Measure the distance s'.	Record the difference between your prediction and the measured values of s'.

c. Explain the difference between predicted and measured values of s' (think of experimental uncertainties and assumptions that we made deriving the lens equation).

22.3.3 Test the lens equation Assemble a convex lens whose focal length is $+10$ cm.

a. Predict where you should hold the lens with respect to a white piece of cardboard placed on your desk to form a focused image of a window onto this piece of cardboard. What mathematical model did you use to make a prediction?

b. Check your prediction and record the result. What is your judgment about the mathematical model that you used to make the prediction?

22.4 Quantitative Reasoning

22.4.1 Regular problem Use ray diagrams and the mirror equation to locate the position, orientation, and type of image formed by an upright object held in front of a concave mirror of focal length $+20$ cm. The object distances are (a) 200 cm, (b) 40 cm, and (c) 10 cm.

22.4.2 Regular problem A large concave mirror of focal length $+3.0$ m stands 20 m in front of you. Describe the changing appearance of your image as you move from 20 m to 1 m from the mirror. Indicate distances from the mirror where the change in appearance is dramatic.

22.4.3 Represent and reason Using a ruler, carefully draw ray diagrams to locate the images of the objects listed below. Measure the image locations on your diagrams and indicate if they are real or virtual, upright or inverted. When you are done, check your work using the lens equation. (Choose a scale so that your drawing fills a significant portion of the width of the paper.)

a. an object that is 12 cm from a concave lens of -5-cm focal length,

b. an object that is 7 cm from the same lens,

c. an object that is 3 cm from the same lens.

22.4.4 Represent and reason A magnifying glass is a convex lens that when held close to an object (slightly closer than the focal length of the lens) allows you to see its enlarged upright virtual image. Draw a ray diagram to explain how a magnifying glass works.

22.4.5 Represent and reason Imagine you have a +20-cm focal-length convex lens. You place an object 15 cm from the lens on the main axis.

a. Fill in the table that follows.

Draw a ray diagram to find the image of the object.	Use the lens equation to calculate the location of the image.

b. Is the calculation consistent with the ray diagram?

c. What is the meaning of the negative sign in the distance of the image?

22.4.6 Represent and reason Imagine that you have a −20-cm focal-length concave lens. You place an object 25 cm from the lens on the main axis.

a. Fill in the table that follows.

Draw a ray diagram to find the image of the object.	Use the lens equation to calculate the location of the image.

b. Is the calculation consistent with the ray diagram?

c. What is the meaning of the negative sign in the distance of the image?

22.4.7 Reason and explain A simple camera system consists of an opening for light, a convex lens, and photographic paper placed where the image of the object is formed.

a. Draw a ray diagram explaining how a camera allows us to produce images of objects.

b. Think of possible technical arrangements of a camera that allow us to make sharp images of objects that are located at different distances from the same lens.

22.4.8 Diagram Jeopardy In the figures below, you see an axis of a lens (the lens itself is not shown) and the location of a shining object and its image. Your task is to find the location and the type of the lens (convex or concave) that could produce this image and find the focal points of the lens. When you think you have found an appropriate lens type and lens location, draw a ray diagram to help justify your choice and show the focal length on the diagram.

a. Object

Image

b. Object

Image

c. Image

Object

22.4.9 Reason and explain An unlabeled sketch of the optical system of a simple camera is shown below (real cameras have multiple lenses).

The eye is similar in some ways. Complete the table to identify the analogous elements in the eye and the camera.

Purpose	Identify the element for the camera.	Identify the element for the eye.
Allows the light from an object to enter the device		
Allows the light to be focused		
A place where the image is captured		
Makes it possible to get sharp images of objects that are at different distances from the place where the image is formed		
Makes it possible to change the amount of light entering the system		

22.4.10 Represent and reason Sketch an eye with its lens and draw a ray diagram showing how the eye forms an image on the retina. Should the eye lens be convex or concave?

22.4.11 Represent and reason The *far point* of the eye is the greatest distance to an object on which an eye can comfortably focus. The *near point* of an eye is the closest distance of an object on which the eye can comfortably focus. A nearsighted person has trouble focusing on distant objects (such as a sign on the highway). A farsighted person has trouble focusing on objects that are near the eye (such as the morning newspaper). Suggest reasons for these defects of vision and possible ways to correct them. Support your answers with ray diagrams.

22.4.12 Regular problem The image distance for the lens of a person's eye is 2.20 cm. Calculate the focal length of the eye's lens system for an object at the following distances:

a. at infinity,

b. 500 cm from the eye,

c. 25 cm from the eye.

22.4.13 Regular problem A farsighted man can see sharp images of objects that are 3.0 m or more from his eyes. He would like to read a book held 30 cm from his eyes. Determine the focal length of the lenses he needs for his glasses.

He will hold the book 30 cm from his glasses. An eyeglass lens should form an image 3.0 m (300 cm) in front of the lens. The optical system of his eye will look at this image. Draw a sketch of the object (book), eyeglass lens, and the image of that object. Enter known information in your sketch.	
Draw a ray diagram for the eyeglass lens system described in the first cell of the table. It's similar to that of a magnifying glass.	
Use the lens equation to calculate the focal length of the desired eyeglass lens.	

22.4.14 Regular problem A nearsighted woman can focus see sharp images of objects that are 2.0 m or less from her eyes. She would like to read road signs while driving on the turnpike. Determine the focal length of the lenses she needs for her glasses.

She will look at distant signs through her glasses (assume an infinite distance). An eyeglass lens should form an image of this distant object that is 2.0 m (200 cm) in front of the lens. The optical system of her eye will look at this image. Draw a sketch of the distant object, the eyeglass lens, and the image of that object. Enter known information in your sketch.	
Draw a ray diagram for the eyeglass lens system described in the first cell of the table.	
Use the lens equation to calculate the focal length of the desired eyeglass lens.	

22.4.15 Evaluate the solution

The problem: A man who can focus only on objects in the range of 1.6 m to 4.0 m wants to buy a pair of nonprescription glasses to wear while reading and another pair to wear while driving.

a. Determine the focal length of the glasses he should buy for reading.

b. Determine the focal length of glasses he should buy for driving.

Proposed solution for part a:

Sketch and translate

A ray diagram of the situation is shown at the right. The image should be 1.6 m from the lenses when the object (the book) is about 0.4 m from the lenses (a comfortable distance to hold a book from the eye).

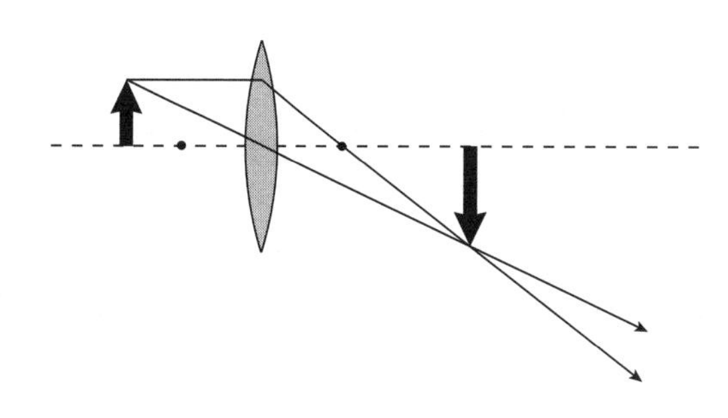

Simplify and diagram

We assumed the book was held 0.40 m from the eye—about 16".

Refer to the ray diagram.

Represent mathematically and solve

Using the lens equation, we can find the focal length.

$$\frac{1}{f_{\text{reading}}} = \frac{1}{0.4} + \frac{1}{1.6} = +2.0\text{ m}$$

Proposed solution for part b:

Sketch and translate

A ray diagram of the situation is shown at the right. The image should be 4.0 m from the lens when the object (a road sign) is far away—an infinite distance from the lenses.

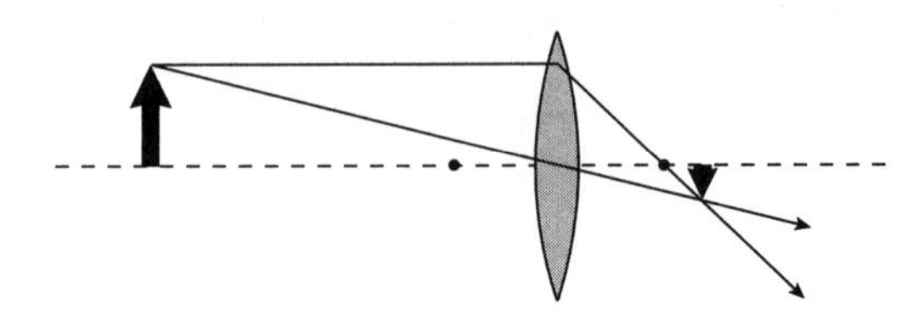

Simplify and diagram

We assume that the object is at infinity.

Refer to the ray diagram.

Represent mathematically and solve

Using the lens equation, we can find the focal length for the glasses used for distance work.

$$\frac{1}{f_{\text{driving}}} = \frac{1}{\infty} + \frac{1}{4.0} = +0.25\text{ m} = 25\text{ cm}$$

a. Identify any errors in the proposed solutions to this problem.

b. Provide corrected solutions if you find errors.

23 Wave Optics

23.1 Qualitative Concept Building and Testing

Competing models of light: According to the *particle model of light,* supported by Isaac Newton, some light phenomena can be explained by assuming that light behaves as a stream of tiny discrete particles (corpuscles) that move in straight lines through space, like tiny bullets. According to the *wave model of light,* supported by Christaan Huygens, some light phenomena can be explained by assuming that light behaves as if it is a wave. Each of these two models can explain some light phenomena.

23.1.1 Observe and explain Descriptions of several experiments follow.

a. Complete the table to explain the result of each experiment using the two previously identified models of light.

Experiment	Describe what you observe and sketch the situation.	Explain your observations using the particle model of light.	Explain your observations using the wave model of light.
Light a candle in a dark room and place the candle 2–3 m from the wall. Hold a pencil close to the wall.			
Direct the light from a laser pointer at a plane, horizontal mirror at a 60°-incident angle relative to the normal line.			
Direct the light from a laser pointer at the surface of water in a large, clear container at a 60°-incident angle relative to the normal line.	The refracted beam in the water makes an angle that is less than 60° relative to the normal line.		

b. Which experiment(s) can be explained by both models of light?

23.1.2 Test your ideas Assemble a candle and a piece of cardboard (about 20 cm long and 10 cm wide). Cut two large slitlike holes in the cardboard, as shown in part a of the figure.

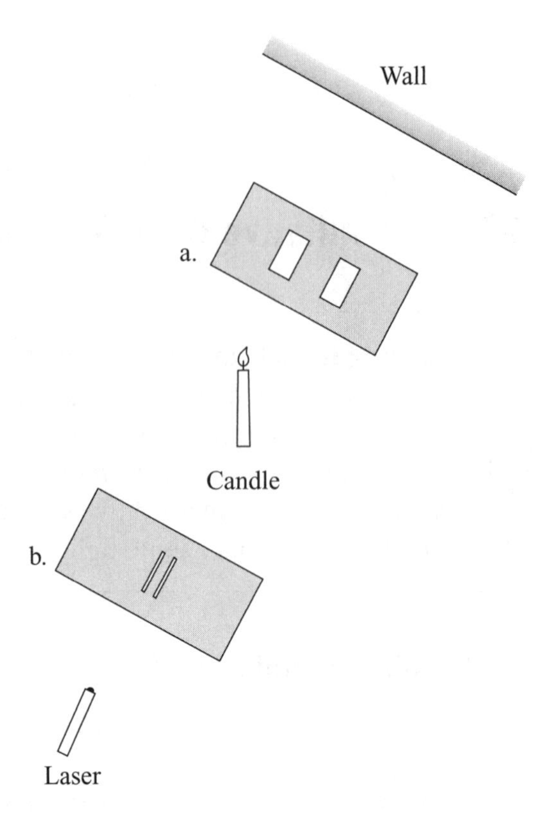

a. Complete the table that follows.

Use the particle model of light to predict what you will see on the wall if you place the cardboard with the large holes between the candle and the wall, situating the cardboard closer to the wall than to the candle.	Use the wave model of light to predict what you will see on the wall if you place the cardboard with the large holes between the candle and the wall, situating the cardboard closer to the wall than to the candle.	Perform the experiment and record the outcome.	Discuss which model gave you a better prediction.

b. Assemble a laser pointer and a plate with two narrow slits situated close to each other. Complete the table that follows.

Use the particle model of light to predict what you will see on the wall if you place the plate with the narrow slits between the laser beam and the wall, situating the plate closer to the laser pointer.	Use the wave model to predict what you will see on the wall if you place the plate with the narrow slits between the laser pointer and the wall, situating the plate closer to the laser pointer. *Note:* Assume that Huygens' wavelets emanate from each slit.	Perform the experiment and record the outcome.	Discuss which model gave you a better prediction.

23.1.3 Explain In Activity 23.1.2b, with the laser pointer and two narrow slits, instead of seeing two thin bright lines (the images of the slits), you saw closely spaced alternating bright and dark narrow bands of light—a bright band at the center with several less-bright bands on the sides. Why were there dark areas between bright areas? *Hint:* Is it possible for two particles to arrive at the same location and cancel each other? Is it possible for two waves to arrive at the same location and cancel each other?

23.2 Conceptual Reasoning

23.2.1 Reason and explain Waves (any type) are incident from the left side on a barrier with two small openings. Consider what happens on the right side of the barrier. According to Huygens' principle, these openings become wave sources. In the illustration, we represent the wave fronts leaving these sources with dark and light circles. The solid dark lines represent wave crests, and the lighter gray lines represent the troughs beyond the slits at one instant of time.

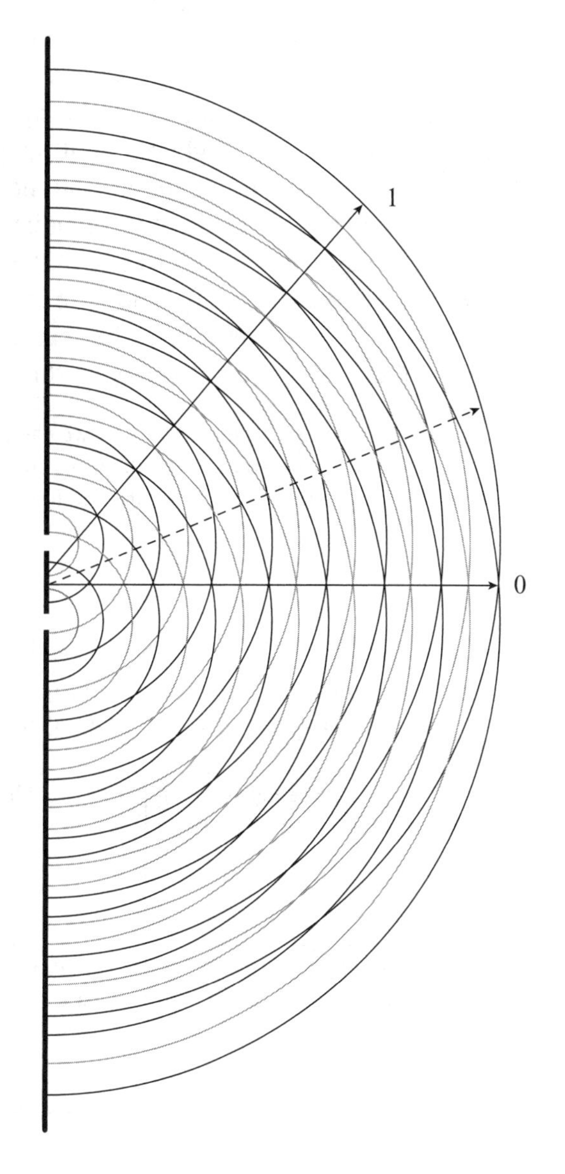

a. Describe what the crests might represent for each of the following types of waves: water waves, sound waves, and light waves.

b. Indicate with the letters "dc" (double crest) places on line 0 that are equal distance from the wave sources (the openings) and where the crests add to form a disturbance that is twice as big as the wave amplitude from one source.

c. Indicate with the letters "dt" (double trough) places on line 0 that are equal distance from the sources and where the wave troughs add to form a negative disturbance whose magnitude is twice the amplitude of a wave trough from one source.

d. The sketch represents the positions of wave crests and troughs at one particular time. Suppose that these are water waves that are now moving and that you stand in the water on the right side of line 0. What would it feel like as the alternating dc and dt points passed you?

e. What if the sketch represents sound waves that are now moving along the line described in part d. What would you hear?

f. What if these were light waves moving along the line described in part d. Would the light be bright or dim? Explain.

g. Would the same effect be observed along line 1 as along line 0? Explain.

h. What would you feel (water waves), hear (sound waves), or see (light waves) if you were located at the end of the dashed line between 0 and 1? Explain.

23.2.2 Reason and explain In Activity 23.1.3, the light and dark bands produced on the wall by laser light passing through two narrow slits are more easily explained using the wave model of light. Consider the wave model and the two-slit phenomenon. Shine the light from a laser pointer onto two closely spaced slits. On a screen several meters to the right of the slits, you observe bright light bands at the positions of the dots shown in the illustration (b_0 at the center, and b_1 and b_2 bands at each side of the center). You see darkness, which we call dark bands, at the positions of the crosses (the d_1 and d_2 bands at each side of the center). *Note:* The separation of the bright and dark bands on the screen is exaggerated in this sketch.

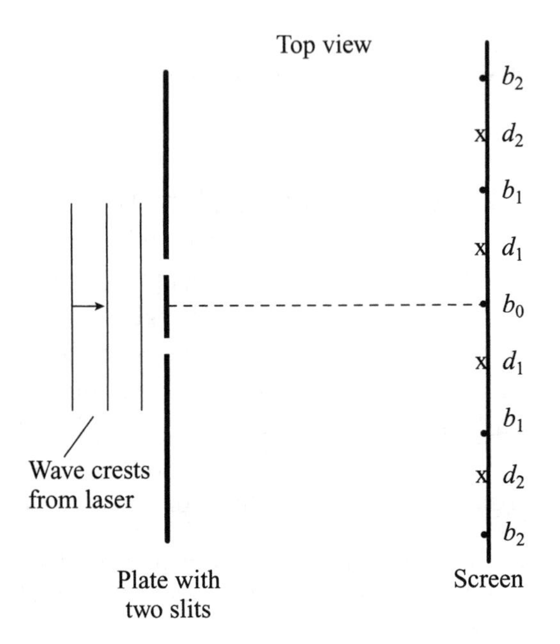

a. Use Huygens' principle to explain why we can assume that the two slits are wave sources and the waves produced by them vibrate synchronously (they are said to be *in phase*).

b. Use the wave model of light to answer the questions in the table that follows. Think about the distances from the two slits to a bright band or a dark band and about superposition of the waves coming from the two slits.

Explain why the center b_0 band is bright.	Explain why b_1 above the center bright band is bright.	Explain why d_1 above the center bright band is dark.
Explain why b_2 above the center bright band is bright.	Explain why d_2 above the center bright band is dark.	

23.2.3 Observe and explain Place a color filter in front of a slide projector so that the beam of light coming out of the projector is of primarily one color. Let the light pass through a single slit whose width can be varied. Place a screen about 1 m beyond the slit. With a 1-cm-wide slit, you see on the screen an image that looks like the slit. Slowly decrease the width of the slit. When the slit width approaches a millimeter (still much wider than the slits in double-slit experiments), a seemingly strange thing happens: The width of the pattern on the screen starts to increase, but the light becomes dimmer. In addition to the widening of the pattern, there are alternating bright and dark fringes. The fringes spread as the slit width decreases. Can you explain these observations with a particle model of light and/or with a wave model of light? *Hint:* You can simplify the situation by using Huygens' principle—that is, by considering the points of the open slit to be tiny light sources emitting half-circle light wavelets in the forward direction. The wavelets can interfere constructively or destructively.

23.3 Quantitative Concept Building and Testing

23.3.1 Represent and reason Imagine that you shine a laser at a screen with two very narrow, closely spaced slits, as in Activity 23.2.2. These two slits can be considered sources of light wavelets of the same wavelength vibrating in phase—produced by the same wave front arriving at the slits from the left. The illustration shows lines going from each of two slits to the second bright band (b_2) above the central bright band (b_0) on a screen. Dots represent the bright spots, and crosses represent dark spots.

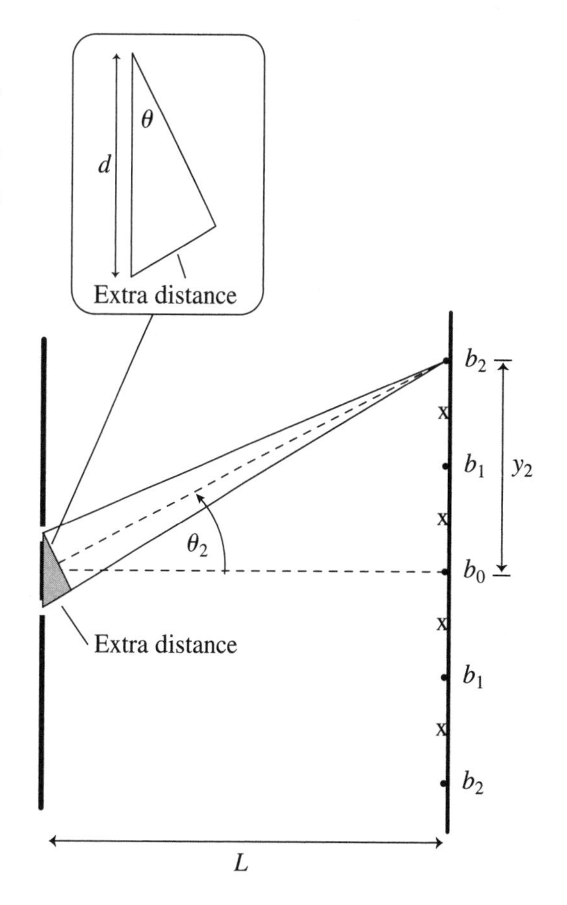

a. Compare the extra distance that light travels from the lower slit to the b_2 bright band and the distance from the upper slit to that bright band. Express this difference in wavelengths of light. Remember that this is the second bright band.

b. How is the angle θ (shown in the inset) related to the angle θ_2 shown in the main part of the figure? Explain.

c. Refer to the triangle insert in the illustration and to the results of Activity 23.2.1 to help you write an expression that relates the extra distance that light travels from the lower slit to the b_2 bright band and the distance from the upper slit to that bright band expressed through the wavelength λ, slit separation d, and the angle θ in the triangle.

d. Write another expression that relates the angle θ_2 to the distance L from the slits to the screen and the distance y_2 from the b_0 central maximum to the position of the b_2 bright spot.

e. Generalize the two expressions developed in parts c and d so that they can be used to determine the angular deflection to the nth bright band. List the assumptions you made when constructing these two expressions.

23.3.2 Test your ideas Gather a laser pointer, a set of double slits of known separations, a screen, and a ruler. Design an experiment to test the relationships you devised in Activity 23.3.1.

a. Complete the table that follows. *Note:* Look at the laser case; it may tell you the wavelength of the light.

Sketch the experimental setup.	Use the expressions devised in Activity 23.3.1c and d to predict the outcome of the experiment.	List additional assumptions you made.	Perform the experiment, record the outcome, and compare it to the prediction.

b. Discuss whether the outcomes of the experiments support the expressions you devised in 23.3.1c and d.

23.3.3 Observe and explain The slit separation for double slits typically used in lecture demonstrations is about 0.5 mm $= 0.5 \times 10^{-3}$ m. An apparatus called a *grating* has about 200 slits in 1 mm.

a. Determine the distance between the centers of the adjacent slits in such a grating.

b. Place the grating about 2 m from a white screen. Shine laser light through the grating and observe a set of bright dots on the screen. These dots are much farther apart than when you use a double-slit apparatus. Draw a diagram of the experimental situation and explain the phenomenon qualitatively.

23.3.4 Reason and explain In Activity 23.3.3, you found a large angular deflection to the first bright band. Suppose you have a five-slit grating with adjacent slits having a small slit separation. The position of the first b_1 bright band to the side of the central b_0 bright band for laser light passing through this grating is shown in the illustration. In terms of light wavelength, how do the distances from second, third, fourth, and fifth slits to the first b_1 bright band compare to the distance from the first (bottom) slit to that bright band? Explain, and be specific.

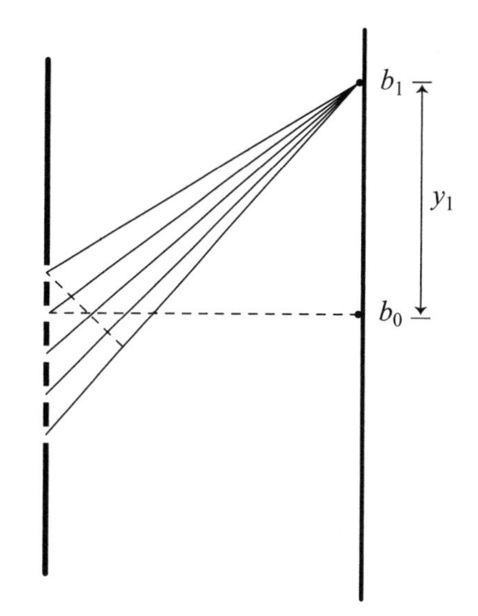

23.3.5 Derive Consider the situation depicted in Activity 23.3.4.

a. Devise a mathematical expression that relates the angular deflection θ_1 to the first bright band, the wavelength λ of the light, and the separation d of adjacent slits. You might want to review what you did in Activity 23.3.1.

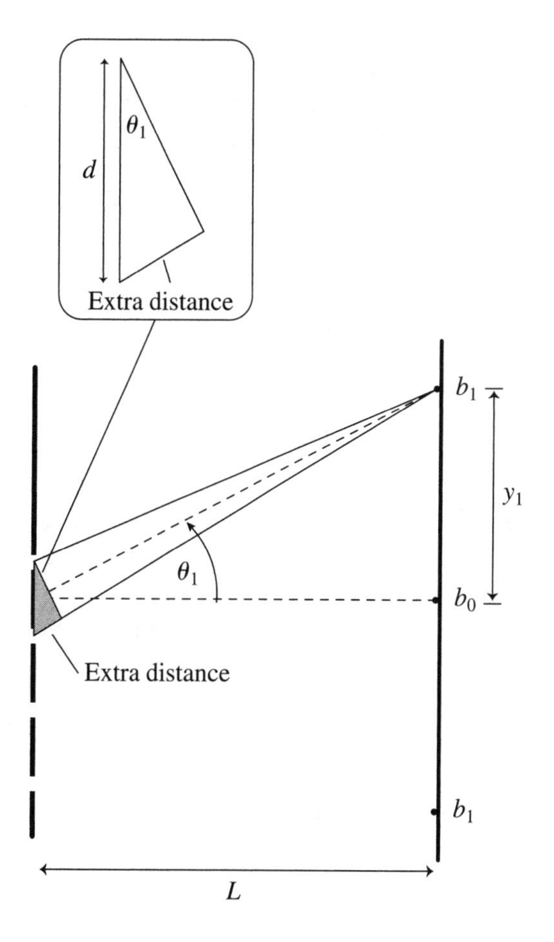

b. Devise a mathematical expression that relates the angular deflection θ_1 to the first bright band on the screen to the distance L of the grating from the screen and the distance y_1 of the first bright band from the central maximum b_0 on the screen.

23.3.6 Test your idea

a. Use the expressions that you devised in Activities 23.3.4 and 23.3.5 to predict how an increase in the number of slits per millimeter in the grating (for example, 400 slits per mm instead of 200) should affect the separation of the bright dots on the screen, assuming you use the same laser as in the previous activities. Perform the experiment and compare the outcome to your prediction.

b. Use the explanations that you devised in Activities 23.3.4 and 23.3.5 to predict how an increase in the total number of slits without changing the number of slits per unit length should affect the separation of the bright dots on the screen, assuming you use the same laser as in the previous activities. Create an experiment that you can perform to check your prediction. Then perform the experiment and record the outcome. Was it consistent with your prediction?

23.3.7 Observe and find a pattern Repeat the experiment in Activity 23.3.3, but this time instead of a laser, use a flashlight and a grating with the greatest number of slits/mm that is available. Complete the table that follows.

Sketch the experimental setup.	Use colored pencils to draw the pattern you observe on the screen.	Describe the pattern in terms of color location.	Write an explanation for the pattern.

23.3.8 Design an experiment Use the apparatus from Activity 23.3.7 to design an experiment to determine the wavelengths of different colored light coming from an incandescent lightbulb.

Sketch the experimental setup.	List the quantities you will measure.	Summarize the mathematical procedure you will use to calculate wavelengths. Then calculate the wavelengths.	List representative calculated wavelengths of the different colored light.

23.3.9 Derive In previous activities involving two or more slits, we used very narrow slits and considered them pointlike wave sources. What happens if an obstacle with a single slit blocks the light? Imagine that a wave approaches a single slit from the left (see the illustration). The opening is about the same width as the wavelength of the wave. We observe on a screen to the right of the slit a bright band of light in the center (b_0) and alternating dark and bright bands on each side of the center bright band (d_1 and b_1 in the sketch—there are usually more than one dark and bright band on the sides).

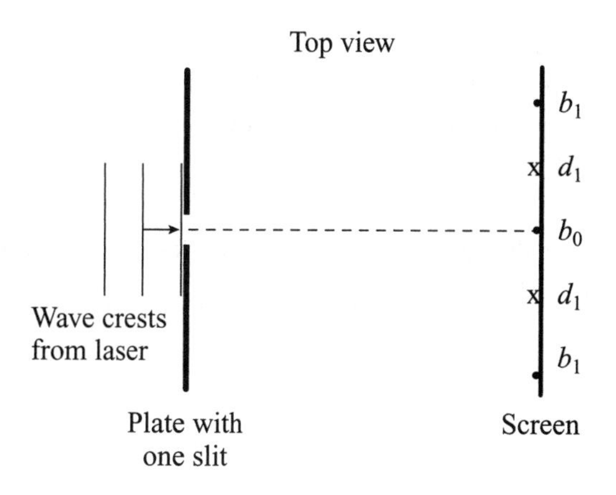

a. To understand the location of the first dark band (d_1) on the screen, divide the slit in half. What condition is necessary for a wavelet from a point at the top of the top half of the slit to interfere *destructively* at d_1 on the screen with a wavelet from the top of the bottom half of the slit opening? Explain. (The sketch should help.)

b. Write an expression that relates the slit width w, the wavelength of the light λ, and the angle θ_1. Refer to the illustration.

c. If you consider a wavelet produced a little lower in the top half of the opening and another a little lower in the bottom half of the opening, will they also interfere destructively? Explain. In fact if the condition in part a is satisfied, does a wavelet from each point in the bottom half of the single slit interfere destructively with a wavelet from each point in the top half of the slit? Explain.

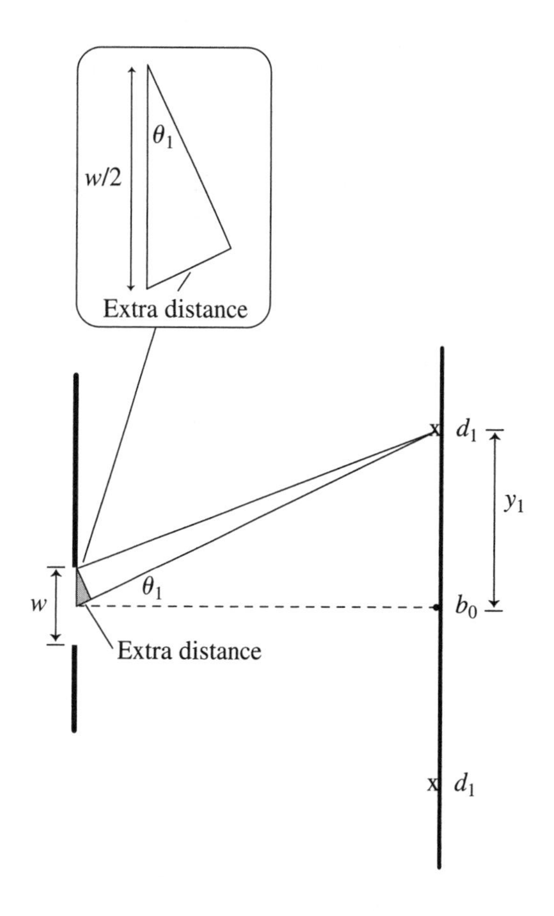

23.3.10 Test your idea Aim a laser pointer toward a single narrow slit that is part of a system of slits with different widths. Use the relationship derived in Activity 23.3.9 to predict the width of the central maximum on a screen placed about 1 m away from the slit. Complete the table that follows.

Sketch the pattern on the screen for the narrowest slit.	Sketch the pattern on the screen for a medium-width slit.	Sketch the pattern on the screen for the widest slit.

Calculate the distance between the dark bands on each side of the central maximum for each slit.	Perform the experiment and compare the results with predicted values.

23.3.11 Test an idea Babinet's principle states that the diffraction pattern of complementary objects is the same. For example, a slit in a screen produces the same diffraction pattern as a screen the same size as the slit; a hair will produce the same diffraction pattern as a slit of the same width as the hair. Use Babinet's principle to predict the difference in the patterns on a screen produced by laser light shining on a strand of your hair and then shining on a thin sewing needle. Explain your prediction. Perform the experiment and record the outcome. Did the prediction match the outcome of the experiment?

23.4 Quantitative Reasoning

23.4.1 Regular problem Light of wavelength 540 nm from a green laser is incident on two slits that are separated by 0.50 mm; the light reaches a square screen 50 cm × 50 cm that is 1 m away from the slits. Describe quantitatively the pattern that you will see on the screen. Complete the table that follows.

Sketch and translate • Visualize the situation and sketch it. • Translate givens into physical quantities.	*Simplify and diagram* • Decide if the small-angle approximation is appropriate. • Represent the situation with a ray diagram showing the path of light waves from the two slits to the screen.
Represent mathematically • Describe the diagram mathematically.	*Solve and evaluate* • Solve the problem and decide if the answer makes sense.

23.4.2 Represent and reason Monochromatic light passes through two slits and then strikes a screen. The distance on the screen between the central maximum and the first bright fringe at the side is 2.0 cm.

a. Sketch the situation. Include symbols of quantities used in the equations.

b. Determine the fringe separation if the slit separation is doubled and everything else remains unchanged.

c. Determine the fringe separation if the wavelength is doubled and everything else remains unchanged.

d. Determine the fringe separation if the screen distance is doubled and everything else remains unchanged.

23.4.3 Evaluate the solution

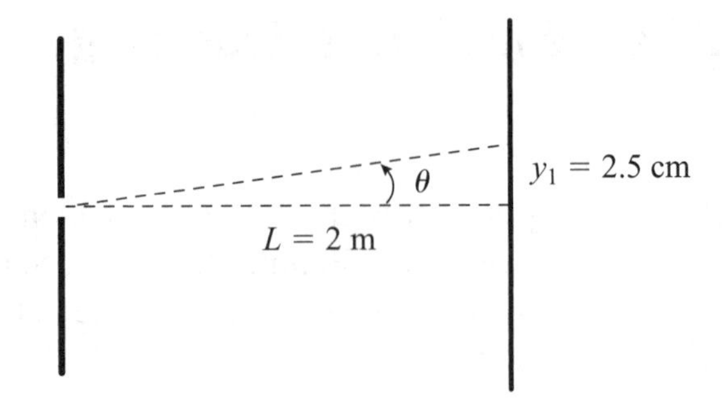

The problem: Determine the width of a hair that when irradiated with laser light of wavelength 630 nm produces a diffraction pattern on a screen with the first minimum 2.5 cm from the central maximum. The screen is 2.0 m from the hair.

Proposed solution: A hair is a very thin obstacle in the path of light; the light bends around it and produces a diffraction pattern on the screen. According to Babinet's principle, the pattern will be similar to that formed by light passing through a single narrow slit whose width is the same as the width of the hair. Thus, we can use the expression for the angular deflection to the first minimum to relate the angular width of the central maximum to the width of the hair. Because the angular deflection is small, the sine and tangent of this angle give the same result:

$$\sin\theta_1 = \tan\theta_1 = y_1/L$$

$$w\sin\theta_1 = n\lambda \quad \text{for} \quad n = 1$$

$$w = \frac{\lambda}{\sin\theta_1} = \frac{\lambda y_1}{L} = \frac{(630 \times 10^{-9})(2)}{(2.5)} = 504 \times 10^{-9}\ \text{m}$$

a. Identify any missing elements or errors in the solution.

b. Provide a corrected solution if there are errors.

23.4.4 Regular problem

A reflection grating reflects light from adjacent lines in the grating instead of allowing the light to pass through slits, as is the case with the so-called transmission gratings we have been studying. Interference between the reflected light waves produces *reflection maxima*. The angular deflection of bright bands, assuming perpendicular incidence, is calculated using the same equation as the angular deflection of transmitted light through a regular grating. White light is incident on the wing of a Morpho butterfly (whose wings act as a reflection grating).

a. Explain why you see different color bands coming from the wings of the butterfly when white light shines on the wings.

b. Red light of wavelength 660 nm is deflected in first order ($n = 1$) at an angle of 1.2°. Determine the angular deflection in first order ($n = 1$) of blue light (460 nm).

c. Determine the angular deflection in third order ($n = 1$) of yellow light (560 nm).

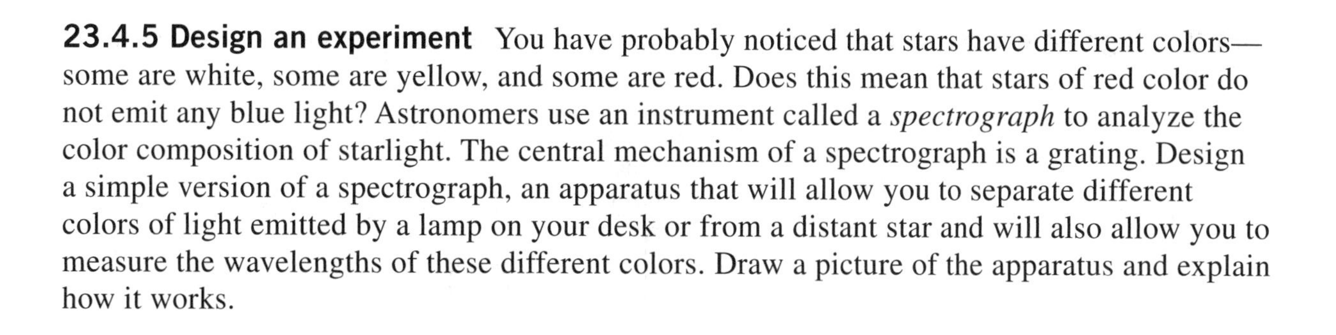

23.4.5 Design an experiment You have probably noticed that stars have different colors—some are white, some are yellow, and some are red. Does this mean that stars of red color do not emit any blue light? Astronomers use an instrument called a *spectrograph* to analyze the color composition of starlight. The central mechanism of a spectrograph is a grating. Design a simple version of a spectrograph, an apparatus that will allow you to separate different colors of light emitted by a lamp on your desk or from a distant star and will also allow you to measure the wavelengths of these different colors. Draw a picture of the apparatus and explain how it works.

23.4.6 Design an experiment Design an experiment to determine the thickness of an individual strand of your hair using a laser pointer, a screen, and a ruler.

a. Describe the design, the procedure, and the assumptions that you make.

b. Perform the experiment, record the measured quantities, and calculate the width of a piece of hair. Then measure the hair strand with a caliper and decide whether the result you obtained with the first method agrees with the second.

23.4.7 Estimate Assume that Earth, its structures, and its inhabitants are all decreased in size by the same factor. *Estimate* the decrease required so that the first-order diffraction dark band of 500-nm light entering a typical room window is at 90° (the central bright band would light most of the room). Explain all aspects of your calculations and the assumptions that you made.

23.4.8 Design an experiment Examine the apparatus in the illustration to answer the questions that follow.

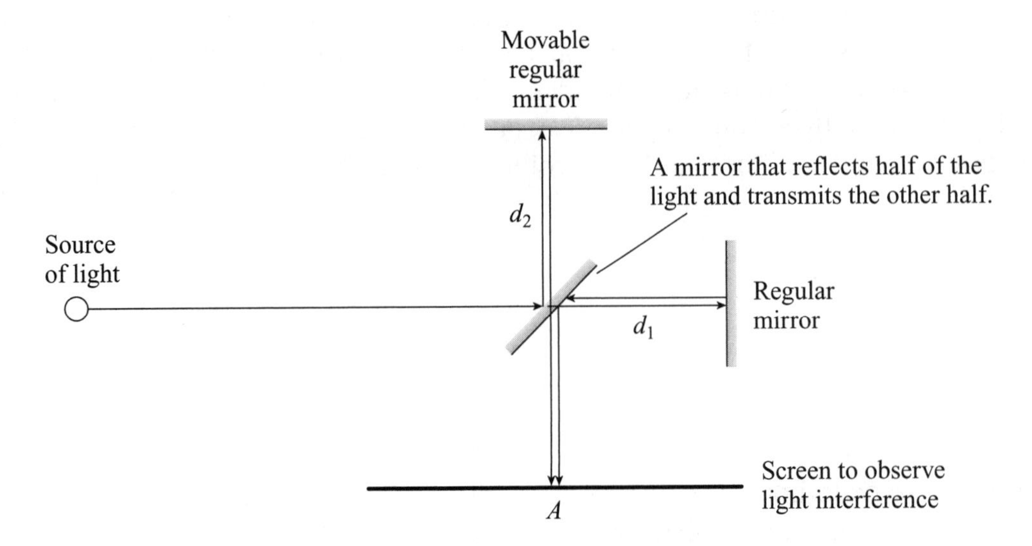

a. Explain why you see an alternating bright light and then a dark spot on the screen as the movable mirror is moved.

b. Explain how this apparatus can be used to determine if the speed of light in one direction is different from the speed of light in another direction.

24 Electromagnetic Waves

24.1 Qualitative Concept Building and Testing

24.1.1 Observe and find a pattern Several experiments involving traveling waves on a Slinky are described below.

a. Construct a sketch for each experiment. The outcome is shown in the right column.

Experiment	Sketch	The outcome
A Slinky rests on its side on a table. One end is held off the edge of the table. The Slinky is shaken left to right, parallel to the table. Make a sketch of the process.		The pulse continues across the table.
The same as above except the Slinky is shaken up and down, perpendicular to the table.		The pulse does not continue across the table.
The same as above except the Slinky is shaken forward and backward, along the direction of the Slinky. Make a sketch of the process.		The pulse continues across the table.
The same as above except part of the Slinky on the table passes through a tube. The Slinky is shaken forward and backward, a longitudinal pulse. Make a sketch of the process.		The pulse continues across the table.

b. Devise a rule or rules that describe the pattern of conditions under which the pulse can continue across the table.

24.1.2 Test an idea Light is either a transverse wave or a longitudinal wave. Several experiments that test these two hypotheses are described below. The experiments make use of a polarizer, a device made from material that prevents light waves from passing through if something in the wave is vibrating perpendicular to the axis of the polarizer.

a. Make predictions based on each of the two hypotheses about the brightness of the light once it has passed through the polarizer(s).

Experiment	Prediction if light is transverse wave	Prediction if light is longitudinal wave	Outcome
Light from a light bulb shines on a polarizer and its brightness is detected on the other side.			The light reaching the other side of the polarizer is significantly dimmer.
The same as above except the polarizer is slowly rotated.			The light reaching the other side of the polarizer is significantly dimmer and does not change as the polarizer is rotated.

Experiment	Prediction if light is transverse wave	Prediction if light is longitudinal wave	Outcome
Light from a light bulb shines on a polarizer. A second polarizer is positioned behind the first one. The second polarizer is slowly rotated relative to the first.			The light is dimmer overall but also fades in and out completely as the second polarizer is rotated.

b. Make a judgment about each of the two hypotheses. Which (if any) of them are disproved by these experiments?

24.2 Conceptual Reasoning

24.2.1 Summarize You learned in a previous chapter that a changing magnetic field produces an electric field. Describe experimental evidence for this.

24.2.2 Represent and reason Antennas are used to produce electromagnetic (EM) waves. This is accomplished by connecting the antenna to a source of an alternating emf. In this activity you will represent the production of an EM wave using field line diagrams. The figures below show the current and charge separation in the antenna at various clock readings.

a. Add appropriate electric and magnetic field lines to these figures. Be sure to include an electric and magnetic field line that passes through the indicated point P.

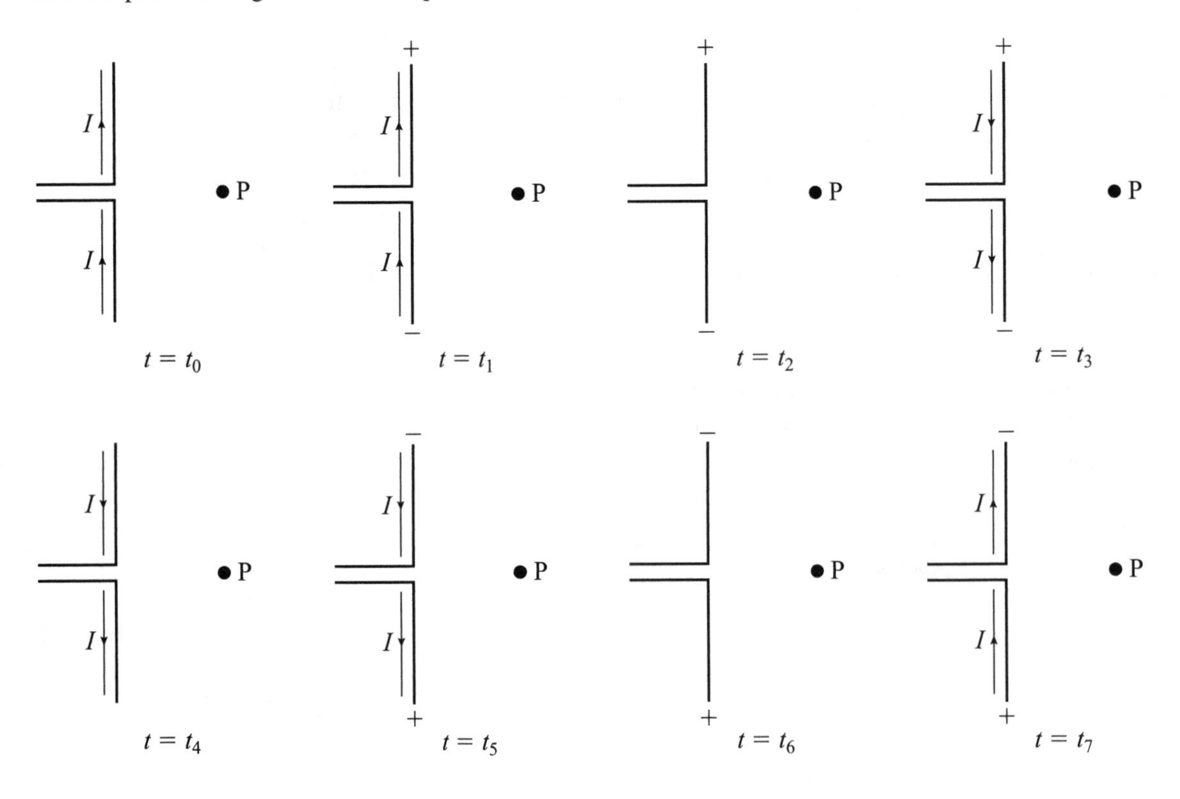

b. For each clock reading t_0 through t_7 draw a vector representing the $\vec{E}$ field at the point P.

c. For each clock reading t_0 through t_7 draw a vector representing the $\vec{B}$ field at the point P.

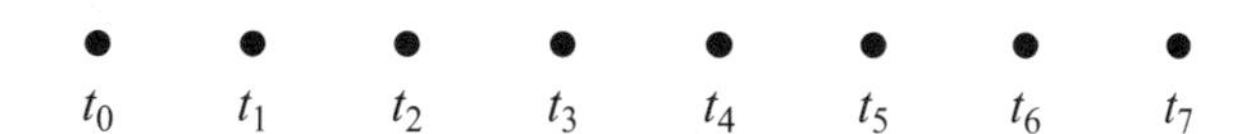

d. Describe the behavior of the $\vec{E}$ field and $\vec{B}$ field at the point P.

24.3 Quantitative Concept Building and Testing

24.3.1 Observe and find a pattern Experiments involving light waves passing through one or more polarizers are described in the table below. The light intensity is measured using a device known as a photometer.

Experiment	Outcome of experiment
a. Light from a light bulb shines on a polarizer. The polarizer is slowly rotated. The light intensity on each side of the polarizer is measured.	I/I_0; 1/2; 180°; θ
b. Light from a light bulb shines on a polarizer. A second polarizer is positioned behind the first one. The second polarizer is slowly rotated relative to the first. The light intensity on each side of the polarizer set is measured.	I/I_0; 1/2; 90°; 180°; θ

Experiment	Outcome of experiment
c. Light that has already passed through a vertically oriented polarizer shines on a second polarizer. The second polarizer is slowly rotated. The light intensity on each side of the polarizer is measured.	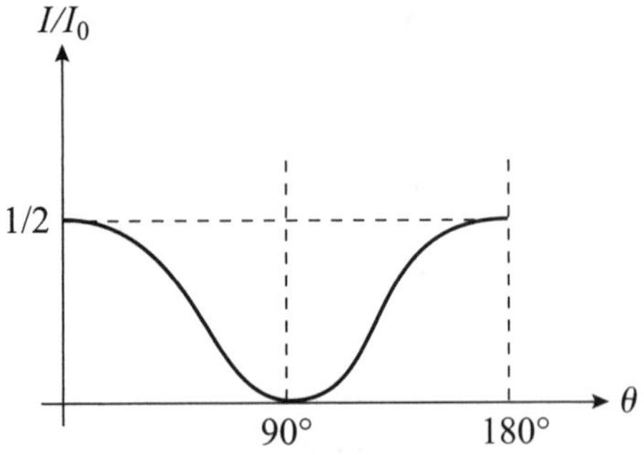 I/I_0; 1/2; 90°; 180°; θ

For each of the above cases, devise a mathematical relationship between the light intensity I_0 before the polarizer(s) and the light intensity I after.

a.

b.

c.

24.3.2 Derive The constant k (Coulomb's constant) and the constant μ_0 (vacuum permeability) have appeared in your study of electric and magnetic phenomenon.

a. By analyzing the units of these two constants, devise a mathematical expression involving these two constants that has units of speed, m/s, and determine its value.

b. What is the significance of this value?

24.3.3 Represent and reason An electromagnetic wave traveling in the positive x direction can be represented by the following wave equations for the $\vec{E}$ field and $\vec{B}$ field.

$$E_y = E_{\text{max}} \cos\left[2\pi\left(\frac{t}{T} - \frac{x}{\lambda}\right)\right] \quad B_z = B_{\text{max}} \cos\left[2\pi\left(\frac{t}{T} - \frac{x}{\lambda}\right)\right]$$

To see if these equations are reasonable, draw graphs of E_y as a function of x at the following clock readings: $t = 0, t = T/4, t = T/2$, and $t = 3T/4$. Assume the wave is a signal from a 2100 MHz 3G cell phone tower.

Graph E_y as a function of x at the given clock readings	
$t = 0$	$t = T/4$
$t = T/2$	$t = 3T/4$

24.3.4 Determine everything you can Below is the graph of the $\vec{B}$ field component of an electromagnetic wave. Write down everything that you can about this wave.

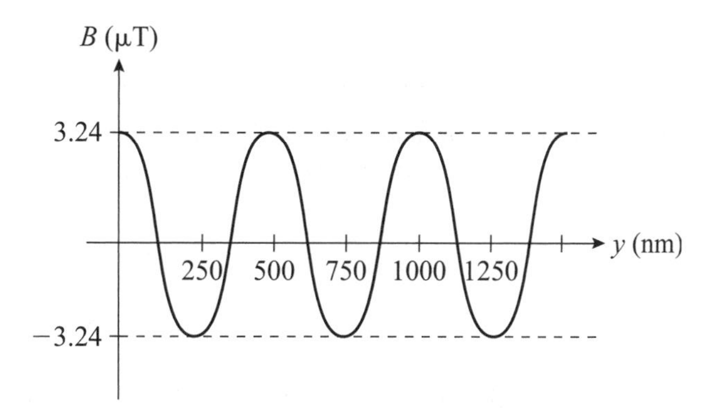

24.3.5 Evaluate The expressions for the electric and magnetic field energy densities are

$$u_E = \frac{1}{2}\varepsilon_0 E^2 \text{ and } u_B = \frac{1}{2\mu_0}B^2$$

Check that the units of these expressions are consistent.

24.3.6 Observe and find a pattern You are working late on a cold winter night. The light from a rising full Moon reflects off a nearby frozen fountain and reaches your eyes. The glare is very bright so you put your sunglasses on. As you do this you notice that as you rotate your sunglasses the reflected light of the Moon passing through the sunglasses varies in brightness. You head outside to investigate this and record your observations shown in the table that follows.

Experiment	Outcome of experiment	
As the Moon rises (decreasing the angle of incidence) you observe the light passing through the rotating sunglasses	θ_i, θ_{refl} (°)	Brightness variation
	35	Moderate
	45	Large
	53	Maximal (brightness goes to zero for some orientations)
	65	Large

a. Analysis of observations.

θ_{refr} (°), from Snell's law	Ray diagram (use a protractor for the angles for accuracy)

b. Devise a rule or rules that describe the conditions under which the reflected light is completely polarized in the plane parallel to the reflecting surface. *Hint: Look at each ray diagram and the corresponding values of the angles of reflection and refraction.*

24.3.7 Derive The pattern discovered in the preceding activity can be expressed more conveniently as an equation for the so-called polarization angle θ_p, the angle of incidence at which the reflected light is completely polarized in the plane of the reflecting surface. Follow these steps to derive this relationship:

1. Begin with Snell's law.
2. Use the third ray diagram above under Activity 24.3.6 (a) to come up with an equation involving angles in the diagram that relates θ_2 and θ_p.
3. Use this equation to substitute for θ_2 in Snell's law.
4. Finally, use the trigonometric identity $\cos\theta = \sin(90° - \theta)$ and simplify the result as much as possible to arrive at what is known as Brewster's law.

24.3.8 Test an idea See the preceding two activities. The next day you are again working late. You see that a piece of Plexiglas® sheet has been placed over the fountain. This is an opportunity to test the idea developed in the preceding activities.

a. Analyze the following experiment:

Experiment	Prediction of reflection angle for completely polarized reflected moonlight off the Plexiglas sheet using equation from Activity 24.3.7.	Outcome of experiment
The light from a rising full moon reflects off the Plexiglas sheet and reaches your eyes.		$54° \pm 3°$

b. Make a judgment about the idea being tested. Has it been disproved by this experiment?

24.4 Quantitative Reasoning

24.4.1 Estimate In 1899 the first transmission of information across the English Channel via radio waves was achieved. Estimate the time interval between when the signal was produced and when it was received.

24.4.2 Equation Jeopardy Come up with a problem that is consistent with the solution below.

$$\frac{I}{I_0} = \frac{1}{4} = \cos^2(\theta)$$

$$\frac{1}{2} = \cos(\theta)$$

$$\theta = \cos^{-1}\left(\frac{1}{2}\right) = 60^\circ$$

24.4.3 Analyze A secret mission needs to be flown by the military into a neighboring country. There are two radar stations that the aircraft must fly between. They are 60 km apart. The radio waves emitted by station 1 have a period of 2.0×10^{-4} s and a pulse width of 12.0 μs. The waves emitted by station 2 have a period of 2.7×10^{-4} s and a pulse width of 10.0 μs. Is it possible for the aircraft to fly undetected into the neighboring country? Justify your answer quantitatively.

24.4.4 Analyze Some mobile phone carriers offer 4G networks in major metropolitan areas. One version of 4G uses a carrier wave frequency of 700 MHz.

a. How does this frequency compare with the frequencies of visible light?

b. What is the wavelength of this carrier wave?

c. How does this wavelength compare with the wavelengths of visible light?

24.4.5 Evaluate the solution Determine the magnetic field between the plates of a parallel plate capacitor. The capacitor is connected to a 12-V source of EMF, which keeps the plates at a constant potential difference of about 12 V. The plates of the capacitor are separated by 1.0 cm. State any assumptions that you make.

Solution:

$$E = \frac{\Delta V}{d} = \frac{12\ \text{V}}{1.0\ \text{cm}} = 12\ \text{N/C}$$

$$B = \frac{E}{c} = \frac{12\ \text{N/C}}{3.0 \times 10^{8}\ \text{m/s}} = 4.0 \times 10^{-8}\ \text{T}$$

This assumes the current in the circuit is constant.

a. Identify any errors in the solution.

b. Provide a corrected solution if there are errors.

24.4.6 Determine everything you can A satellite broadcasting a satellite television signal orbits at an altitude of 22,300 miles above Earth's surface. It broadcasts radio waves with a total power output of 200 W per channel. Receivers designed to tune in to this signal are dish-like in shape and have a radius of about 40 cm. Determine as many physics quantities relevant to this situation as you can.

25 Special Relativity

25.1 Qualitative Concept Building and Testing

25.1.1 Observe and find a pattern In the following experiments the speed of physical objects and of light when bouncing off objects is investigated.

Experiment	Sketch	Outcome of experiment
A cart moving to the right at 2 m/s on an air track bounces off a barrier at the end.	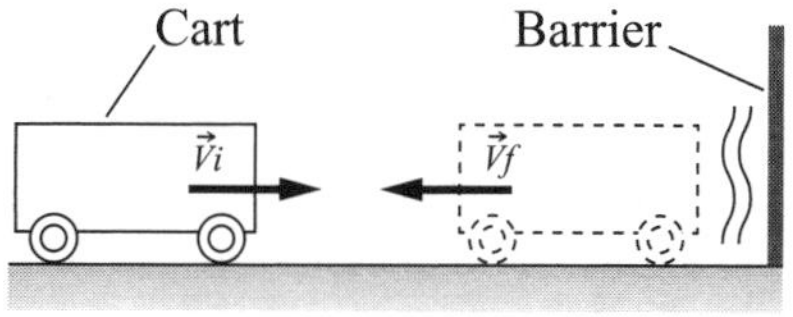 Cart, Barrier, $\vec{v}_i$, $\vec{v}_f$	The cart now travels to the left at 2 m/s.
As above but the barrier now moves to the left at 1 m/s relative to the track.	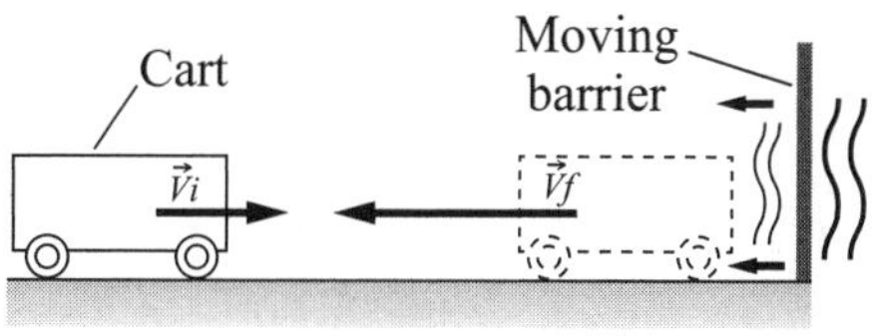 Cart, Moving barrier, $\vec{v}_i$, $\vec{v}_f$	The cart travels to the left at 4 m/s after bouncing (the cart was traveling at a speed of 3 m/s relative to the barrier both before and after the collision).
A laser is fired toward a mirror 10 m away.	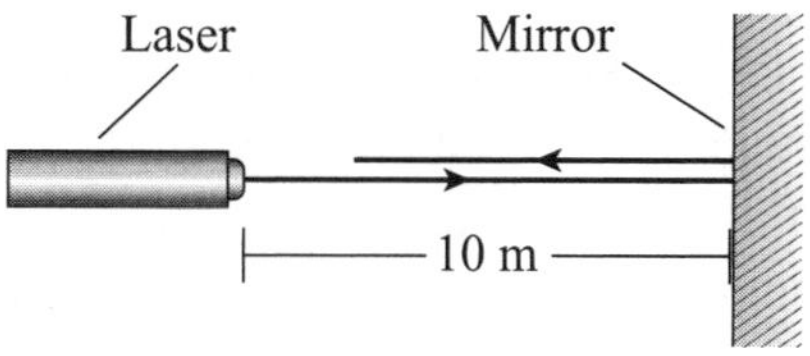 Laser, Mirror, 10 m	The laser light returns 6.673283×10^{-8} s later.
As above, but the mirror is being spun so that it is moving toward the laser at 100 m/s.	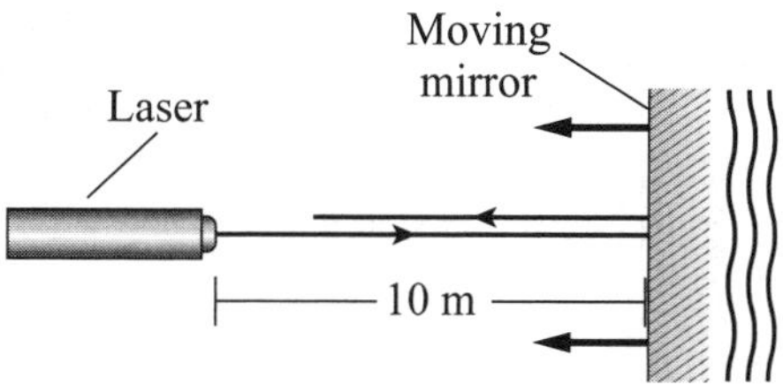 Laser, Moving mirror, 10 m	The laser light returns 6.673283×10^{-8} s later.

Analysis Draw a conclusion about the speed of light. *Hint:* Make an analogy between the cart and the laser light. Also, remember that the index of refraction of air is 1.0003 which has an effect on the speed of light.

25.1.2 Test an idea The experiment below tests the idea that the speed of light is independent of reference frame (it has a value of 299,792,458 m/s in vacuum). The experiment involves an unstable elementary particle that emits electromagnetic waves when it decays.

a. Analyze the experiment and make predictions using the idea under test.

Experiment	Prediction of EM wave travel time to the detector (based on speed of light being independent of reference frame)	Prediction of EM wave travel time to the detector (based on speed of light being relative to the emitting object)	Outcome of experiment
A beam of unstable particles moves to the right at 10^8 m/s. The particles decay 5 m to the right of an EM wave detector.			The average travel time of the EM waves is $(1.7 \pm 0.4) \times 10^{-8}$ s

b. Based on the results of this testing experiment, what is your judgment about these two hypotheses? Explain your reasoning.

25.2 Conceptual Reasoning

25.2.1 Reason The table below describes various phenomena being observed from several different reference frames. For each phenomenon, determine which reference frame(s) is the proper reference frame. Then, for each of the other reference frames determine whether the time interval between the events will be longer, shorter, or the same as the proper time interval.

Phenomena	Events	Reference frames		
1. A cheetah runs across the savannah, its heart beating.	Successive beats of the cheetah's heart.	**a.** A tourist standing on the ground watching.	**b.** The cheetah.	**c.** A hippopotamus floating in the water where the cheetah stops for a drink.
2. Two people are playing catch. One person throws the ball at a 45 degree angle. The ball is caught by the other person.	The ball being thrown, and the ball being caught.	**a.** The ball.	**b.** The person throwing the ball.	**c.** A person standing on the spot below where the ball reaches its greatest height.
3. You drive from San Francisco to Los Angeles, using your GPS to find your way.	You leave San Francisco; you arrive at Los Angeles	**a.** The GPS satellites.	**b.** You.	**c.** The road you are driving on.

Answers:

Phenomenon	Proper reference frame	Time interval from other reference frames (specify which frame)	
(1)			
(2)			
(3)			

25.2.2 Reason Aaron is able to throw a baseball at 80 mph. If Aaron were to do this from the back of a pickup truck moving at 50 mph with respect to the ground, then the baseball would be moving at 130 mph with respect to the ground. This seems very reasonable. But, what if the truck were moving at $0.5c$ and Aaron could throw the baseball at $0.6c$? Going further, what if Aaron were shooting a laser pointer rather than throwing a baseball? Discuss the relevant issues here and suggest ideas that might resolve them.

25.2.3 Reason A deep space probe is positioned halfway between two stars, A and B, each 1 light-year away. These stars are near the end of their life cycle and are each due to become supernovae soon. A scientist in a spacecraft is traveling from star A to star B. Just as the scientist passes the probe he sees both stars turn supernova at precisely the same time! "What a coincidence! Each star must have exploded precisely one year ago." He also decides to download the probe's data. To his astonishment, the probe reports that the two stars did not explode simultaneously. Use your understanding of relativity to carefully explain how this is possible.

25.2.4 Reason When you studied magnetism you learned that two parallel wires with electric currents in the same direction would exert an attractive magnetic force on each other. One current produces a magnetic field, which in turn exerts a magnetic force on the other current. This explanation is being made from a reference frame at rest with respect to the wires. What about explaining it from a reference frame at rest with respect to the current? Because the charged objects that make up the current are at rest in this reference frame, all magnetic forces exerted on them will be zero!

To resolve this, choose the system of interest to be a single free electron in one of the wires. Assume that the other wire is made of both positively and negatively charged objects. Consider (1) what the motion of these objects is relative to the system, and (2) the phenomenon of length contraction.

25.3 Quantitative Concept Building and Testing

25.3.1 Derive Muons produced in the upper atmosphere by cosmic rays reach Earth's surface because their lifetimes are extended in Earth's reference frame by time dilation. But, how can this be explained in a muon's reference frame where the muon's lifetime is not extended? Is it possible that the distance the muon must travel to reach Earth's surface is *less* in the muon's reference frame? Let's examine the situation.

a. First, write a kinematics equation describing the motion of the muon in Earth's reference frame. Assume the muon moves in a straight line with constant velocity.

b. Next, write a corresponding equation describing the motion of Earth's surface in the muon's reference frame.

c. The velocities in each of these equations have the same magnitude. Use this to combine the equations.

d. The two time intervals in the equation are related through time dilation. Use this to arrive at a new equation relating the distance the muon travels in Earth's reference frame to the distance Earth's surface travels in the muon's reference frame.

You've just used the ideas of relativity to learn that distances depend on reference frame, a phenomenon known as length contraction.

25.3.2 Analyze The one-dimensional classical velocity transformation equation is:

$$v_{\mathrm{OS},x} = v_{\mathrm{OS}',x} + v_{\mathrm{S'S},x}$$

The one-dimensional relativistic velocity transformation equation is:

$$v_{\mathrm{OS},x} = \frac{v_{\mathrm{OS}',x} + v_{\mathrm{S'S},x}}{1 + \dfrac{v_{\mathrm{OS}',x}v_{\mathrm{S'S},x}}{c^2}}$$

The goal of this task is to determine when it is reasonable to use the simpler classical equation, and when the relativistic equation should be used.

a. First, devise a situation where velocity transformation is relevant. Describe this situation.

b. Now, use a limiting case analysis to investigate the circumstances under which the classical and relativistic equations produce significantly different results (a 5% or greater difference, just to be specific).

25.3.3 Design an experiment Your goal is to design an experiment that will test the following hypothesis:

Doubling the momentum of an object doubles its velocity.

a. Describe your experimental design. Include the prediction and any assumptions made.

b. Explain why your experiment is likely to succeed.

25.3.4 Represent and reason In this activity, momentum is considered as a function of the speed of an object.

a. On a single graph, plot the magnitude of the momentum of an object as a function of its speed 1) using the classical equation and 2) using the relativistic equation. The range of speeds covered by the graph should go *beyond* the speed of light.

b. Discuss consistencies/inconsistencies between the two functions. Discuss any interesting features they have.

c. Repeat part (a), but for the kinetic energy of an object.

d. Repeat part (b) but for the kinetic energy of an object.

25.4 Quantitative Reasoning

25.4.1 Estimate During his working career, David commutes 1 hour to work and 1 hour back. Anya, however, works from home. Each is given a very precise watch that they carry with them throughout their careers. After they retire, will Anya find David's watch to be behind hers, ahead of hers, or synchronized with hers? Explain. If David's watch is not synchronized with hers, estimate the difference between the clock readings of the two watches.

25.4.2 Equation Jeopardy Come up with a problem that is consistent with the solution below.

$$24\text{ hr} = \frac{\Delta t_0}{\sqrt{1 - ((2.8 \times 10^8\text{m/s})/c)^2}}$$

$$\Rightarrow \Delta t_0 = (24\text{ hr})\sqrt{1 - ((2.8 \times 10^8\text{m/s})/c)^2}$$

$$\Rightarrow \Delta t_0 = 8.6\text{ hr}$$

25.4.3 Equation Jeopardy Come up with a problem that is consistent with the solution below.

$$2(2.4 \times 10^{-28}\,\text{kg})c^2 = \frac{2(9.1 \times 10^{-31}\,\text{kg})c^2}{\sqrt{1-(v/c)^2}}$$

$$\Rightarrow \sqrt{1-(v/c)^2} = \frac{9.1 \times 10^{-31}\,\text{kg}}{2.4 \times 10^{-28}\,\text{kg}}$$

$$\Rightarrow (v/c)^2 = 1 - \left(\frac{9.1 \times 10^{-31}\,\text{kg}}{2.4 \times 10^{-28}\,\text{kg}}\right)^2$$

$$\Rightarrow v = 0.999993c$$

25.4.4 Analyze At what speed must a proton be traveling so that its kinetic energy is equal to its rest energy? What about an electron? Explain the relationship between your two answers.

25.4.5 Regular problem A futuristic starship is going to be used to travel to Alpha Centauri (4 light-years away from Earth). The starship has two stages. The first stage accelerates the ship to a speed of $0.5c$ relative to Earth. The first stage then separates from the rest of the ship and the second stage activates. The second stage accelerates the ship to a speed of $0.9c$ relative to the first stage.

a. Determine how much time passes on Earth during the journey, assuming the acceleration phases of the ship take only a small fraction of the total travel time.

b. How much time passes on the ship?

c. What distance does the ship travel in its reference frame?

25.4.6 Evaluate the solution A starship moves away from Earth at 40% of light speed. After traveling for 1 month (Earth time) the starship launches a probe back towards Earth at 50% of light speed with respect to the ship. How fast is the probe traveling with respect to Earth?

Solution:

$v_{PE} = v_{ES} + v_{PS} = 0.5c + 0.4c = 0.9c$

a. Identify any errors in the solution.

b. Provide a corrected solution if there are errors.

25.4.7 Analyze A galaxy that emits primarily blue light appears primarily red according to detectors here on Earth. Explain how this is possible and describe in as much quantitative detail as you can the motion of the galaxy that would result in this happening.

25.4.8 Regular problem The sun has a total power output of about 4×10^{26} W. How much does its mass decrease each year? What percentage of its total mass will it lose in the next 4.5 billion years (the approximate remainder of its lifespan)?

25.4.9 Determine everything you can Two stars form a binary star system. Each has a mass of 10^{30} kg and travels in a circular path of radius 10^7 m as shown. Each star emits primarily yellow light (550 nm wavelength). Astronomers on Earth (located far to the left of the figure) analyze the light coming from this system over time. Determine the value of as many relevant physical quantities as you can.

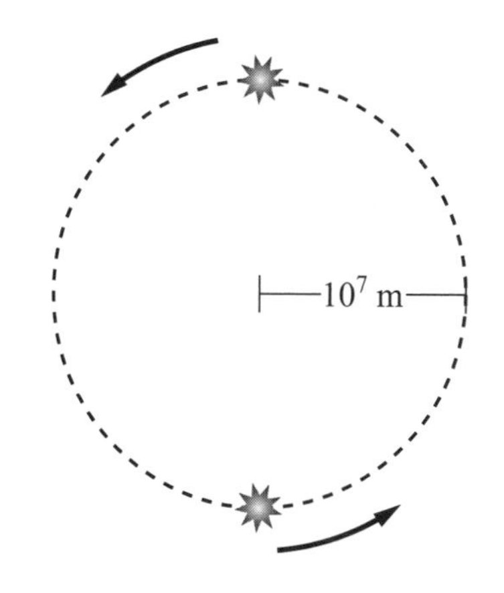

26 Quantum Optics

26.1 Qualitative Concept Building and Testing

26.1.1 Observe and explain Rub the top of an electroscope (see the sketch at right; rubbing is necessary to "transfer" more charge from the charged object to the electroscope) with differently charged objects. Note what happens to the electroscope leaves in each case. Provide an explanation for the outcome of each experiment.

The experiment	Outcome of experiment	Explanation
a. Rub with a negatively-charged foam pipe. Then remove the pipe. (The foam pipe is charged negatively if rubbed with wool.)	The leaves remain deflected for a long time.	
b. Rub with a positively-charged foam pipe. Then remove the pipe. (The foam pipe is charged positively if rubbed with plastic wrap.)	The leaves remain deflected for a long time.	
c. Rub with a negatively-charged foam pipe. Remove the pipe. Then shine a flashlight on the metal disc on the top.	The leaves remain deflected for a long time.	
d. Rub with a negatively-charged foam pipe. Remove the pipe. Then shine an ultraviolet (UV) light on the metal disc.	The electroscope discharges immediately (as indicated by the leaves moving together).	
e. Repeat (c) and (d), only this time use a positively charged foam pipe.	The leaves remain deflected for a long time.	

The discharge of the negatively charged electroscope due to exposure to light is called a *photo-electric effect.*

26.1.2 Explain Assume you know that free electrons inside metals can move and positively charged ions cannot. Assume that light is an electromagnetic wave in which $\vec{E}$ and $\vec{B}$ fields oscillate periodically.

a. Use your knowledge of the effects of the electric field on charged particles to explain the effect of the light on the negatively charged metal surface in Activity 26.1.1d (think microscopically).

b. Repeat for Activity 26.1.1e.

26.1.3 Observe Physicists use an evacuated glass container such as the one shown to study the photoelectric effect. Light of different frequencies can shine through a quartz window onto a metal plate connected to the negative pole of the battery. Such a plate is called the cathode container. When no UV light shines on the cathode, the ammeter does not register any current.

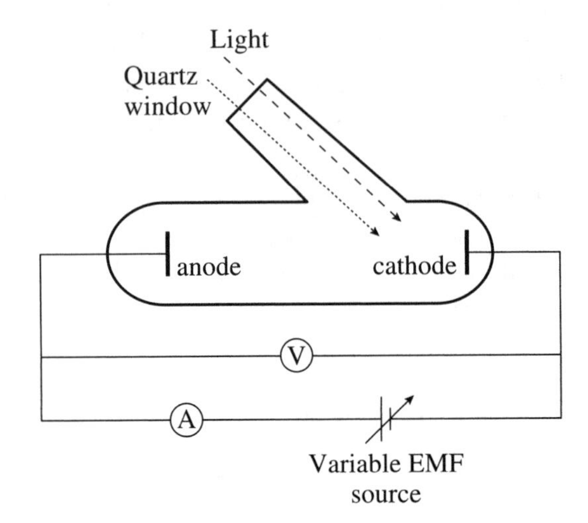

a. When a UV light shines on the cathode, the ammeter registers a current in the circuit. Explain how UV light can cause the current. Note: A voltmeter has very high electric resistance.

b. When the battery is turned off but the UV light still shines on the cathode, the ammeter registers a small current—much smaller than in case (a). Explain.

c. If the polarity of the battery is reversed, then the plate on which the light shines is at a higher potential than the plate on the left side. When this reversed potential difference reaches a certain value, the ammeter stops registering any current. Explain why. (This potential difference is called a stopping potential difference, ΔV_s.)

26.1.4 Observe and explain Referring to the apparatus in Activity 26.1.3, the current is stopped when the metal cathode on which the light shines is at a positive potential relative to the more negative potential of the plate on the left side.

a. Surprisingly, you find that the potential difference that stops the current (see Activity 26.1.3c) does *not* depend on the intensity of light. The electric current induced by high-intensity light is stopped as easily as electric current induced by low-intensity light. Explain this observation.

b. While the stopping potential does not depend on the intensity of light, it does depend on the color: the higher light frequency (UV versus visible, violet versus red), the higher the stopping potential. How can you explain this observation?

26.1.5 Explain You observed in Activity 26.1.1 that visible light does not discharge a negatively charged electroscope. The increase of the intensity of visible light does not make a difference—no current is observed. However, even at very low intensity UV light produces electric current. Explain.

26.1.6 Represent and reason Analyze and represent the following two historical findings:

a. In 1902 the German physicist Phillip Lenard suggested an explanation for the photoelectric effect. He proposed that light being an electromagnetic wave knocked out electrons from the surface of the cathode by continuously exerting force on the electrons. These electrons were then accelerated by the electric field of the battery inside the glass tube, reached the opposite electrode and closed the circuit. He reasoned that if the energy of interaction between electrons and the lattice is negative and equal to $-\phi$, and light had the energy E_{light} larger than ϕ, then the leftover energy of light would be given to the electrons in the form of kinetic energy K_f. Draw a new energy bar chart that represents this energy exchange process between light and electron-lattice system during the photoelectric effect.

$U_{qi} + E_{light} = K_{ef}$

0

b. In 1905 A. Einstein suggested that the photoelectric effect can be explained assuming that light is a stream of bundles of energy (photons), which are individually absorbed by electrons in the metals. The energy of each photon is determined by the frequency of light ($E = hf$, where $h = 6.63 \times 10^{-34}\ \text{J} \cdot \text{s}$); the higher the frequency, the higher the photon energy. An electron is bound to the crystal lattice, and the energy of the interaction of one electron with the lattice is $-\phi$. An electron can absorb only the energy of one photon. Draw a new energy bar chart that represents the energy exchange process between a photon and an electron-lattice during the photoelectric effect.

$U_{qi} + E_{light} = K_{ef}$

0

26.1.7 Explain Use the photon model of light to explain why light exerts pressure on a surface on which it shines. On what surface would the same photon exerts a greater pressure: a shiny one or a black one? Explain. *Hint:* Think about elastic collisions and inelastic collisions.

26.2 Conceptual Reasoning

26.2.1 Represent and reason Draw an energy bar chart to represent the following process: a photon of light hits a metal and gets absorbed. The energy of the photon is exactly equal to the magnitude of the negative electric potential energy of the interaction between the electron and the lattice.

26.2.2 Represent and reason Draw an energy bar chart to represent the following process: a photon of light hits a metal and ejects an electron with zero kinetic energy.

$U_{qi} + E_{light} = K_{ef}$

0

26.2.3 Represent and reason Draw an energy bar chart to represent the following process: a photon of light hits a metal and ejects a fast-moving electron.

$U_{qi} + E_{light} = K_{ef}$

0

26.3 Quantitative Concept Building and Testing

26.3.1 Observe and explain: In the table below try to use the *wave model* of light to explain each of the experimental results involving the apparatus shown in Activity 26.1.3.

The experiment	Result	Explain using wave theory.
a. As the light intensity increases, the electric current changes as shown in the graph.	Current 0 Intensity	
b. The dependence of the electric current on the potential difference across the electrodes is shown. Explain the steady part of the graph. The intensity of light remains constant during the experiment.	Current V_s 0 ΔV Note: ΔV is positive when the left metal plate (the cathode) connects to the positive battery terminal.	

The experiment	Result	Explain using wave theory.
c. Use the wave model to try to explain why the current decreases to zero when there is a negative stopping potential difference ΔV_s in (b).		
d. You repeat the previous experiment for increasing intensity light. The stopping potential difference ΔV_s does not change.		
e. The potential difference ΔV_s needed to stop the electric current depends on the light frequency—see the graph.	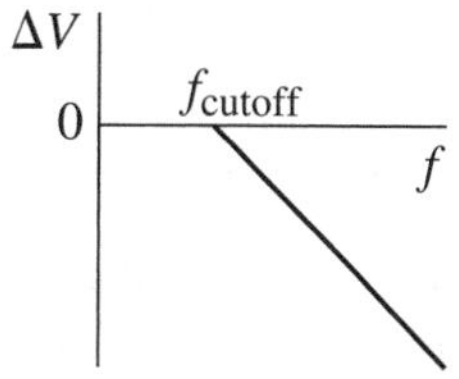 ΔV, 0, f_{cutoff}, f	

26.3.2 Observe and explain: Use a photon model to try and explain the results of the five experiments in Activity 26.3.1. Before starting, carefully describe the new model.

The experiment	Result	Explain using the photon model.
a. As the light intensity increases, the electric current changes as shown in the graph.	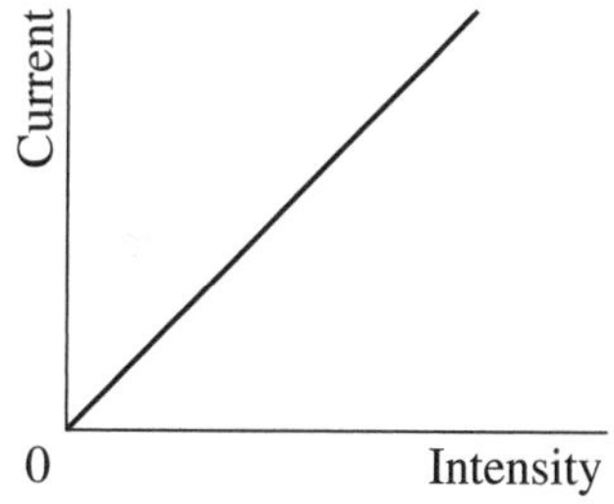 Current, 0, Intensity	
b. The dependence of the electric current on the potential difference across the electrodes is shown (measured by the voltmeter) when the intensity of light remains constant. Explain the steady part of the graph.	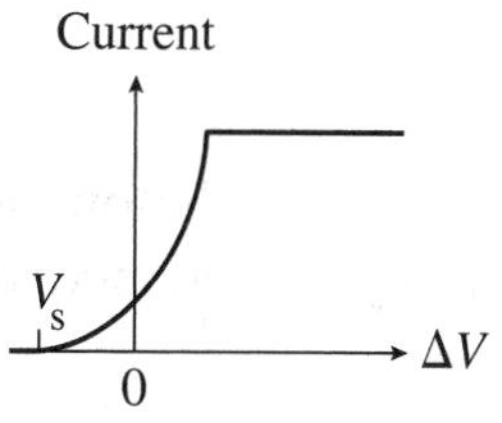 Current, V_s, 0, ΔV Note: ΔV is positive when the left metal plate (the cathode) connects to the positive battery terminal.	

(continued)

The experiment	Result	Explain using the photon model.
c. Use the photon model to explain why the current decreases to zero when there is a negative stopping potential difference ΔV_s in (b).		
d. You repeat the previous experiment for increasing intensity light. The stopping potential difference ΔV_s does not change.		
e. The potential difference ΔV_s needed to stop the electric current depends on the light frequency—see the graph.	ΔV vs. f graph: 0, f_{cutoff}, f	

26.3.3 Test your ideas: Light passes through double slits and illuminates a screen producing the double-slit interference pattern observed and explained in Chapter 23 using a wave model of light. However, we now have experiments that can only be explained if we consider light to be a stream of photons—a photon model for light.

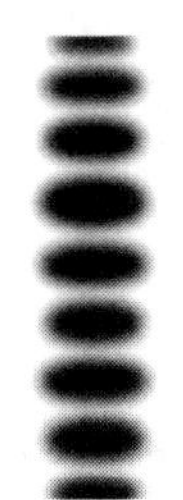

a. First use the wave model of light to predict what you expect to observe on the screen if you illuminate the double slits with very low intensity light. Then use the photon model of light to make a prediction.

b. Reconcile these two models of light with the outcome of this experiment, which is described in the textbook in Section 26.4.

26.3.4 Evaluate Your friend Mark says that the photon model of light is not a new model, but just an old particle model of light. How can you convince Mark that his opinion is not correct?

26.3.5 Derive We found that light has a particle-like behavior when interacting with matter. If a photon is particle-like, it should have momentum.

a. Write an expression for the energy of a photon and set it equal to the relativistic energy of a particle with mass m moving at speed c. (Massive particles do not move at the speed of light, c. Here we are assuming that the photon is a particle moving at speed c.) Use this to determine an expression for the equivalent mass of a photon.

b. Write an expression for the momentum of a photon—it moves at speed c and has the equivalent mass m derived in part (a). You should now have an expression for the momentum of a photon in terms of its frequency.

c. Rewrite this expression in terms of the wavelength of the photon.

d. Compare a photon to a classical particle—such as a billiard ball. What are the properties that are similar? What are the properties that are different?

26.3.6 Represent and reason: Find v_2 when a moving billiard ball collides elastically with a stationary ball.

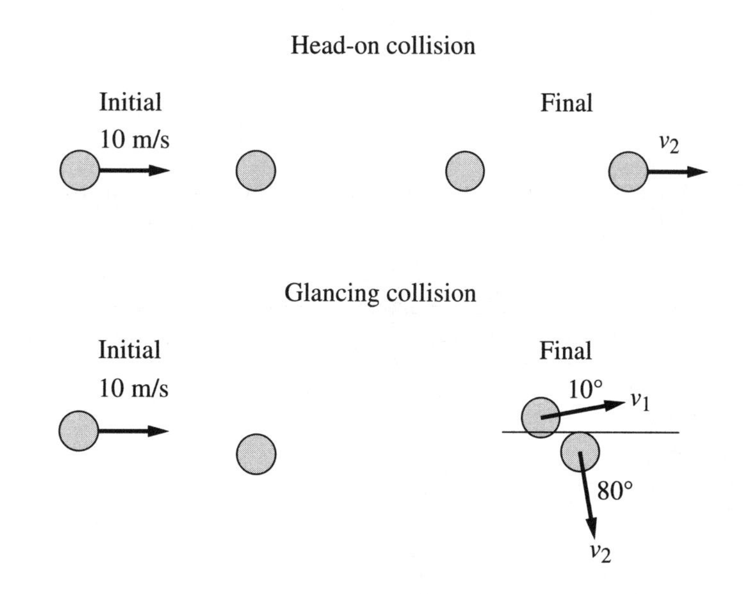

26.3.7 Test your idea Use concept of the momentum of a photon to analyze the following process. A photon of wavelength λ scatters off a particle of mass m that is at rest. The photon moves off at angle θ and the particle at angle α. In addition to the change in the direction of travel, what else should happen to the photon? Use momentum bar charts to analyze the collision; remember that momentum is a vector quantity and you need to make two component bar charts.

26.4 Quantitative Reasoning

26.4.1 Observe and explain: The experiment described in Activities 26.3.1(e) and 26.3.2(e) is repeated for different types of metal. The results are shown at the right. The work functions of several metals are given in the table below. Use the photon model of light to explain these observations.

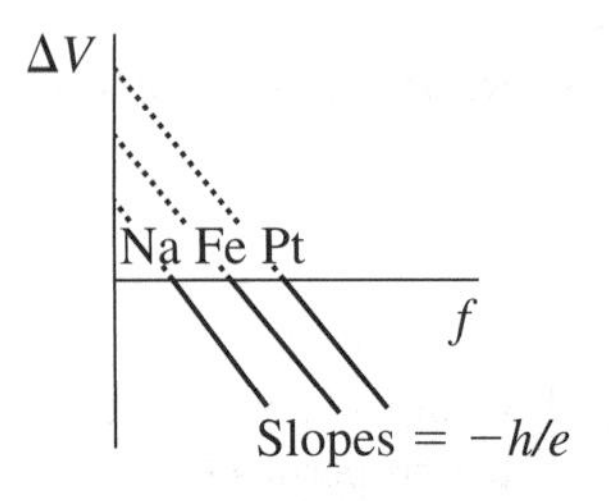

Metal	Aluminum (Al)	Copper (Cu)	Iron (Fe)	Sodium (Na)	Platinum (Pt)	Zinc (Zn)
Work function (eV)	4.1	4.7	4.7	2.3	6.4	4.3

26.4.2 Practice Use some of the information from Activity 26.4.1 to determine the minimum frequency light that will cause a photoelectric emission (a) from sodium and (b) from iron.

26.4.3 Represent and reason A 200-nm light source shines on a sodium surface. Represent with a bar chart and an equation each of the processes described below.

Word description of a process	Energy bar chart description of the process	Mathematical description of the process
a. The process starts with a 200-nm photon and ends just after the ejection from the sodium of an electron that moves at maximum speed.		
b. The same as above, only now the electron has traveled across the photoelectric tube. A potential difference has stopped the electron just before reaching the metal collector electrode.		

26.4.4 Represent and reason: Draw an energy bar chart for each of the three photoelectric effect processes described below. Then write a mathematical description for each process (if it can occur).

Word description of a process	Energy bar chart description of the process	Mathematical description of the process
a. The process starts with a photon whose energy is $hf > \phi$ and ends just after the ejection of an electron from a metal surface while the electron moves at maximum speed.		
b. The same as above only now the energy of the photon is $hf = \phi$.		
c. The same as above only now the energy of the photon is $hf < \phi$.		

26.4.5 Evaluate Your friend Tamara is working on physics problems. She provides the following answer to the problem below. Evaluate the answer to see if you agree. If not, correct the answer.

The problem:

a. What is maximum speed of electrons leaving the copper cathode of work function 4.7 eV? The stopping potential difference is 3.0 V.

b. What is the wavelength of this light?

Proposed solution:

a. The electron's kinetic energy must have been enough to traverse a region with a $(4.7 + 3.0)$ V potential difference. Thus, using energy conservation we get:

$$\frac{1}{2}mv^2 = e\Delta V$$

or

$$v = \left(\frac{2e\Delta V}{m}\right)^{1/2} = \left(\frac{2(1.6 \times 10^{-19}\,\text{C})(7.7\,\text{V})}{(9.11 \times 10^{-31}\,\text{kg})}\right)^{1/2} = 1.64 \times 10^6\ \text{m/s}.$$

The photon's energy equals the kinetic energy that the electron acquired from it, which equals the stopping energy of the electric potential difference:

$$hf = \frac{1}{2}mv^2 = (-e)(-V_s)$$

b. Note that $f = c/\lambda$. Thus,

$$\lambda = \frac{hc}{eV_s} = \frac{(6.63 \times 10^{-34}\,\mathrm{J\cdot s})(3.0 \times 10^8\,\mathrm{m/s})}{(1.6 \times 10^{-19}\,\mathrm{C})(3.0\,\mathrm{J/C})} = 414 \times 10^{-9}\,\mathrm{m} = 414\,\mathrm{nm}.$$

26.4.6 Explain You have a laser pointer. Remember that lasers emit monochromatic light—light having a single frequency.

a. How is the color of the laser beam related to the energy of the photons?

b. How is the intensity of the light (energy/time) related to the number of photons per second?

c. How is the intensity of the light (energy/time) related to the frequency of the photons?

27 Atomic Physics

27.1 Qualitative Concept Building and Testing

27.1.1 Observe and explain In the late 19th century, several observational facts had been established that any model of the atom had to explain. One included observations of light emitted by low-density gases. The experimental set up for these observations looks as follows: A glass tube filled with gas has two metal electrodes in it. The electrodes are connected to the poles of a high-voltage power supply. When the power is on, the gas in the tube glows. Different gases glow with different colors.

a. Use a spectroscope (or a simple grating) to examine light emitted by a tube filled with hydrogen. Describe your observations.

b. Compare your observations to observations of the light of a bright lightbulb filament when viewed through the spectroscope.

27.1.2 Observe and explain To help determine atomic structure, Philip Lenard shot cathode rays (electrons) at a thin sheet of aluminum. He observed that electrons moved through the foil without any deflection. At this time physicists already knew that atoms are electrically neutral and contain several electrons. Electrons were known to have a very small mass, much smaller than the mass of the atom. An atomic model developed by J. J. Thomson included positive charge equally distributed within the atom with electrons embedded in it—like plums in a positively-charged plum pudding. Explain how this model accounted for Lenard's experiments.

27.1.3 Observe and explain Alpha particles are elementary particles emitted by some materials, such as uranium; an alpha particle has an electric charge of $+2e$ and the mass of a helium atom.

To help determine atomic structure, Rutherford and his graduate students shot alpha particles at a thin sheet of mica or metals. They looked at places where the particles hit a fluorescent screen after passing through the thin sheets. They also looked for alpha particles that possibly bounced backward off the thin sheet (on the same side as the incoming alpha particles). They observed that most alpha particles moved through the sheet with minor deflection. However, a small fraction of the alpha particles bounced backward. At this time physicists already knew that atoms are electrically neutral and contain several electrons. Electrons were known to have a very small mass, much smaller than the mass of the atom. What could Rutherford suggest about the distribution of the positive charge in the atom to explain his experiments?

27.1.4 Design an experiment Three solid discs are secured below and hidden under a board above. The discs of diameter *d* are spread randomly across a distance *L*, as shown at the right. You obtain BBs (ammunition for BB guns) that have a diameter much smaller than diameter *d*. You roll these small BBs through the opening under the boards and observe that some of the BBs are scattered to the sides and even backwards. Many BBs move straight ahead with no scattering. Design an experiment and carefully describe how you can estimate the diameter of the hidden discs by this BB scattering experiment. You can roll the BBs many times.

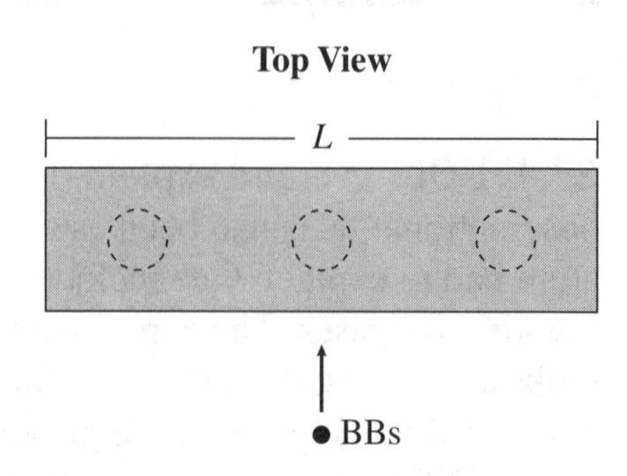

27.1.5 Explain Using similar ideas Rutherford estimated the size of a positively charged atomic nucleus to be about 10^{-15} m. At that time people already knew the approximate size of atoms to be about 10^{-10} m. How could he reconcile these two numbers? What possible model of the atom could he suggest that would explain Thomson's cathode ray experiments, Lenard's experiments, the experiments of his colleagues, and the fact that atoms are neutral and stable?

27.1.6 Explain Does the planetary atomic model proposed by Rutherford and information about the radiation of the electromagnetic waves that you learned in Chapter 24 (any accelerated electrically charged particle emits electromagnetic radiation):

a. Produce a stable model for the hydrogen atom with one electron circling around the nucleus? Explain.

b. Explain the lines in a spectrum of hydrogen (see Activity 27.1.1).

27.1.7 Evaluate In 1913 Niels Bohr devised a solution to save Rutherford's planetary model from collapsing and to explain the line spectrum of hydrogen. He suggested that electrons are charged particles, which when bound in an atom behave in this way: they obey Newton's laws, interact with the nucleus via Coulomb forces, and do not radiate energy when the atoms are in the preferred energy states (called stable energy states) even though electrons in those atoms are moving in a circle. But as a trade off, atoms can only have energies corresponding to the energies of the stable states. To change its state, the atom needs either to emit a photon (when the energy decreases), or to absorb a photon (when the energy increases). Discuss how this model helps to explain observations of gas spectra and observations conducted by Rutherford and his colleagues.

27.1.8 Observe and explain Imagine a cathode ray tube with a very hot cathode. Electrons in the metal wire have considerable kinetic energy and escape the cathode and accelerate toward the positively charged anode. Some of these electrons traveling at about the same speed pass through an opening in the anode and then reach two closely-spaced narrow slits. A fluorescent screen beyond the slits indicates places where electrons hit the screen. The pattern of electrons that hit the screen is shown below—first only a few electrons, and later after many have hit it. Develop two or more explanations for this observed pattern.

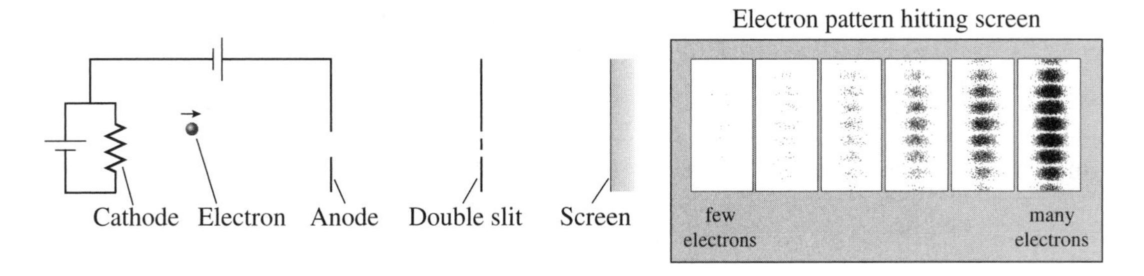

27.1.9 Observe and explain You repeat the experiment described in Activity 27.1.8 only now you increase the electric potential difference from the cathode to the anode so that the electrons are moving faster when they reach the two slits. You find that the pattern on the screen changes in a systematic way—the faster the moving electrons, the closer the spacing of the bands of electrons hitting the screen. Try to account for this observation.

27.2 Conceptual Reasoning

27.2.1 Reason Discuss different aspects of the energy of an electron-nucleus system in a hydrogen atom in Bohr's model. Fill in the table that follows.

Does the system posses kinetic energy? Explain.	**Does the system posses electric potential energy? Explain.**	**Does the system possess gravitational potential energy? Explain.**	**Draw an energy bar chart representing one energy state of the system. Decide on the scale and the direction of the bars.**

27.2.2 Explain Explain the observations of light emitted by gases in tubes using the photon model of light. What do the colors of the lines correspond to? What does the brightness of the lines correspond to?

27.2.3 Reason In Bohr's model of the atom, the electron revolves around the nucleus similar to the planets revolving around the Sun (the solar system is an analogy for the atomic system). To understand the model better, answer the questions about the analogy in the table below.

Find the analogous objects in two systems.		Find analogous interactions in two systems.		Find the aspects of the atomic model that do not have analogies in the solar system model.
Atomic model	**Solar system**	**Atomic model**	**Solar system**	**Atomic model only**

27.2.4 Test your idea An electron beam such as described in Activity 26.1.6 passes through a grating instead of double slits.

a. You keep the speed of the electrons fixed but use different gratings that have first few slits per centimeter and then other gratings with increasing numbers of slits per centimeter. Predict how the pattern of electrons hitting the screen beyond the gratings varies. Give reasons for your predictions.

b. Next you keep the grating fixed (using the one with the most slits per centimeter) and now vary the electric potential from the cathode to the anode. Qualitatively, predict how you think the pattern of the electrons hitting the screen changes and explain carefully the reasons for your predictions.

c. Compare and contrast a photon beam with an electron beam. What are the similar properties of both beams? What are the properties that only one beam possesses?

27.3 Quantitative Concept Building and Testing

27.3.1 Reason Bohr's postulates suggest that the frequency of the emitted photons can be found using the following mathematical expression: $U_{a2} - U_{a1} = hf$, where U_{a2} is the energy of the atom in state 2, U_{a1} is the energy of the atom in state 1, and hf is the energy of the emitted photon.

a. Does this relation make sense in terms of units? Explain.

b. Does it make sense in terms of the observations of light emitted by low-density gases? Explain.

c. Does this expression make sense in terms of light emitted by the lightbulb filament? Explain.

27.3.2 Reason Bohr's postulates suggest that an the atom can only be in the states where the product of the electron's mass, speed, and distance from the nucleus, mvr (its orbital angular momentum) equals a positive integer times $h/2\pi$.

a. Does this expression make sense in terms of units? Explain.

b. What does this expression imply about the velocity of the electron in different energy states?

27.3.3 Derive Use Bohr's postulates and your knowledge of electrostatic interactions and circular motion to find the value of the smallest electron orbit in the hydrogen atom.

a. What is the electric potential energy of the electron–nucleus system? Is it positive or negative? Assume that the atom is neutral.

b. What is the kinetic energy of the system?

c. What is the total energy of the system?

d. How is the velocity of the electron related to the distance between the electron and the nucleus? Use your knowledge of circular motion and Bohr's third postulate to answer this question.

e. Combine the results of (a) through (d) to determine the smallest-radius electron orbit in a hydrogen atom (the atom is said to be in the ground state).

f. Evaluate your result.

27.3.4 Derive Using Bohr's postulates and the relationships that are assumed to be valid to describe the hydrogen atom (namely Coulomb interactions between the electron and the nucleus, the circular motion of the electron, and Newton's second law), develop a plan to determine the wavelengths of bright lines that you can see when looking at the hydrogen tube through a spectrometer. Do not calculate anything yet. Remember that a human eye is sensitive to the wavelengths of 300–800 nm.

27.3.5 Test your idea Use the plan that you outlined in Activity 27.3.4 to predict the wavelengths of visible light emitted by hydrogen gas.

a. What energy transitions are occurring to produce these wavelengths?

b. Compare the spectrometer lines to your predictions. Were the predictions matched?

27.3.6 Reason Imagine that a hydrogen atom is in its ground state.

a. What is the total energy of the atomic system?

b. In part (a) you obtained a negative value. Does it make sense? Explain.

c. What is the minimum energy of the photon that a ground-state hydrogen atom needs to absorb for the electron to become free? Explain.

d. Express the value that you obtained in part (c) in the units of electron volts. One electron volt is the energy that an electron acquires when it passes through a potential difference of 1 V. The magnitude of the charge of the electron is 1.6×10^{-19} C.

27.3.7 Observe and explain Using a variety of detectors one can observe spectral lines from hydrogen atoms in the visible, ultraviolet and infrared parts of the electromagnetic spectrum. The series of four lines in the visible part were called the Balmer series with wavelengths of 656.21 nm; 486.07 nm; 434.01 nm, and 410.12 nm. A series of lines in the infrared part of the spectrum (3 lines) were called the Paschen series. And finally, the most recently discovered series in the UV part of electromagnetic wavelengths (4 lines) were called the Lyman series. The names of the series are the names of the physicists who found the lines. They also found that the wavelengths of light for the lines in all series can be empirically described by the formula

$$1/\lambda = R_H\left(\frac{1}{n_f} - \frac{1}{n_i}\right).$$

a. To understand the meaning of the symbols in the empirical formula, fill in the table.

Determine n_i and n_f for the Balmer series.	Determine the value of the constant R_H using the knowledge of the wavelengths of the Balmer series.	Determine n_i and n_f for the Lyman series.	Determine the wavelengths of the 4 lines in the Lyman series.	Determine n_i and n_f for the Paschen series.

b. Derive the empirical formula using the knowledge of Bohr's model of the atom.

27.3.8 Represent and reason In the figure at the right the hydrogen atom is represented as a series of spheres corresponding to the possible orbits of the electron corresponding to the allowed energies. The orbit with $n = 1$ is called a ground state. Use this representation to answer the following questions:

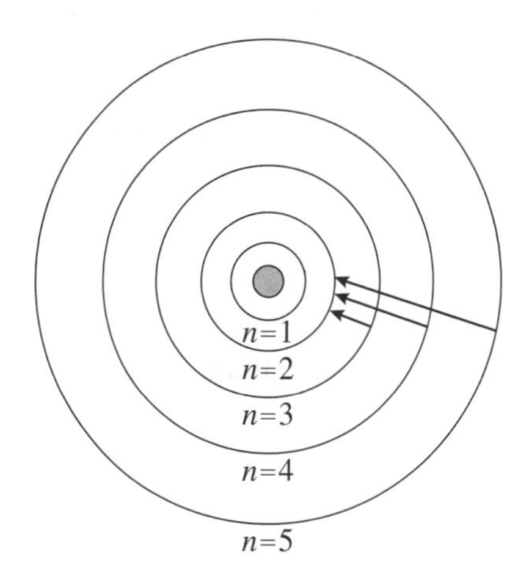

a. What is the approximate scale of the picture, i.e., how large is the radius of the $n = 1$ sphere?

b. Where in this picture is the electron when the atom has the smallest energy? The largest energy?

c. Which series of spectral lines do the arrows represent? How do you know?

d. Draw similar arrows corresponding to the other two series.

27.3.9 Observe and explain Imagine using a set up as in the Activity 27.1.6. Only this time light emitted by a hot lightbulb filament passes through a container with cold atomic hydrogen. A person observes the light though a spectrometer. She sees a colored band in which red color slowly turns into orange, then into yellow, then into green, into blue, and finally into violet (continuous spectrum) with dark lines. The wavelengths at which the dark lines appear are equal to: 656.21 nm; 486.07 nm; 434.01 nm, and 410.12 nm. Explain why a person sees dark lines at these particular wavelengths. Assume that the lightbulb radiates light of all wavelengths.

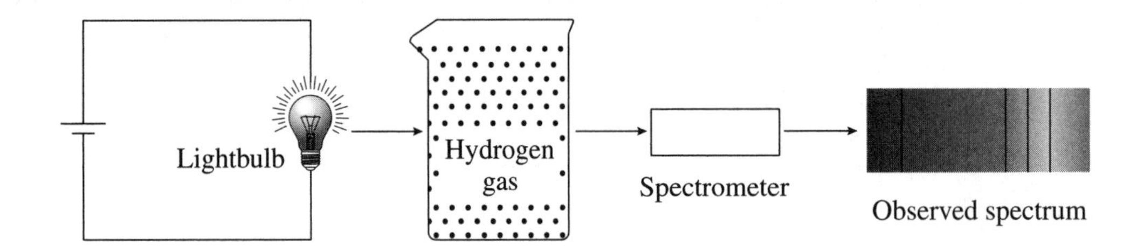

27.3.10 Reason Electrons could be described as having wave-like properties. Their wavelength can be determined using the relationship $p = \frac{h}{\lambda}$. Then one can suggest that if an electron has a wavelength that allows it to form a standing wave while orbiting the nucleus, then this state will be a stable state.

a. Show how this idea can explain Bohr's third postulate. *Hint:* Think of what wavelength an electron's standing wave can have when it occupies a certain circular orbit.

b. Discuss how this idea changes Bohr's model of the atom.

27.3.11 Derive A small atomic-size particle (like an electron or proton) of mass m is moving with speed v.

a. Write an expression for the momentum of this particle. Set this expression equal to the expression for the momentum of a photon (the expression in terms of the wavelength associated with the photon).

b. Solve this equation for an expression for the wavelength of the moving atomic-size particle.

27.4 Quantitative reasoning

27.4.1 Reason Determine the temperature of the hydrogen gas at which most of its atoms are ionized. Indicate all assumptions that you made.

27.4.2 Represent and reason Use the knowledge of allowed energy states of a hydrogen atom to draw a scaled energy diagram for the hydrogen atom.

27.4.3 Represent and reason Complete the energy diagrams to represent each of the following processes:

An atom absorbs a 15 eV photon.	An atom in the state with $n = 5$ changes into the state with $n = 2$.	An atom absorbs a photon and changes from the state with $n = 1$ to the state with $n = 4$ and then after 10^{-8} s changes to the state with $n = 2$.	Hydrogen gas is placed in the electric field so the potential difference between the ends of the container slowly changes.	Hydrogen gas surrounds a hot object whose temperature is above 10^6 K.
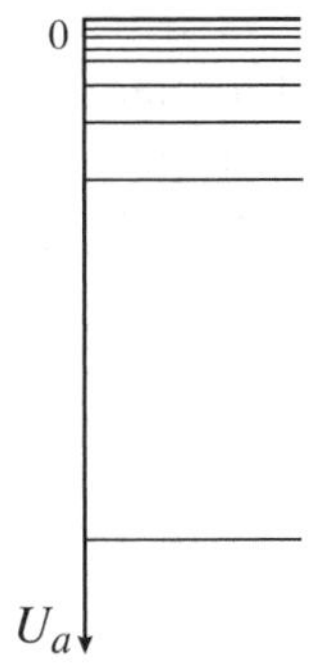 0 U_a	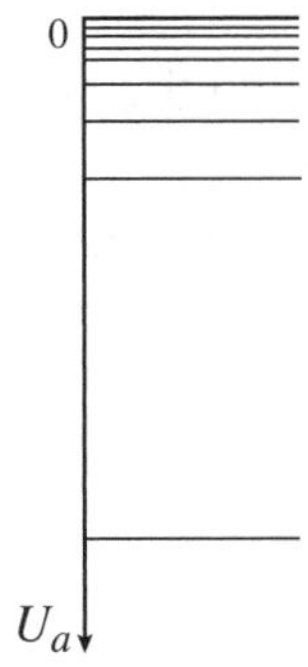 0 U_a	0 U_a	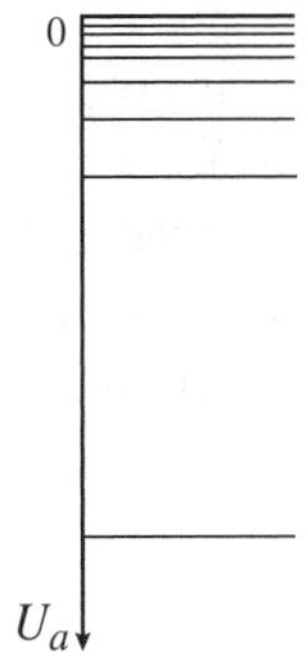 0 U_a	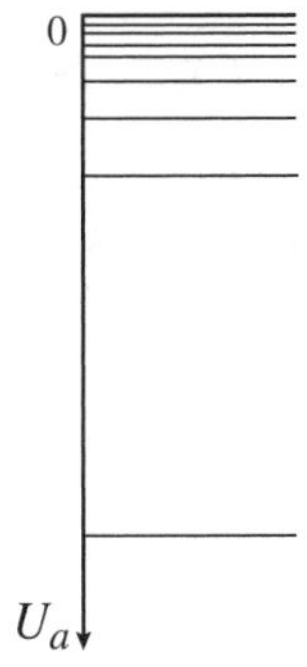 0 U_a

27.4.4 Reason Consider the following process: a hydrogen atom absorbs a photon and changes its energy from the second state to the fifth state. Complete the table that follows.

What is the energy of the atom in the second state?	What is the energy of the atom in the fifth state?	What is the energy difference between the states?	What is the energy of a photon that the atom must absorb to undergo this change?	What is the frequency of the photon that the atom must absorb to undergo this change?

27.4.5 Reason Imagine that you have an atom of He that lost one electron and is now a positively charged ion with the charge of the nucleus equal to +2 (He^{+}).

a. Determine the radii and energy of the $n = 1, 2, 3$ energy states in this ion.

b. Construct an energy-level diagram for this ion.

27.4.6 Reason In Bohr's model of the hydrogen atom the electron revolves around the nucleus like a planet revolves around the Sun. Determine the speed and frequency of the revolution of an electron around the first Bohr orbit in hydrogen. According to classical physics, the atom should emit electromagnetic radiation at this frequency (because circular motion described mathematically is very similar to vibrational motion). In what portion of the electromagnetic spectrum is this frequency?

27.4.7 Reason Niels Bohr proposed the postulates that described the behavior of the electron in the hydrogen atom as different from electrons that are not bound to nuclei. However, he also stated that "At sufficiently large *n*, the new model should predict the same behavior of the electron as classical physics." Discuss what this statement means and whether Bohr's model of the atom behaves according to this principle.

27.4.8 Regular problem A hydrogen atom changes its state from $n = 15$ to $n = 5$. Complete the table that follows.

Represent the process with an energy diagram.	Discuss whether the energy of the system increases or decreases.	Discuss whether the atom absorbs or emits a photon to undergo this process.	Calculate the energy, the frequency, and the wavelength of the emitted photon.

27.4.9 Reason The average thermal energy of the random translational motion of a hydrogen atom at room temperature is $(3/2)kT$ where k is the Boltzmann constant. Would a typical collision between two hydrogen atoms be likely to transfer enough energy to one of the atoms to cause a transition from the $n = 1$ energy state to the $n = 2$ state?

a. Draw an energy bar chart for a process during which two hydrogen atoms collide and one of them changes state from $n = 1$ to $n = 2$.

b. Would a typical collision between two hydrogen atoms in Earth's atmosphere be likely to transfer enough energy to one of the atoms to change the energy state from $n = 1$ to $n = 2$? Explain your answer. (*Note*: Actually, free hydrogen on Earth is in the molecular form H_2. However, the above reasoning still explains why a gas composed of hydrogen molecules at room temperature is seldom excited to the point of emitting light.)

c. Find the temperature at which the collisions between hydrogen atoms can lead to this change of state.

27.4.10 Regular problem A gas composed of hydrogen atoms in a tube is excited by collisions with free electrons. If the maximum excitation energy gained by an atom is 12.5 eV, determine the wavelengths of light emitted from the tube as atoms return to the ground state.

27.4.11 Regular problem Determine the de Broglie wavelength of an electron that has been accelerated across a potential difference of 200 V. The electron is initially at rest and has a mass of 9.1×10^{-31} kg.

27.4.12 Regular problem

a. Determine the states of the four electrons in the ground state of beryllium.

b. Determine the electron configuration of the twelve electrons in the ground state of magnesium.

27.4.13 Design an experiment Design and describe an experiment to determine whether gas collected in a cave contains carbon monoxide (CO).

28 Nuclear Physics

28.1 Qualitative Concept Building and Testing

28.1.1 Observe and explain Photographic plates are glass plates covered with material that undergoes a chemical reaction when light shines on it – this is called "exposure". After the photographic plate undergoes another chemical process called "development" the places on the plate exposed to light change color. In 1896 Henri Becquerel experimented with uranyl crystals and found that when placed on top of a photographic plate the crystals could expose the plate, even when the materials were stored in a dark box. Suggest an explanation for this observation.

Hint: Some phosphorescent or fluorescent materials, such as barium sulfide, zinc sulfide, etc., can expose photographic paper after they have been exposed to light. However, uranyl salts do this without exposure to light.

28.1.2 Test your idea One if the explanations of Becquerel's experiments is that uranyl salts emit some kind of charged particles that affect the photosensitive paper in a way similar to light. How can you test this explanation?

28.1.3 Observe and explain Pierre Curie invented an electrometer that could be used to measure how much charge the electrometer loses per unit time (the current). Marie Curie used this electrometer to measure the amount of electric current produced when the electrometer was placed near uranium salts.

- She found that the amount of current was proportional to the amount of uranium present.
- She found that the intensity of the current was independent of the identity of the uranium salt, its wetness, temperature, physical appearance, or the amount of light shining on it.

a. Explain why the electrometer would lose its charge when a sample of uranium salt was placed nearby.

b. If you were Marie Curie, with a strong background in chemistry, what could you conclude about how the uranium rays are produced? (*Hint:* Sometimes in science you can determine what a phenomenon *is not* long before you have an idea about what it *is*.)

28.1.4 Observe and explain In 1899 Ernest Rutherford and his colleagues investigated the ability of uranium salts to ionize air. He set up two parallel plates, with a potential difference between them. When a uranium sample was placed between the plates, ions created by the radiation would be pulled to the plates before they could recombine. This caused a detectable current. Covering the uranium sample with thin aluminum sheets decreased the amount of current observed, but only up to a point. After this point, no further decrease in current was observed, even with the addition of more aluminum plates. Propose an explanation of why the current decreased with more aluminum shielding, but only to a point.

28.1.5 Explain In 1903 Rutherford placed his radioactive sample in a magnetic field in an apparatus such as shown below. He and his assistants used a scintillating screen, which glowed when a charged particle hit the surface (similar to the screen of an old-fashioned TV that had a cathode-ray tube inside). In the second experiment they used photographic paper and found that it was exposed around point O. Describe below the cause of each exposure.

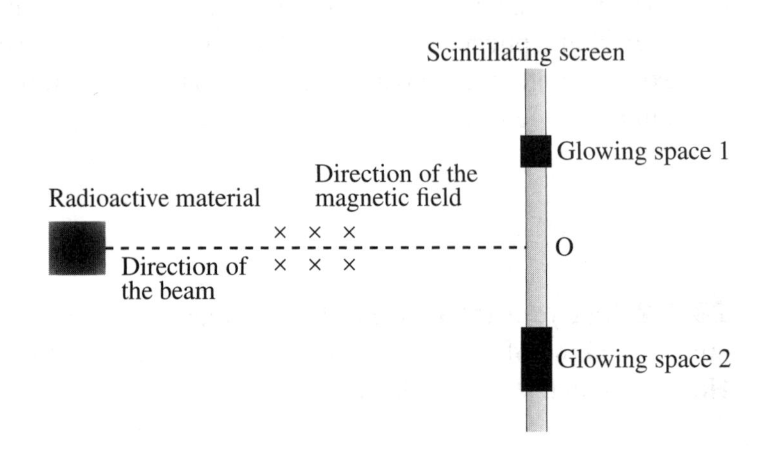

a. Describe everything you can about what caused the glowing screen at space 1.

b. Describe everything you can about what caused the glowing screen at space 2.

c. Describe everything you can about what caused the photographic paper to be exposed at O.

28.1.6 Evaluate reasoning Based on the experiments such as described above, scientists proposed that the nucleus of an atom is made of positively charged alpha particles and negatively charged electrons. Their electrostatic attraction holds them together. When a nucleus has a lot of alpha particles, they start repelling each other and are likely to leave the nucleus (alpha decay). This leaves too many electrons inside that repel each other and thus electrons are emitted (beta decay). After each transformation the nucleus is left in an excited state and emits a high-energy photon—a gamma ray (gamma decay). Describe how this proposal is consistent with the experiments in Activity 28.1.5.

28.1.7 Observe and find a pattern Frederic Soddy (1913) collected the following data related to the radioactive transformation of uranium. The first product that appears in the sample is thorium, then protoactinium, then another isotope of uranium, and so on. Examine the series of the transformation found by Soddy and explain it using your knowledge of alpha and beta decays. Discuss what quantities are conserved in each process. In the series presented below the left subscript indicates the positive charge of the nucleus in the units of the electron charge and left superscript indicates the mass of the nucleus in the atomic units. Using this notation system hydrogen can be written as ${}^{1}_{1}\text{H}$.

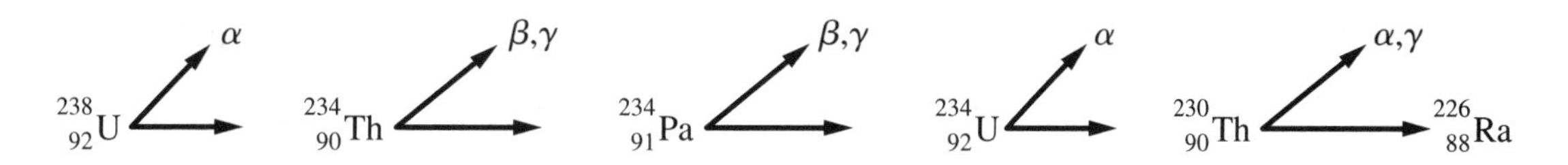

28.1.8 Explain According to Heisenberg's uncertainty principle, the uncertainties of the position Δx and momentum Δp_x of an atomic-size particle can be known no better than $\Delta x \cdot \Delta p_x \geq h/4\pi$, where h, called Planck's constant, equals $6.63 \times 10^{-34}\ \text{J} \cdot \text{s}$. Experiments by Rutherford's colleagues led to the estimation of the size of atomic nucleus to be about 10^{-15} m. Explain how the application of the uncertainty principle indicates why electrons could not be inside a nucleus.

28.1.9 Explain In 1920 Rutherford suggested that experimental evidence about nuclei at that time indicated the need for a particle that has no electric charge and has a mass slightly higher than the proton nucleus of hydrogen. Irène Joliot-Curie (daughter of Marie Curie and Pierre Curie) and her husband, Frédéric Joliot-Curie, performed experiments in which they found that fast moving alpha particles passing through and being stopped by targets made of light elements (for example, beryllium) created radiation that could penetrate materials better than gamma rays. James Chadwick explained the results of their experiments by using the idea of the particle predicted by Rutherford—a neutron. Use this information to answer the following questions.

a. Use the concepts of protons and neutrons to explain the atomic number and the mass of nuclei.

b. Use the concepts of protons and neutrons to explain alpha decay.

c. Use the ideas of protons and neutrons to explain the beta decay.

d. Suggest an explanation for the fact that positively charged protons and electrically neutral neutrons stay together in a nucleus.

28.1.10 Terminology Determine the number of protons and neutrons in each of the following nuclei:

$^{11}_{5}B$	$^{19}_{9}F$	$^{39}_{19}K$	$^{63}_{29}Cu$	$^{138}_{56}Ba$	$^{208}_{82}Pb$
protons: neutrons:	protons: neutrons:	protons: neutrons:	protons: neutrons:	protons: neutrons:	protons: neutrons:

28.1.11 Observe and find a pattern Devise one or more rules that seem necessary in order for the nuclear reactions below to occur.

a. $^{1}_{0}n + ^{235}_{92}U \rightarrow ^{147}_{56}Ba + ^{86}_{36}Kr + 3\,^{1}_{0}n$ + energy

b. $^{239}_{94}Pu \rightarrow ^{235}_{92}U + ^{4}_{2}He$ + energy

c. $^{191}_{76}Os \rightarrow ^{191}_{77}Ir + ^{0}_{-1}e$ + energy

28.2 Conceptual Reasoning

28.2.1 Estimate Use what you have learned to estimate the following values.

a. Estimate the density of a nucleus. Indicate any assumptions you made.

b. Estimate Earth's radius if it had this same density. Indicate any assumptions you made.

c. Estimate the radius of a sphere that would hold all of your body mass if its density equaled the density of a nucleus. Indicate any assumptions you made.

28.2.2 Estimate Use what you have learned to estimate the following quantities.

a. Estimate the total number of nucleons (protons and neutrons) in your body.

b. Estimate the total number of electrons in your body.

c. Indicate roughly the volume in cm^3 occupied by these nucleons.

28.2.3 Reason Insert the missing symbol in the following reactions.

a. ${}^{4}_{2}\text{He} + {}^{12}_{6}\text{C} \rightarrow {}^{15}_{7}\text{N} + ?$

b. ${}^{2}_{1}\text{H} + {}^{3}_{1}\text{H} \rightarrow {}^{4}_{2}\text{He} + ?$

c. ${}^{1}_{0}n + {}^{235}_{92}\text{U} \rightarrow {}^{140}_{54}\text{Xe} + ? + 2\,{}^{1}_{0}n$

d. ${}^{2}_{1}\text{H} \rightarrow ? + {}^{0}_{-1}e$

28.2.4 Reason The following nuclei produced in a nuclear reactor each undergo radioactive decay. Write the decay reaction equation.

a. ${}^{239}_{94}\text{Pu}$ alpha decay:

b. ${}^{144}_{58}\text{Cs}$ β^- decay:

c. ${}^{65}_{30}\text{Zn}$ β^+ decay:

28.3 Quantitative Concept Building and Testing

28.3.1 Observe and find a pattern Rutherford found that by blowing air across a sample of thorium, a radioactive gas could be collected. This gas lost its radioactivity rapidly, as shown in the table. Is there a mathematical trend in Rutherford's data? Summarize it in your own words.

Time (min)	Portion of time-zero radioactivity lost
0	0
1	~50 %
2	~75 %
3	~87 %
4	~93 %
5	~97 %
10	Undetectable $\approx$ 100 %

28.3.2 Observe and explain Measurements with a mass spectrometer indicate the following particle masses: ${}^{4}_{2}\text{He}$ (4.002604 u), ${}^{1}_{1}\text{H}$ (1.007825 u), and ${}^{1}_{0}n$ (1.008665 u). Compare the mass of the helium atom to the mass of the particles of which it is made. What do you conclude? *Note:* 1 u is 1/12 the mass of a carbon-12 atom and equals $1.660566 \times 10^{-27}\text{ kg} = 931.5\text{ MeV}/c^2$.

28.3.3 Reason in Activity 28.3.2 you found that the mass of a helium nucleus is less than the mass of the nucleons inside it.

a. Explain how this observation led scientists to the idea that it is possible to convert hydrogen into helium to produce thermal energy.

b. Does this process mean that energy is not conserved? Explain.

c. Represent the process with an energy bar chart. What is the binding energy on this chart?

d. Explain why very high temperatures and pressures are needed for this reaction.

e. Estimate the temperature at which two protons will join together due to their nuclear attraction. Remember, that nuclear forces are effective at distances less than or equal to about 10^{-15} meters. Hint: Use an energy approach and not a force approach.

28.3.4 Explain Explain why the following reaction does not occur spontaneously: ${}^{4}_{2}\text{He} \rightarrow {}^{3}_{1}\text{H} + {}^{1}_{1}\text{H}$.

28.3.5 Reason In the 1940s Lise Meitner, Otto Hahn, and Fritz Strassmann irradiated uranium with neutrons. They found that instead of getting a heavier uranium isotope, the reaction produced lighter nuclei, like isotopes of Ba (barium). How can you explain their findings?

28.3.6 Reason Lise Meitner asked her nephew Otto Robert Frisch to help with the explanation. They thought about Bohr's liquid drop model; heavy nuclei behaved like a drop of water with a kind of "surface tension" holding it together. The only problem was that the nucleus had an electric charge that would counteract the effect of the surface tension, especially if the nucleus was not very spherical. The model suggested that the nucleus could elongate and divide into two smaller pieces. This meant that the uranium nucleus would be very unstable and ready to divide with the slightest provocation. This could happen when a neutron hit it.

a. Explain how Meitner's and Frisch's reasoning could explain the findings in Activity 28.3.5.

b. Use the liquid drop model to predict another product or products of uranium disintegration. If a chemist finds such products in the mixture resulted from the irradiation of uranium with neutrons, then the water drop model is a productive model to explain the observations in Activity 28.3.5.

c. Explain how the model of Meitner and Frisch can be used to explain why uranium can be used as a source of thermal energy. Additional information: the products of the irradiation of uranium with neutrons included not only chemical elements in the middle of the periodic table but also extra neutrons.

28.4 Quantitative Reasoning

Note that $1\text{ u} = 1.6606 \times 10^{-27}\text{ kg} = 931.5\text{ MeV}/c^2$.

28.4.1 Reason Determine the binding energies per nucleon for $^{238}_{92}\text{U}$ and for $^{120}_{50}\text{Sn}$. Based on these numbers, which nucleus is more stable? Explain.

28.4.2 Reason One part of the carbon-nitrogen cycle that provides energy for the Sun is the reaction $^{12}_{6}\text{C} + ^{1}_{1}\text{H} \rightarrow ^{13}_{7}\text{N} + 1.943\text{ MeV}$. Using the known masses of ^{12}C and ^{1}H and the results of this reaction, determine the mass of ^{13}N.

28.4.3 Reason Determine the missing nucleus in the following reaction and the energy released in the reaction: $^{232}_{92}\text{U} \rightarrow ? + ^{4}_{2}\text{He} + \text{energy}$.

28.4.4 Reason Radon-222 ($^{222}_{86}\text{Rn}$) is released into the air during uranium mining and undergoes alpha decay to form $^{218}_{84}\text{Po}$ of mass 218.0089 u. Determine the energy released by the decay reaction. Most of this energy is in the form of α particle kinetic energy.

28.4.5 Reason Cesium-137 is a waste product of a nuclear reactor. Determine the fraction of ^{137}Cs remaining in a reactor fuel rod:

a. 120 y after it is removed from the reactor.

b. 240 y after it is removed.

c. 1000 y after it is removed.

Half-Lives and Decay Constants of Some Common Nuclei

Isotope	Half-life	Decay constant (s^{-1})
$^{87}_{37}Rb$	4.88×10^{10} yr	4.50×10^{-19}
$^{238}_{92}U$	4.5×10^{9} yr	4.9×10^{-18}
$^{40}_{19}K$	1.28×10^{9} yr	1.72×10^{-17}
$^{239}_{94}Pu$	2.44×10^{4} yr	9.00×10^{-13}
$^{14}_{6}C$	5730 yr	3.84×10^{-12}
$^{226}_{88}Ra$	1602 yr	1.37×10^{-11}
$^{137}_{55}Cs$	30.0 yr	7.32×10^{-10}
$^{90}_{38}Sr$	28.1 yr	7.82×10^{-10}
$^{3}_{1}H$	12.4 yr	1.77×10^{-9}
$^{60}_{27}Co$	5.26 yr	4.18×10^{-9}
$^{131}_{53}I$	8.05 day	9.96×10^{-7}
$^{11}_{6}C$	21 min	5.5×10^{-4}

28.4.6 Regular problem A sample of radioactive technetium-99 of half-life 6 h is to be used in a clinical examination. The sample is delayed 15 h before arriving at the lab for use. Use two methods to determine the fraction that remains.

28.4.7 Regular problem To estimate the number of ants in a nest, 100 ants are removed and fed sugar made from radioactive carbon of long half-life. The ants are returned to the nest. Several days later, it is found that of 200 ants tested, only 5 are radioactive. Roughly, how many ants are in the nest? Explain your estimation technique.

28.4.8 Regular problem A sample from a tree uprooted and buried during the Wisconsin glaciation contains 50 g of carbon when it is discovered.

a. If 1 in 10^{12} carbon atoms in a *fresh* tree sample were carbon-14, how many carbon-14 atoms would be in 50 g of carbon from a fresh tree?

b. Determine the carbon-14 activity of this fresh sample.

c. Determine the age of the buried tree if its 50 g of carbon has an activity of $-2.2\ \text{s}^{-1}$.

28.4.9 Regular problem Determine the energy released in the following fission reaction:

$$ {}_{0}^{1}n^{-} + {}_{92}^{235}\text{U} \rightarrow {}_{56}^{141}\text{Ba} + {}_{36}^{92}\text{Kr} + 3\,{}_{0}^{1}n. $$

The masses of the nuclei are: $m_n = 1.0087$ u, $m_{\text{U}} = 235.0439$ u, $m_{\text{Ba}} = 140.9141$ u, and $m_{\text{Kr}} = 91.8981$ u.

28.4.10 Energy from gasoline compared to uranium Use the data provided below to compare these two energy sources.

a. Determine the energy release in MeV of gasoline per molecule of n-heptane burned. The molecular mass of n-heptane is 100 u, and it releases energy at a rate of 4.8×10^7 J/kg.

b. Determine the ratio of energy released by one uranium-nucleus fission (approximately 200 MeV) and one n-heptane molecule combustion.

29 Particle Physics

29.1 Qualitative Concept Building and Testing

29.1.1 Summarize In your own words, describe the models and reasoning that resulted in the prediction of the existence of antimatter.

29.1.2 Observe and analyze The figure shows the path of a charged particle that was produced when a cosmic ray interacted with the atmosphere. The particle is shown passing through a sheet of lead. Assuming the $\vec{B}$-field points into the page, determine the direction this particle is traveling and the sign of its electric charge.

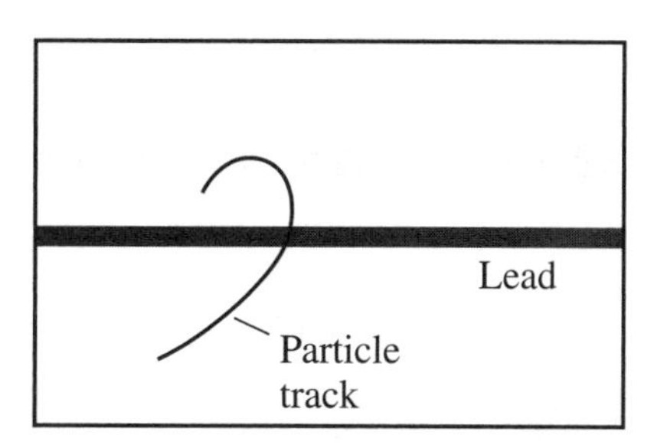

It is determined that the mass of this particle (within experimental uncertainty) is equal to the mass of the electron. Explain why this is interesting.

29.1.3 Observe and find a pattern The table below contains energy and momentum data just before and after a photon undergoes electron-positron pair production.

	Photon	Electron	Positron
Rest energy (MeV)	0	0.511	0.511
Kinetic energy (MeV)	1.62	0.300	0.300
Momentum, x-component (kg · m/s)	8.65×10^{-22}	4.32×10^{-22}	4.33×10^{-22}
Momentum, y-component (kg · m/s)	0	2.50×10^{-22}	-2.50×10^{-22}

Sketch this process.

Find as many patterns as you can in this data. Are they consistent with your understanding of energy and momentum?

29.1.4 Predict and test Describe as many predictions of the Standard Model as you can. For each prediction, was the result of the corresponding experiment consistent or inconsistent with the prediction, or is the result as yet unresolved?

29.1.5 Observe Describe as much evidence as you can in support of the Big Bang model.

29.1.6 Observe Describe as much evidence as you can for the existence of dark matter.

29.1.7 Explain Describe as may hypotheses as you can that explain the nature of dark matter.

29.1.8 Observe Describe as much evidence as you can for the existence of dark energy.

29.1.9 Explain Describe as may hypotheses as you can that explain the nature of dark energy.

29.2 Conceptual Reasoning

29.2.1 Explain Positron emission tomography can be used in pharmacology to determine if a drug is being delivered to the appropriate places in an organism. Explain how this could work.

29.2.2 Observe and analyze A detector at the Fermi National Accelerator Center (Fermilab) records the collision between a proton and an antiproton. Each gray line in the figure represents the path of a particle created in the collision. Assuming the $\vec{B}$ field points out of the page, determine the sign of the electric charge of each particle. Explain your reasoning.

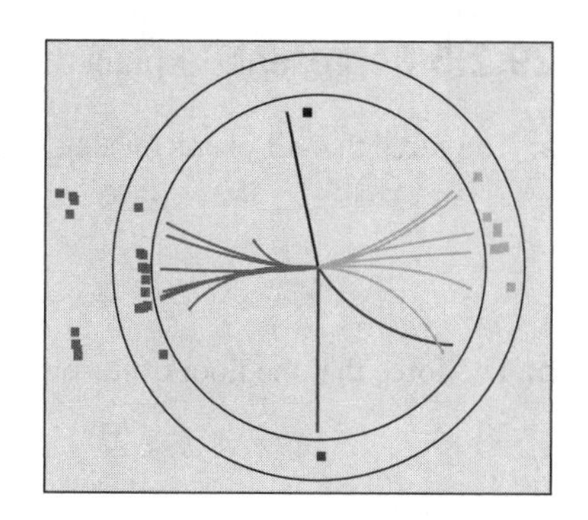

29.2.3 Represent An electron and a positron collide head-on, annihilate, and produce a muon and an antimuon. Sketch a valid version of this process. What assumptions did you make?

Sketch an invalid version of this process. Explain why it is invalid.

29.2.4 Represent Explain why someone might think electron-positron annihilation into a single photon (shown in the figure) is correct.

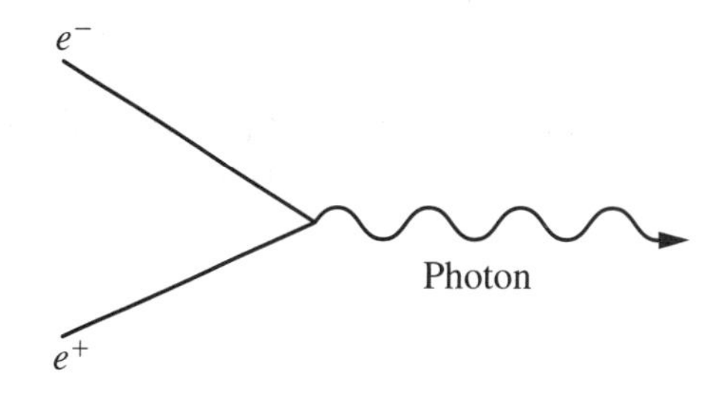

There is a problem with this process, however. Go to the reference frame where the electron and positron are heading toward each other at equal speeds. Draw a sketch of the process from this reference frame.

Is it possible for a single photon to be produced by this annihilation process? Explain your reasoning. Would adding an additional photon help? Explain why or why not.

29.2.5 Represent Explain the following non-fundamental interactions in terms of fundamental ones.

a. The force that air exerts on a moving car.

b. The force that the floor exerts on you while you are standing.

c. The force that one section of a stretched bungee cord exerts on an adjacent section.

d. The force that water exerts on a submerged submarine.

29.2.6 Explain How do we know that the strong interaction has an extremely short range?

29.2.7 Explain Consider the following process:

$$2p \rightarrow 3p + \bar{p}$$

Here the p represents a proton and the $\bar{p}$ represents an antiproton. How is this process possible? Why might someone argue that it is not possible?

29.2.8 Reason Four baryons have been discovered and are known as the delta particles: Δ^-, Δ^0, Δ^+, and Δ^{++}. The superscripts indicate the electric charge of these particles. The delta particles are composed of different combinations of up and down quarks. Determine these combinations.

Given the quark content of the Δ^+, suggest an explanation for what distinguishes it from a proton. *Hint:* Try to relate what you have learned about the hydrogen atom.

29.2.9 Explain Consider the following process:

$$2p \rightarrow p + n + \pi^+$$

Represent this process in terms of the quark content of the particles. Then, explain why this process is an example of confinement.

29.2.10 Explain According to the Big Bang model, how old are the protons and neutrons in your body?

29.2.11 Explain How does the idea of supersymmetry help in explaining dark matter? Dark energy?

29.3 Quantitative Concept Building and Testing

29.3.1 Analyze How do we know that the electric interaction is much stronger than the gravitational interaction? Why might someone believe the opposite to be true?

Use the interactions between the proton and electron in the hydrogen atom to argue quantitatively why the electric interaction is many orders of magnitude stronger than the gravitational interaction.

29.3.2 Explain You've learned that the electrostatic interaction is weaker the farther apart the two interacting objects are. Here, you will consider how the particle interaction mechanism of fundamental interactions is consistent with this. First, explain this particle interaction mechanism.

Consider two electrons that are very close to each other (say, 10 times the diameter of an atom). Use the uncertainty principle to estimate the energy and momentum of the virtual photons.

Now consider two electrons that are much farther away from each other (say, 1 million times the diameter of an atom). Estimate the energy and momentum of these virtual photons.

Use your above estimates to decide whether the particle interaction mechanism is consistent with your understanding of the electrostatic interaction. Explain your reasoning.

29.4 Quantitative Reasoning

29.4.1 Reason A photon (gamma ray) of wavelength 1.15×10^{-12} m is produced in a supernova. Is it possible for this photon to produce an electron-positron pair?

Explain your reasoning.

29.4.2 Reason Neutrons are unstable with a half-life of about 15 minutes. They decay into a proton, an electron, and an electron antineutrino. Explain this process in terms of energy, quantitatively.

On the other hand, protons are not observed to spontaneously decay into neutrons, positrons, and electron neutrinos. Explain why.

29.4.3 Regular problem At the Large Hadron Collider protons are collided with each other at extremely high speeds. What is the minimum speed each of two protons must have in order for them to produce one W^+ and one W^- weak interaction mediator? *Hint:* The two protons do not annihilate since they are both protons rather than one proton and one antiproton.